Urushi

Urushi

漆

Proceedings of
the Urushi Study Group
June 10–27, 1985
Tokyo

N. S. Brommelle and Perry Smith, *Editors*

The Getty Conservation Institute

Cover:
Food Container, *detail*
Japan, Momoyama Period, 1568–1615
Black lacquer with gold makie
The Los Angeles County Museum of Art
M.87.202 a.b
Gift of Donald & Iris Blackmore

Coordination: Irina Averkieff, The Getty Conservation Institute,
Marina del Rey, California
Design: Joe Molloy, Mondo Typo, Inc., Santa Monica, California
Typography: Ventura/Adobe PostScript Stone Serif
and Berthold Sabon and Univers
Printing: Dai Nippon, Tokyo, Japan

Printed in Japan

Library of Congress Cataloging-in-Publication Data

Urushi Study Group (1985: Tokyo, Japan)
Urushi: Proceedings of the 1985 Urushi Study Group/
N.S. Brommelle and Perry Smith, editors.
p. cm.
Bibliography: p.
ISBN 0-89236-096-8: $40.00
1. Lacquer and lacquering—East Asia—Congresses.
2. Lacquer and lacquering—Conservation and restoration—
East Asia—Congresses. I. Brommelle, N.S. II. Smith, Perry.
III. Title.
NK9900.7.E15078 1985 745.7'2-dc19 85-21394 CIP

The Getty Conservation Institute

The Getty Conservation Institute (GCI), an operating program of the J. Paul Getty Trust, was created in 1982 to enhance the quality of conservation practice in the world today. Based on the belief that the best approach to conservation is interdisciplinary, the Institute brings together the knowledge of conservators, scientists, and art historians. Through a combination of in-house activities and collaborative ventures with other organizations, the Institute plays a catalytic role that contributes substantially to the conservation of our cultural heritage. The Institute aims to further scientific research, to increase conservation training opportunities, and to strengthen communication among specialists.

Contents

Prologue ix

Map x

Chronology xi

History

Hirokazu Arakawa — On the Chinese *Kyushitsu* Method Based on a Study of *Kyushoku-roku* Documents 3

Motoo Yoshimura — Innovations in Kodaiji *Makie* 7

George Kuwayama — Chinese *Guri* Lacquers 13

Beatrix von Ragué — *Inro* Research: Some Proposed Future Steps 23

Akio Haino — Sano Chokan, the Urushi Master, Studied through His Work 31

Norimitsu Kimura — *Heidatsu* and *Hyomon* in the Nara Period 37

Yoshiyuki Kuraku — Origins of the Use of Urushi in Japan and Its Development 45

Jongseok Lee — Lacquerware in the Unified Silla Period, with Special Reference to the Finds at Anap-chi 51

Yoshio Hasegawa — Urushi Coating and Color Painting Applied to Japanese Architectural Cultural Monuments 57

Conservation and Techniques

Zhou Bao Zhong — The Protection of Ancient Chinese Lacquerware 71

Shin Yagihashi — The Preservation and Handing Down of Traditional Urushi Art Techniques in Japan 79

Toshie Nakajima — Conservation of Chinese Urushi: Methods and Difficulties 87

Shogyo Ohba	The *Kyushitsu* Technique Demonstrated on a *Natsume*	91
W. T. Chase	Lacquer Examination and Treatment at the Freer Gallery of Art: Some Case Histories	95
Kenichi Kitamura	Some Thoughts about Conserving Urushi Art Objects in Japan, and an Example of Conservation Work	113
Susanne Barchalia	Apprenticeship and Conservation	121
Kyotaro Nishikawa	Dry Lacquer Statues of Japan	127
Alena Skálová	Chinese Lacquer	133
Yoshikuni Taguchi	Essay on Shells That Became Cherry Blossoms	139
Toshikatsu Nakasato	Urushi Technique in the Prehistoric and Antique Periods in Japan	147

Science

Toshiko Kenjo	Scientific Approach to Traditional Lacquer Art	155
A. Burmester	Technical Studies of Chinese Lacquer	163
Yumin Du	The Production and Use of Chinese Raw Urushi and the Present State of Research	189
Billie Milam and Hélène Gillette	X-ray Radiography in the Study of Oriental Lacquerware Substructures	199
M. Derrick, C. Druzik and F. Preusser	FTIR Analysis of Authentic and Simulated Black Lacquer Finishes on Eighteenth Century Furniture	227
Rosamond Westmoreland	Solvent Testing Method for Identification of Oriental Lacquer Used in European Furniture	235
Ju Kumanotani	The Chemistry of Oriental Lacquer (*Rhus Verniciflua*)	243
	Glossary	253

Oval Tray
China, Yuan Dynasty, 13–14th Century
Carved red lacquer
The Los Angeles County Museum of Art
M.81.125.1
Gift of Mr. & Mrs. John H. Nessley

Prologue

漆

In 1985, the Getty Conservation Institute (GCI) organized an international meeting on the study of Oriental lacquerware—urushi—in collaboration with the Tokyo National Research Institute of Cultural Properties in Japan. Held on June 10–27, the meeting included specialists in the history and techniques of urushi in an effort to stimulate an interdisciplinary discussion and approach to the conservation of this ancient and contemporary art form. The GCI wishes to express its gratitude for the opportunity to cooperate with the Tokyo National Research Institute of Cultural Properties in this important endeavor. Special appreciation is extended to Dr. Nobuo Ito, Director of the Institute at the time of the meeting, and to his colleagues who, through the organization of visits to museums, temples, urushi workshops, conservation studios, and manufacturers, facilitated a cultural and technical understanding of urushi objects.

The papers presented here are the proceedings of the meeting, which have been edited through the good offices of the International Institute for Conservation. We are very grateful to Mr. Norman Brommelle and Miss Perry Smith who undertook the editorial responsibility. Appreciation is also expressed to Miss Barbara Roberts, Decorative Arts Conservator, The J. Paul Getty Museum, who served as coordinator of the meeting, and to the many participants who travelled to Japan from other countries with the hope of gaining new scientific, historical, and cultural insights into the manufacture and conservation of urushi objects.

Luis Monreal
Director
The Getty Conservation Institute

HEILONGJIANG
JILIN
GANSU
LIAONING
QINGHAI
Beijing
NORTH KOREA
HEBEI
Seoul
SHANXI
SOUTH KOREA
SHANDONG
SHAANXI
HENAN
JIANGSU
CHINA
ANHUI
SICHUAN
HUBEI
ZHEJIANG
HUNAN
JIANGXI
GUIZHOU
FUJIAN
YUNNAN
GUANGXI ZHUANG
GUANGDONG
TAIWAN
Aomori
Yamagata
Niigata
Nikko
Wajima
JAPAN
Tokyo
Izumo
Kyoto
Ise
Hiroshima
Nagasaki
Ryukyu Islands

Chronology

China

Neolithic Period	*c.* 7000–*c.* 1600 B.C.
Shang	*c.* 1600–1027 B.C.
Zhou	1027–256 B.C.
Western Zhou	1027–771 B.C.
Eastern Zhou	770–256 B.C.
Warring States Period	475–221 B.C.
Qin	221–206 B.C.
Han	A.D. 220
Western Han	206 B.C.–A.D. 8
Eastern Han	A.D. 25–220
Period of the Three Kingdoms and Six Dynasties	A.D. 221–581
Sui	A.D. 581–618
Tang	A.D. 618–906
Five Dynasties Period	A.D. 907–960
Liao	907–1125
Song	960–1279
Northern Song	960–1126
Southern Song	1127–1279
Yuan	1280–1368
Ming	1368–1644
Qing	1644–1912

Japan

Jomon	*c.* 200 B.C.
Yayoi	*c.* 200 B.C.–*c.* A.D. 250
Kofun (Tumulus)	*c.* A.D. 250–552
Asuka	A.D. 552–646
Nara	A.D. 646–794
Heian	794–1185
Kamakura	1185–1392
Muromachi	1392–1568
Momoyama	1568–1600
Edo	1600–1868
Meiji	1868–1912

Korea

Three Kingdoms Period	37 B.C.–A.D. 935
Kingdom of Koguryo	37 B.C.–A.D. 668
Kingdom of Paekche	18 B.C.–A.D. 660
Kingdom of Silla	57 B.C.–A.D. 935
Unified Silla	A.D. 668–918
Koryo	918–1392
Choson	1392–1910

This chronology is intended to give the reader a general overview of the periods discussed in the text and not as a definitive reference.

Toshogu *(main shrine)* honsha *and* honden.

On the Chinese *Kyushitsu* Method, Based on a Study of *Kyushoku-roku*

Hirokazu Arakawa
Tokyo National Museum

The documents on the art of urushi in China rarely tell us about the detailed process of manufacture. To the best of my knowledge, there are only a few examples. One of these is *Kyushoku-roku*, which records methods of *kyushitsu*, the process and techniques for the application of urushi from the ground coating to the surface coating.

Kyushoku-roku is a document for specialists on the art of urushi. It covers almost all the techniques, with a proper explanation of each. It was written by Huang Wei, an urushi craftsman, in the Longqing period (1567–1572), but the original is lost. The edition with the stamp "*Kenka-do Zosho-in*" is thought to be the most reliable of the manuscripts still in existence. (All references in this paper to *Kyushoku-roku* refer to this edition; Arakawa 1963.) It consists of eighteen chapters, divided into two volumes: *Ken* and *Kon*. *Ken* describes tools, materials, and various points to note in making urushi; *Kon* classifies each technique and describes it in detail. A note to this edition says that the preface was written by Yan Min, but does not give any explanation about his background. Presumably he was an urushi craftsman of the Tianqi period, since the preface was written in the fifth year of Tianqi (1625).

Other important documents on *kyushitsu* include:

1. A volume of *Doten Sei-roku*, which was compiled by Tsuao Chi Hu in the Song dynasty (960–1279). He explains how to differentiate old vessels; the explanation of *kyushitsu* is included in the chapter, *Kokinben*.

2. *Tekko-roku*, which was compiled in thirty volumes by Tao Son I in the Ming dynasty (1368–1644). It is actually a book on legal systems and disturbances in the Yuan period (1280–1368), but the author also deals with paintings and calligraphic works. The last volume, *Kyuki*, describes the art of urushi.

3. Fourteen volumes of *Kinkyo*, which were compiled by Tsuan Ta Mi in the Ming dynasty. This was revised by Sun In, U Ie Shi, and Wan Mao Tao, and published in the thirty-seventh year of Wanli (1609). Included is a detailed description of *kyushitsu* for the seven-string *koto*. (It is interesting to find comments on *kyushitsu* in documents on *kotos*; I have already made a detailed study of this subject (Arakawa 1981a, 1981b, 1981c) and the reader is referred to these publications for a fuller account.)

***Documents on* kyushitsu**

Kyushitsu is explained in principle in volume 17 of *Kyushoku-roku*, titled *Shippo*. This volume mentions seven processes from the wooden substrate to *nakanuri* (intermediate coating). The types of *uwanuri* (final coating), according to color, are explained in volume 3, *Shisshoku*. Eight types are named: black, vermilion, yellow, green, purple, brown, "oil coating," and *kinshitsu*. Oil coating is a technique of applying oil on the lacquerware surface, and *kinshitsu* is ornamenting with gold filings and gold paint. In this paper we shall concentrate on *kokukyo* (black urushi).

The processes of *kyushitsu* explained in *Kyushoku-roku* and other documents are as follows:

1. *Kenso*. *Ken* is defined in *Gyokuhen*[1] as a bent work, a vessel made of curved wood. *So* means *kiji*, *shitaji*, and *sozai*, all substrates. *Kenso*, then, means *magemono*, or a bent-work substrate. In *Kyushoku-roku*, however, the terms *haitai* and *kikotsu* are used instead of *kenso*. The notes by Yan mention that materials such as bamboo, metal, and others were used as substrate. We should therefore interpret *kenso* as substrate (Arakawa 1977).

1. *Gyokuhen* is a kind of dictionary ("Yupian" in Chinese) published in the Six Dynasties in A.D. 543.

Yan's notes list bamboo, rattan, bronze, tin, ceramic, *nerimono* (a molding paste), cloth and paper, and layers of cloth as examples of substrates. In addition, wood is mentioned as being used for square boxes, bent works, and turned works. For example, paulownia and other woods were used as materials for zithers.

2. *Goho*. *Goho* means joining each surface or attaching legs. A craftsman joins surfaces with *hoshitsu*, fastens them with strings, and wedges them. *Kinkyo* explains *hoshitsu* as a mixture of raw urushi and *hoan min* (animal glue).

3. *Soto*. This is the filling with *kokuso* (a thick paste) of connecting parts, missing areas, and knots in the *kiji* (wooden substrate). Yan's notes explain the method of filling in deep gaps using a mixture of *hoshitsu* and wood powder or cotton dust.

4. *Fushitsu*. This is a method of using cloth as a reinforcement. Hemp cloth is carefully pasted on with *hoshitsu* so that there are no irregularities on the surface. *Hoshitsu* is applied to the edges and connecting parts because they are liable to exfoliate. According to Yan's notes, hemp fiber or thick paper was used during the Tianqi period instead of leather, which had been used previously. It should be noted that *fushitsu* is not done for *koto*s.

5. *Kanshitsu, kaishitsu*. These are types of urushi priming. *Kyushoku-roku* mentions the following materials mixed with urushi for priming: *tsunoko* (calcined deerhorn) and fine porcelain powder are first-class materials, bone powder and clam-shell powder are second-class materials, and ceramic powder and *tonoko* (finely ground baked clay) are third-class materials. These types of powder are sieved into coarse, fine, and intermediate grades, then applied in five separate stages.

First, coarse *kaishitsu* is applied thinly many times. Next, slightly finer *kaishitsu* is applied evenly in a thick coat. Third, a still finer *kaishitsu* is used to make angles and edges and fill in dents. Fourth, very fine *kaishitsu* is applied, neither thin nor thick. Finally, edges, angles, and border lines are "sharpened."

Yan's notes say that there are other *shitaji* (ground layers), known as *shitaji* substitutes. One is made of a mixture of ceramic powder and charcoal powder, to which starch, pig's blood, lotus juice, and glue are added; it was used by lower-ranking craftsmen in the Tianqi period and considered worthless. Another is called *mansui*, made from oil and lime; a detailed description is given in the section on mansui in *Tekko-roku*.

According to *Tekko-roku*, the *kai* of *kaishitsu* is ground ceramic powder. It is sieved into coarse, medium, and fine powder, and combined with urushi mixed with *nikawa* (animal glue). The *shitaji* substitute made from pig's blood, starch, etc., as opposed to urushi mixed with *nikawa*, is regarded as easily breakable. The chapter on *kaishitsu* in *Kinkyo* defines deer-horn powder as the best, bull-horn powder as second best, and copper and brass powder as "unique." It divides the method of applying *shitaji* into four parts and includes grinding processes—coarse grinding, water grinding, and oil grinding—between each layer.

Doten Sei-roku says that fine powders of gold, copper, and porcelain are used in making zithers. In addition the document mentions *happokai* (mixed powder), rokkakukai (deer-horn powder), and *mumeii-kai* (limonite powder), noting that mixtures of plaster, copper alloy, and ceramic were used for zither *shitaji*. *Happokai* is the ground powder of various gemstones; *mumeii-kai* is mixed with brown iron-ore powder.

6. *Soshitsu*. *Soshitsu* is *nakanuri* (intermediate coating). *So* means unmilled rice. *Soshitsu* is interpreted as *aranuri* (rough coating), or as urushi for *aranuri*. *Kushoku-roku* says that this smooths the surface of the *shitaji* and improves the density of the coating. There are three stages in the process of *nakanuri*: the first, *kaiso*, should be thick and smooth; the second, *kiurushi-so*, should be thin and even; and the third, *senso*, should be applied without wrinkles. Notes by Yan say that these three processes are old methods used for the *kyushitsu* of zithers. Vermillion and yellow are sometimes applied as *nakanuri*. It is known that raw urushi and fine powder are used for *suri-urushi*.

The chapter on *Soho* in *Kinkyo* also mentions three stages but adds a grinding process to the first stage in which high-quality raw urushi is used. It is interesting that the author mentions the *senso* method in detail. The process of making urushi is complete when the final coating and decoration are applied after the processes described above.

7. *Hoshitsu*. This term is used for *uwanuri*, the final coating. *Ho* means applying urushi in multiple layers. It also has the meaning of "dark-red urushi." The volume on *Shippo* in *Kyushoku-roku* does not say anything about *hoshitsu*, but it is recorded in the chapter on *Soshitsu* that *hoshitsu* is applied over the *nakanuri*, and *homen* (a surface prepared by laminating cloth using urushi as the binder) is mentioned in the chapter on *fushitsu* (cloth reinforcement). The book *Ken* lists some weaknesses of *uwanuri*, including cracking, wrinkling, and brush marks. As mentioned earlier, six kinds of *uwanuri* are given in volume 3, *Shisshoku*. Four methods for *kawari-nuri* (crackle pattern) are described in volume 4, *Monho*.

Kyushoku-roku explains *shissai* in the chapter following *hoshitsu*. This is a special method of applying urushi only on important points such as connecting parts and edges. It does not seem necessary to go into this in detail since it has no direct connection with the processes described above.

Afterword

During the meeting of the Urushi Study Group in Japan, there seemed to be great interest in the repair and preservation of *shikki* (lacquerware), but a lack of fundamental knowledge of the techniques involved. Only a few members of the Group were aware of the existence of old records on the process of *kyushitsu*. I have had a particular interest for some time in the differences and similarities in *kyushitsu* techniques from different parts of Asia: Japan, China, Korea, Ryukyu (Okinawa), and southeast Asia. On this occasion, the records on Chinese *kyushitsu* technique have been discussed, in order to introduce one aspect of this study.

[The translation of this paper has resulted in a variation of the author's original text. Ed.]

Bibliography

Arakawa, Hirokazu

1963 "Min-Shin no urushi-kogei to *Kyushoku-roku*" (The Kyushoku-roku and lacquer art of the Ming and Qing periods). *Museum* 151:16–21. Tokyo.

1977 "Urushi kogeishi no ni-san no mondai: kensoko" (Some problems on the historical development of lacquer craft). *Museum* 318:13–24. Tokyo.

1981 "Kin no kyushitsu ni tsuite" (Lacquer work executed on kin zither). In *History of Lacquer Art* (*Shikkoshi*). Vol. 4. Tokyo

1981 "Kin no kyushitsu ni tsuite" (Lacquer work executed on kin zither). Lecture for Academy of Lacquer Research. Unpublished manuscript.

1981 "Kin no kyushitsu" (Lacquer decoration of kin zithers). *Museum* 369:11–33. Tokyo

Innovations in Kodaiji *Makie*

Motoo Yoshimura
Kansei Gakuin University

From the sixteenth to the seventeenth centuries, *makie* furniture, the so-called Kodaiji *makie*, was very popular. Since I have already written about the dates of production and the manufacturers, this paper will concentrate on the innovations in Kodaiji *makie*, with specific reference to actual works (Yoshimura 1971).

First, it must be stressed that Kodaiji *makie* (a style of lacquer associated with the Kodaiji temple in Kyoto) does not employ the *okime* method. This very fact accounts for the peculiar beauty of Kodaiji *makie*. The *okime* method consists of drawing a rough sketch on paper with *yaki-urushi,* which is *e-urushi* (thickened raw lacquer and red ocher) made by putting it on a paper and scorching it with fire. By placing this rough sketch face down on the surface to be decorated, and rubbing lightly with either fingers or brush, the sketch is transferred. Today, almost all the rough sketches for *makie* are prepared in this way. However, in Kodaiji *makie,* there is evidence that this method was not employed.

Figure 1 shows the *asagao makie jikiro* (food container). Next to the upper right edge of the morning glory leaf in the center, there is a trace of a vermilion line. Similar vermilion lines are also found on the leaf to the left and on the upper side of the leaf done in *enashiji* in the left corner. It is difficult to see these lines in the picture, but they are easily observed on the actual work because they are conspicuous on the *kuronuri* (black) background. On this food container, it may be said that preliminary lines were made in vermilion and a rough sketch was drawn before the process of *makie* (sprinkling metallic powder on a wet lacquer surface to form a picture or design). In the final stage of the rough sketch, the total effect of the design was reconsidered and the preliminary lines were slightly revised. Usually, after the completion of *makie,* these vermilion lines are erased completely, but in this particular work this procedure was forgotten and hence the rough sketch is revealed.

Drawing a rough sketch in such an impromptu manner is a very risky proce-

Figure 1. Asagao makie jikiro *(food container).*

dure. Therefore, except in the work of extremely skillful *makie* artists, such a method may result in an "unrefined" product. In order to minimize this risk, the following techniques were employed in Kodaiji *makie*. First of all, comparatively large flowers and leaves, such as those of chrysanthemums, Chinese bellflowers, and the morning glory seen in the food container under discussion, as well as sprays of chrysanthemum and paulownia, were expressed by the repetition of several patterns. The designs were prevented from becoming monotonous by the diversities of detail and the arrangement of the pattern of branches and stems, fronds of Japanese pampas grass, and small bush-clover flowers, which were described with lines and points. The details that give life to the design were probably drawn by the most skilled *makie* artist at the final stage. This means that there might have been a studio of artisans who drew the same patterns repeatedly and a *makie* artist who took the leading role. This division of labor would have been the ideal method for the manufacture of Kodaiji *makie* furniture, which was then mass-produced.

Figure 2a. Kiku-kiri-mon makie oigata gusoku-bitsu *(box for storing armor) of Inuyamajo.*

Next, let me introduce an example of a work using a compass, which enabled the rough sketch to be drawn easily and precisely: the *kiku-kiri-mon makie oigata gusoku-bitsu* (a box for storing armor) of Inuyamajo (Fig. 2a). In this work there are two types of chrysanthemum design, single and double, and both large and small designs. For the paulownia design, a five- and seven-flowered pattern is used, and here, too, large and small sizes are found. A careful study of the details reveals a small hole at the center of the chrysanthemum design (Fig. 2b), and a similar small hole is found at the point where the paulownia leaf and flower stalk meet (Fig. 2c). These small holes were without doubt made by the point of a compass.

The following method demonstrates a most effective use of the compass in

Figure 2b. Detail showing the hole in the center of the chrysanthemum design made by the point of a compass.
Figure 2c. Detail showing the hole made by the point of a compass where the paulownia leaf and flower stalk meet.

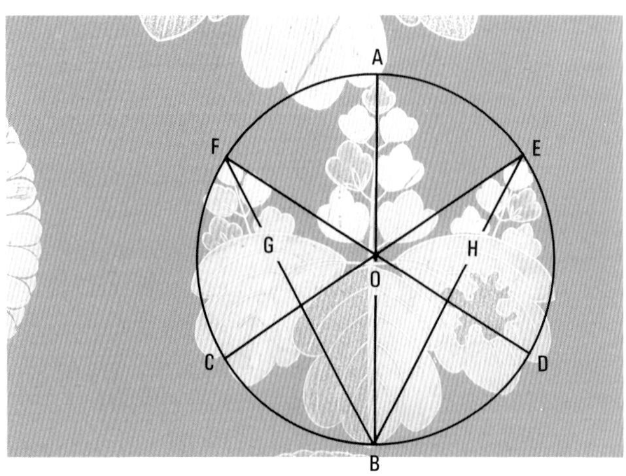

Figure 2d. Diagram showing the composition of the paulownia design using a compass.

drawing the paulownia design. First, a circle of a certain size is made with a compass and the figure drawn so that the center of the lower stalk and the central leaf meet at the center of the circle (Fig. 2d). Obviously in this figure the lengths of the central flower stalk and central leaf are the same: each is equal to the diameter of the circle. Then the circumference is divided with the compass into six sections. The points of contact, A through F, are placed on the circumference and leaves are added to the left and right of the central leaf, having their tips on points C and D respectively. Line OB becomes the mid-rib of the central leaf; the arc touching COD becomes the mid-rib of the leaf on each side. Next, points B and F and B and E are joined; the points of contact of this line with the left and right leaves G and H are marked; then the lines FG and EH become the stalks of flowers on the left and right. This forms a figure similar to the paulownia design. Though we may not be able to say that all the drawings of the chrysanthemum and paulownia designs on Kodaiji *makie* were made using a compass, it is interesting to note that the forms of chrysanthemum and paulownia designs of the Keicho period (1596–1614) resemble designs made in this way.

It is also a well-known fact that Kodaiji *makie* did not employ *togidashi makie* (design flush with surface) or *takamakie* (high relief), or use methods of intricate decoration such as *kanagai* (gold leaf), *kirigane* (gold flakes), or *raden* (shell inlay); only *hiramakie* (low relief) was used. Yet the use of *hiramakie* alone would tend to make a work monotonous in color. Since the design is formed by a repetition of set patterns, diversity in color is necessary to enliven the design. *Makie* furniture had to be decorative because it was surrounded by brilliantly colored gilded screens and used by people dressed beautifully in costumes embellished with Tsujigahana dyes and gold embroidery. In order to bring about this ornamental appearance, various methods were used, such as sprinkling *aokin* (an alloy of gold and silver) while applying the gold filings, or using *nerigaki* (a mixture of urushi and gold filings) or *enashiji* (gold and silver powders sprinkled to make a definite design).

Furthermore, many more effects were created by applying *enashiji* on red ocher coatings or on black urushi. The combined use of *gin-makie* and *gin-enashiji* (techniques using silver) is seen in the *makie* in the *shaden* (building complex) of the Tsukubusuma shrine and in the *sumi-ire* (charcoal basket) of Daigoji temple.

Now let us discuss the compositional characteristics. In Kodaiji *makie* there are several motifs, such as pine trees, bamboo, cherry trees, and scattered musical instruments. But the most popular are the autumn plants, especially those that are depicted realistically. These realistic depictions of autumn plants did not appear suddenly in Kodaiji *makie*. There were already some works made between the fifteenth century and the first half of the sixteenth century that exhibited this tendency. For example, the *kiku-tsuru-kame makie tebako* (cosmetic box; Fig. 3) depicts chrysanthemums blooming on a mound with flowing water, as well as a crane and tortoise design, which are traditional symbols of good fortune. The diagonal depiction of the chrysanthemum on the side is also very traditional. However the chrysanthemum is much larger than the crane; the actual relative size of the motifs is ignored.

Figure 3. Kiku-tsuru-kame makie tebako *(cosmetic box).*

In works between the late fifteenth century and the first half of the sixteenth century, there is an increasing number of designs lacking a sense of balance. For example, as Professor von Ragué (1967) notes, in the *ogura-yama makie suzuribako* (writing box) in the Suntory Museum, the bushes growing on a distant mountain peak are drawn much larger than the nearby trees, ignoring perspective. Similarly, on the back of the lid of the *kasugayama makie suzuribako* (writing box) in the Nezu Art Museum, the autumn plants blooming in the distance are depicted much larger than the hut in the foreground. Such concentration gives a strong impression of the autumnal season to the viewer.

The design of the *akikusa tsuru-kame makie bundai* (low table; Fig. 4) is probably modeled after the previously mentioned cosmetic box (Fig. 3). However, with

Figure 4. Akikusa tsuru-kame makie bundai *(low table).*

Figure 5. Bunko *(document box) from Osakajo.*
Figure 6. The autumn plant design typical of Kodaiji makie.
Figure 7. Susuki-kuzu makie choshi (sake *container) owned by Gonroku Matsuda.*

the addition of Chinese bellflowers, gentian, and Japanese pampas grass to the chrysanthemum design, the symbolism of good fortune is not as obvious. On this table there is also a crescent moon in the upper right corner of the picture. This at first reminds the viewer of the *baigetsu makie bundai* in the Tokyo National Museum and other *bundai* designs of the Muromachi period (1392–1568), yet the autumn plants on the left are too large and gorgeous compared with those on the right where there is more empty space. This destroys the lyrical and lingering impression that is created by the diagonal composition.

Designs like these, which seem to break with tradition, appear one after another from the Momoyama period. The *bunko* (document box; Fig. 5) in Osakajo, which presents an excessively large maple tree and autumn plants with male and female deer standing below, is one example. Compared with traditional designs that would have depicted the motifs of deer, maple tree, and autumn plants with equal balance in order to portray the seasonal atmosphere, in this work the artist's attention is completely on the maple tree and the autumn plants, and the deer are merely supplementary. As this example reveals, plants, especially autumn plants, which had been only one of the motifs in the scenic pictures, seasonal pictures, or symbolic designs of good fortune since the Heian and Kamakura periods, became the focus of interest in the Momoyama period. Hence, in Kodaiji *makie* furniture, supplementary motifs such as cranes, deer, crescent moons, and flowing water disappeared, and the so-called "autumn plant design" of Kodaiji *makie* was established (Fig. 6).

The *susuki-kuzu makie choshi* (*sake* container; Fig. 7) is a work that manifests the spirit of the Momoyama period. Its accurate depiction of autumn plants and the design of pampas grass and arrowroot spread over the comparatively small pitcher gives a forceful impression that this work is much larger than it actually is. Both pampas grass and arrowroot are wild plants growing at the foot of mountains in the wilderness; neither often attracts attention. These plants, which are described as almost sprouting from the surface of the pitcher, do not evoke the feeling of "pity" that the Japanese feel for frail autumn plants whose lives are terminated by the coming of winter. Instead people are affected by the purely artistic nature of the color and shape of these plants.

In addition to realistic depictions of autumn plants, in Kodaiji *makie* there are designs contrasting various contradictory elements. On the surface of the lid and the four sides of the main body of the *jubako* (stacking boxes) in Figure 8, pampas grass is depicted as fine lines, and drops of dew, chrysanthemums, and paulownia are scattered. The contrast between the delicately patterned lines of pampas grass and the distinctly solid shape of the chrysanthemum and paulownia design suggests the relationship of background and foreground. The design on the *sumiaka tebako* (cosmetic box) in Figure 9 contrasts the straight line of the cypress fence and the controlled curve of the paulownia arabesque design. There are also designs such as those of the *kobon* (tray; Fig. 10), in which fan-shaped designs are scattered over pampas grass, creating a sense of diversity. This design undoubtedly follows the tradition of *senmen chirashi* from the Kamakura period. However, what was important in the traditional *senmen chirashi* was the arrangement of various fans—fully open, half open, completely closed, and slightly open—scattered over the surface. In this *kobon,* large pampas grasses are used as background and elaborately decorated fan-shaped designs are scattered over them, thus creating a contrast between boldness and delicacy. This contrast is further emphasized because of the designs on the fan face, which are classical motifs of *sumiyoshi*—pine trees, arched bridge, *torii* (gateway), and shrine building—and *wakanoura* patterns—*katawaguruma* (waterwheel) in a reed bush and strips of paper and cranes. These traditional motifs introduced within the frame of a picture make the contrast all the more vivid.

Figure 8. Jubako *(stacking boxes) showing contradictory design elements.*
Figure 9. Sumiaka tebako *(cosmetic box) from the Suntory Museum.*

Among these designs of contrasts, the most effective is *katamigawari,* where the two halves of the object are patterned differently. This design was originally developed for costumes; however, in the Momoyama period, when gorgeousness and splendor were preferred, the design was popular even in *makie* and in pottery such as *oribe-yaki*. In "Renchu kyuki" it is written:

> Both *katakata-nui* and *katakata-haku* have no definite seasonal restrictions. There may be objects belonging to various seasons in one and the same design. It is not unnatural to have a classical picture design on one *katakata* (half) and something of seasonal import on the other *katakata.* (Ise n.d.)

Originally, in dyed fabrics, either contrary designs possessing seasonal atmosphere were drawn with *nui* on one side and *haku* on the other, or sometimes classical pictorial designs without any seasonal significance on one side were contrasted with designs of the four seasons on the other side.

Figure 10. Kobon *(tray) with a design of fans and pampas grass.*

Figure 11. Suzuribako *(writing box) from Shoshuraigoji in Sakamoto, Shiga prefecture.*

Often, in *makie*, different colors were used: *nashiji* (gold) on one side and *kuronuriji* (black) on the other. On the gold, sprays were scattered and on the black autumn plants were depicted. The *suzuribako* (writing box; Fig. 11) of Shoshuraigoji in Sakamoto, Shiga prefecture, somewhat altered the conventional form. On one side of the diagonally divided section is a screen design characterized by straight lines; on the other, the curved forms of paulownia and spider's web are contrasted. Moreover, within both sections, lines and solid shapes are contrasted.

Conclusion

As has been seen, the technique and the design of Kodaiji *makie* made mass-production possible, but at the same time exhibited splendid effects not inferior to individually produced pieces. When the origin of such designs and techniques is considered, both go far beyond the Momoyama period. Kodaiji *makie*, however, did not follow the main trend of tradition. Rather it employed indirect techniques and designs and, by skillfully combining them, created *makie* adaptable to the modern age. Since Kodaiji *makie* had to be mass-produced, it had to avoid using very intricate techniques like *takamakie* (high relief). However, the mere employment of very simple techniques such as those of *hiramakie* (low relief) and *maki-hanashi* (unpolished *makie*) would hardly compete with the gorgeousness of *takamakie* pieces that are decorated with *kanagai* (gold leaf) and *kirigane* (gold flakes). In order to compensate for this difference between Kodaiji *makie* and *takamakie*, special techniques were employed in sprinkling *makie-fun* (metal powders), *enashiji* (pictorial work) was diversified, and original and creative designs that were pleasing to the eye were added. The Momoyama period was a time when the development of these innovations was possible.

Bibliography

Ise, Sadamutsu

n.d. "Renchu kyuki." In *Gunsho-ruiju*. Buke.

von Ragué, Beatrix

1967 *Geschichte der Japanischen Lackkunst*. Berlin.

Yoshimura, Motoo

1971 *Kodaiji Makie*. Kyoto National Museum.

Chinese *Guri* Lacquers

George Kuwayama
Los Angeles County Museum of Art

Of all the techniques of embellishing lacquerware, the art of carved lacquer is quintessentially Chinese. Typically these works occur in two basic forms: pictorial lacquers carved with illustrative or narrative themes, and *qulun*, commonly known as *guri*, which are sculpted scroll patterns arranged geometrically. *Guri* are simple spirals compounded into a complex pattern of curving and counter-curving scrolls, or into concentric rows of repeated sword-pommel patterns. It is an enduring theme in Chinese art, arresting in its geometric order and understated beauty. Dating Chinese lacquers with carved *guri* patterns is a problem that this paper will try to address.

The earliest Chinese specimen of scroll patterns in lacquer reproduces the *lei-wen* spiral, that ubiquitous background design on Shang bronzes. In a fragmentary example excavated from a Shang tomb at Taixicun, Gaochang Xian, Hebei province, *lei-wen* (thunder pattern) and other traditional bronze designs were carved into the body of the wood and then coated with red and black lacquer (Hebei sheng bowuguan 1974). Other materials unearthed there date the site to the late Anyang period. The Zhou dynasty site of Shangcunling, Henan, attributed to about 700 B.C., also yielded carved wood painted with lacquer.

Molded lacquers have been made since the Han dynasty (206 B.C.–A.D. 220) and possibly as early as the Warring States period (475–221 B.C.; Garner 1979:65). The most notable example of molded lacquer occurs in the ogival tracery of the lacquer casket dated 1042 from the Huiguang pagoda, Zhejiang (Zhejiang Provincial Museum 1973).

True carved lacquers, with simple geometric scroll patterns cut into a lacquer ground, came to be made in the Tang period (A.D. 618–906). The numerous layers of lacquer, totaling several dozen or more, are often colored in different hues creating a succession of contrasting bands. The earliest lacquers with *guri* carvings are armor plates excavated at Fort Miran in East Turkestan by Sir Marc Aurel Stein in 1906 (Fig. 1; Whitfield 1985). These British Museum pieces, dated by style and archaeological context to the Tang dynasty, consist of seven alternating layers of brownish-red and black bands of lacquer applied to a leather core. Crescents, commas, and S-shaped designs are cut at a sloping angle into the lacquer layers. The many applications of lacquer on leather armor plates were protective in their thickness, and the desire for adornment probably led to simple relief carvings that ultimately produced *guri* lacquers.

Figure 1. Armor plates excavated at Fort Miran, East Turkestan: earliest examples of lacquer with guri *carving; Tang dynasty; British Museum.*

Aside from these Fort Miran pieces there are no other *guri* lacquers that can be attributed with confidence to the Tang period. However, several ancient Chinese textual references support their manufacture during this era, particularly the ninth century *Yinhualu* by Zhao Lin (quoted in Garner 1979:70–72,275–277).

Textual sources also provide a basis for assuming the extensive manufacture of *guri* lacquers during the Song dynasty (A.D. 960–1279). Cheng Dachang's *Yanfanlu* of 1181 (Garner 1979:70–71,277) and the *Gegu Yaolun* edition of 1388 by Cao Zhao (David 1971:144–145) refer to carved lacquers. All three of the aforementioned texts refer to *xipi*, a term with as many interpretations as there are writers. However, from its contextual usage a general definition may be ventured. *Xipi* refers to lacquer layers of different colors with carved designs through these layers. Prior to Song it may have referred to flat marbled lacquers, which the Japanese call *tsugaru*.

A sizable corpus of *guri* lacquers from the Song period can be assembled from extant pieces that can be stylistically related to materials from datable archaeological finds or to similar Song objects in other media.

The 1977–1978 excavation of a mortuary site at Wujin Xian, Jiangsu, yielded a handled mirror box with carved *guri* patterns (Fig. 2; Chen 1979). The contents of the tomb indicate a date in the second half of the Southern Song period. The mirror box has a black surface and the beveled carving proceeds through layers of black, yellow, and vermilion to a base of dark brown. The design is an agglomeration of sword-pommel scrolls arranged around a central rosette of four dotted half circles. The upper surface of the *guri* relief is smooth and rounded.

In the summer of 1982, the tomb of Yu Gongzhu and his wife, who died in 1226 and 1199 respectively, was discovered in Pengxian, Sichuan (CPAM Sichuan Province 1985). The pillaged tomb yielded a red carved lacquer-covered box with five sword-pommel scrolls and a pentagonal rosette on its cover (Fig. 3). Its sides were embellished with severed half *guri* scrolls arranged in a horizontal band. This datable *guri* work has sword-pommel scrolls with a central filling that does not descend between the two curvilinear spirals, thus identifying a *guri* form that can be firmly dated to Southern Song (1127–1279). The continuance of this *guri* form into the Yuan period (1280–1368) is indicated by a Jizhou ceramic vase in the de Santos collection (Medley 1974:Pl. 120).

Figure 2. Handled mirror box with carved guri *patterns excavated at Wujin Xian, Jingsu; Southern Song period.*
Figure 3. Cover of carved lacquer box excavated at Pengxian, Sichuan, showing sword-pommel scrolls and a pentagonal rosette; Southern Song period.

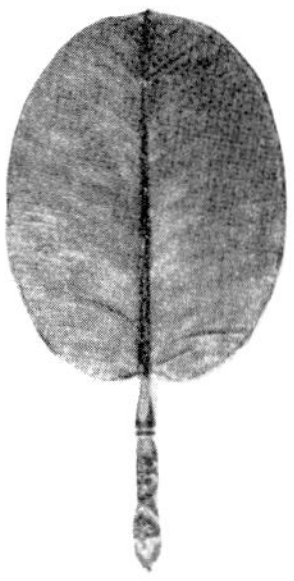

Figure 4. Pottery model excavated from a Liao kiln site in Beijing, showing a simple double-spiral pattern; Liao dynasty.
Figure 5. Fan handle carved with three elliptical groupings of facing heart-shaped scrolls; Jintan Xian, Jiangsu; Southern Song period.
Figure 6. Silver-lined cup with rows of curvilinear sword-pommel scrolls, discovered at Shazhou Xian, Jiangsu; Song dynasty.

An early type of simple double-spiral pattern was discovered on a pottery model excavated from a Liao (907–1125) kiln site in Beijing (Fig. 4; Lu 1978). It is without an upper arc and without the interior filling common to later sword-pommel scrolls.

Several lacquer fan handles adorned with carved *guri* scrolls have been excavated from a Southern Song tomb datable to around 1249. The interred remains and belongings of Zhou Yu were found at Jintan Xian, Jiangsu (Zhejiang City Museum 1977). One of the fan handles was carved with three elliptical groupings of facing heart-shaped scrolls (Fig. 5). The multiple layers of reddish-brown lacquer are divided into over ten segments by hairlines of black lacquer.

A silver-lined lacquer cup with rows of curvilinear sword-pommel scrolls cut through layers of black and red was discovered in a tomb at Shazhou Xian, Jiangsu (Fig. 6; Shazhou County Cultural Office 1981). This bamboo-bodied cup comes from a funerary site attributed by its excavators to the Song period.

One of the *lohan* paintings from a group of sixteen at the Seiryoji, Kyoto, illustrates an attendant holding a lacquer incense case with carved *guri* designs (Mainichi Newspapers 1976). This set of Japanese National Treasures is believed to have been brought from China by the Todaiji monk, Chonen. Japanese archival sources relate that Chonen returned to Japan in 987 with many Buddhist relics including the *Sixteen Lohans*. Chonen's paintings, however, are now believed to have been lost, and the existing Seiryoji *lohan*s are widely accepted as Southern Song Chinese Buddhist paintings.

A good number of Song silver objects with scroll designs provide intriguing analogies to Song *guri* lacquers. The recent archaeological excavation of Tomb 201 at Laoheshan, Hangzhou, Zhejiang, uncovered a silver box with rows of sword-pommel scrolls on its cover and body (Jiang 1957). The find can be dated by a bowl inscribed with the date 1162.

A silver bottle was among the finds from a Song hoard at Deyang, Sichuan (Shen 1961). This bottle was adorned with two facing S-shapes configured into a heart motif. This identical pattern occurs along the border of a box cover found in the tomb of Zhu Tan, who was buried in 1389 at Jiu Xian, Shandong, attesting to the continuation of Song styles into Yuan and early Ming (Shandong sheng bowuguan 1972).

Another Southern Song tomb at Huangyueling, Jiangpu Xian, Jiangsu, yielded silver vases and round boxes with undulating repoussé designs that enlarge the known vocabulary of *guri* patterns popular during the Song period (Nanjing shi bowuguan 1973). A round lead box bearing similar scroll designs was recovered from another Song tomb located in Shanghai (Huang 1962).

Useful stylistic parallels can also be drawn from Song ceramics. Sword-pommel and *rui* (scepter) scroll designs painted on Jizhou stoneware are identical to those on many carved *guri* lacquers (Medley 1974:133, Pl. 117B). Stonewares with spiral patterns chiseled on their bodies before firing reproduce the effect of carved lacquerware (Medley 1974:Pl. 120). The proximity of kilns and lacquer workshops in Zhejiang and Jiangxi may account for this similarity.

All of the aforementioned *guri* lacquers, metalwork, and ceramics are from datable finds establishing reference points for stylistic comparison. There are, in addition, several groups of *guri* lacquers that relate closely to these stylistic landmarks, enlarging the corpus of Song *guri* lacquers. They reflect the popularity of relief scroll designs on Southern Song metalwork, lacquer, and ceramics. All of these pieces are very light in weight with thin wooden bodies, a characteristic shared by Song ceramics of the Ding and Qingbai variety. There is an amazing variety in the design of the scrolls, which are handsomely composed into geometric patterns. The coats of lacquer are usually applied in bands of various hues, creating parallel linear contours in colors that are revealed by the wide, shallow angle of carving. The knife cuts are sharp, creating a ridge where the surface meets the edge of the bevel. There are twelve lacquer pieces in Japanese and American collections that share these characteristics:

1. Circular box and cover, height 11.8 cm, diameter 22.7 cm, carved brown lacquer in *guri* decoration, Tokugawa Art Museum, Nagoya (Tokugawa Art Museum 1984:No. 1).

2. Circular tray, height 4.6 cm, diameter 11.4 cm, carved brown lacquer in *guri* decoration, Tokugawa Art Museum, Nagoya (Tokugawa Art Museum 1984:No. 2).

3. Bottle, height 16.3 cm, diameter 11.4 cm, carved brown lacquer in *guri* decoration, private collection (Tokugawa Art Museum 1984:No. 3).

4. Circular incense container, height 2.8 cm, diameter 7.1 cm, carved brown lacquer in *guri* decoration, Tokugawa Art Museum, Nagoya (Tokugawa Art Museum 1984:No. 4).

5. Circular tray, height 3.6 cm, diameter 22.8 cm, carved brown lacquer in *guri* decoration, private collection (Tokugawa Art Museum 1984:No. 5).

6. Circular incense container, height 3.9 cm, diameter 7.8 cm, carved brown lacquer in *guri* decoration, Tokyo University of Arts (Tokugawa Art Museum 1984:No. 6).

7. Oblong tray, height 3.4 cm, length 27.7 cm, width 13.9 cm, carved brown lacquer in *guri* decoration, Tokugawa Art Museum, Nagoya (Tokugawa Art Museum 1984:No. 7).

8. Tray, height 3.5 cm, diameter 19.2 cm, lacquer carved with *guri* decoration, Tokyo National Museum (Tokyo National Museum 1977:No. 431).

9. Incense container, height 3 cm, diameter 7.2 cm, lacquer carved with *guri* decoration, private collection (Tokyo National Museum 1977:No. 437).

10. Tray for scrolls, 15.8 x 30.8 x 3.2 cm, lacquer carved with *guri* decoration, private collection (Tokyo National Museum 1977:No. 439).

11. Circular dish (Fig. 7), diameter 28 cm, brownish-black, red, and yellow lacquer carved with *guri* decoration, Sackler Collection, Tokyo National Museum (Lee 1972:100, Fig.39).

12. Circular dish, height 2.5 cm, diameter 19 cm, brownish-black, red, and yellow lacquer carved with *guri* decoration, Sackler Collection, Freer Gallery of Art (Lee 1972:102–103, Fig. 41).

The section on *xipi* in the "Ancient Lacquer" chapter of the *Gegu Yaolun* has this revealing sentence: "The pieces which used to be made at Fuzhou have a burnished yellow ground and circular patterns. They are known as Fuxi. They are solid but thin and are also difficult to come by" (David 1971:144).

A black lacquer tray with a foliate rim in the Itsuo Art Museum relates stylistically to the above group (Tokugawa Art Museum 1984:No. 8). Although the swirling curvilinear composition of the interconnecting spirals is vigorous in its movement, it lacks the ordered symmetry and geometric balance of the other twelve.

Figure 7. Circular dish carved with guri *decoration; Song dynasty; Tokyo National Museum.*

Two bowl-stands, one in the British Museum (Garner 1979:75, Fig. 27) and another in the Los Angeles County Museum of Art (Kuwayama 1982:59, No. 7), may be added to this Song corpus. The slower rhythmic movement of the design and the increased stylization suggest a date subsequent to the group of twelve.

There are numerous *guri* lacquers that can be assigned with confidence to the Yuan period. Textual references such as the 1388 edition of the *Gegu Yaolun*, archaeological excavations, and the large number of extant pieces attributable to Yuan enable us to assemble an impressive corpus.

Technical developments in the manufacture of lacquerware may account for this larger corpus of Yuan lacquer objects. The thin, light, Song *guri* pieces with sharply-cut, wide, beveled grooves evolve into the heavier Yuan type with sturdier bodies, more numerous layers of lacquer, and grooves more deeply cut at a steeper angle with their surfaces rounded and smoothed. The wooden core is usually assembled so that the grain at one side is perpendicular to the rest, giving structural strength and rigidity to the body. The cavetto core is a series of wooden strips laid on top of one another and planed to the desired shape, giving stability against warpage.

Guri patterns of the sword-pommel variety now cover the entire surface of a piece in ordered rows. The elaborate and elegant tracery of Song *guri* is gone and the creative energies of the Chinese lacquer-carver turn instead to a new interest in depicting narrative scenes or pictorial elements sculpted in relief. Two round trays, one in the Okayama Art Museum (Tokyo National Museum 1977:No. 468) and the other in the collection of the Los Angeles County Museum of Art (Kuwayama 1982:64–65, No. ll), typify this style of carved black lacquer with *guri* designs on the upper face as well as under the cavetto. Concentric rings of sword-pommel scrolls in rows of six and twelve surround a symmetrical central medallion of roof-shaped arcs and pointed spirals with bifurcating tips. Under the cavetto there is a typical Yuan pattern of a shorn row of sword-pommel scrolls. *Guri* scrolls also occur as subsidiary embellishments under the cavetto of platters carved with pictorial themes. These cavetto scrolls are a characteristic feature of many fourteenth century Chinese lacquers and appear as severed sword-pommel designs whose connecting arcs have been removed leaving two facing spirals. Identical in design to these pieces, but differing in its red color and in an eight-lobed outer rim, is another tray in a Japanese private collection (Tokyo National Museum 1977:No. 463).

There is another group of *guri* lacquers that may be attributed to the Yuan period. They have shorn sword-pommel scrolls under the cavetto and concentric rings of *guri* scrolls on the face. However, their central roundels differ in design from the previously cited group. Alternating pointed arrows and mushroom lobes surround a central circle, forming a pentagonal or hexagonal shape. The group is exemplified by a red lacquer tray in the Tokyo National Museum (Tokyo National Museum 1977:No. 464) and another in black lacquer in the Okayama Art Museum (Tokyo National Museum 1977:No. 467).

A fourteenth-century Buddhist painting of the *Eleventh Lohan* in the Fogg Art Museum illustrates a round lacquer box with *guri* scrolls.[1] Rows of severed sword-pommel designs embellish this box, resembling those under the cavetto of Yuan platters.

1. The *Eleventh Lohan*, fourteenth century, ink and color on silk, 126 x 62.2 cm, Fogg Art Museum, Cambridge, Mass., purchase 851:1920.

The archaeological excavation of a Yuan tomb at Wuxi, Jiangsu, unearthed a matched pair of round silver boxes with *guri* patterns (Wuxi shi bowuguan 1964). Simple sword-pommel scrolls are arranged concentrically in multiples of five around a pentagonal center.

In the absence of dated monuments and solely on the basis of stylistic projection, the Yuan group of *guri* lacquers can be extended to early Ming, up to the end of the fourteenth century. It is difficult to attribute *guri* works to the first half of the fifteenth century (Yongle, 1403–1424 and Xuande, 1426–1435), which is the apogee of pictorially carved lacquers. The under-cavetto *guri* designs on carved pictorial lacquers attributable to the fourteenth century can no longer be found.

Attributions to the later fifteenth century may be made on the basis of a *guri* tray (private collection, Tokyo) inscribed with the date of 1488. This tray continues the *guri* styles of the fourteenth century, but in slightly shallower relief and in somewhat wider proportions. Dated fifteenth-century lacquers with *guri* designs are extremely rare, thus establishing the uniqueness of this tray.

During late Ming, carved lacquer objects with *guri* scrolls revived in popularity. A tray (private collection, Tokyo) dated 1549 fortunately provides a stylistic monument. A large number of existing lacquerworks share stylistic characteristics with this dated tray. Its carving is shallow and flat, particularly in comparison with the deeper and heavier early Ming examples. Most of the late Ming *guri* pieces are black, in contrast to the red-on-black works of Yuan and early Ming. None of the late Ming *guri* carvings is polychromed, despite the enormous production during Jiajing (1522–1566) and Wanli (1573–1619) of multicolored lacquers bearing symbolic and narrative elements in carved relief.

Figure 8. Porcelain beaker with a lacquered surface carved with rows of sword-pommel scrolls and S-shaped guri *designs; the central medallion is the character "shou" meaning "long life;" Kangxi period; Palace Museum, Beijing.*
Figure 9. Square lidded box decorated with horizontal rows of guri *sword-pommel scrolls as well as the simpler severed form; Qianlong mark and period; Palace Museum, Beijing.*

Carved *guri* lacquers continued to be produced during Qing (1644–1912) although in smaller numbers than previously. A tall porcelain beaker with a lacquered surface in the Palace Museum, Beijing, is carved with rows of sword-pommel scrolls and S-shaped *guri* designs in addition to a central medallion with the emblematic character *shou* (long life; Fig. 8; Gugong bowuyuan 1985:No. 269). A Kangxi attribution for this vase may be ventured, since the shape is identical to contemporary porcelain beakers.

Several *guri* works in the Beijing Palace Collection, which bear Qianlong (1736–1795) marks, form a corpus of eighteenth century lacquer works. A typical example is a square lidded box decorated with horizontal rows of *guri* sword-pommel

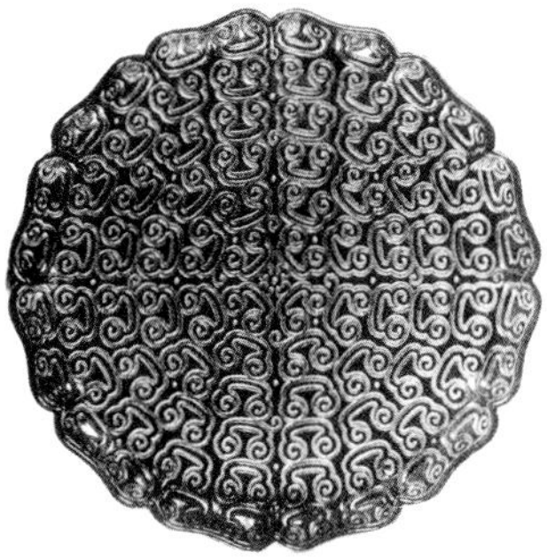

Figure 10. Platter with foliated rim and six concentric rows of guri *patterns clearly divided into quarters; Qianlong mark and period; Palace Museum, Beijing.*

Figure 11. A Yuan platter showing a harmonious design that contrasts with the Quianlong platter in Figure 10; Yuan dynasty; Yamato Bunkakan.

scrolls as well as the simpler severed form (Fig. 9; Gugong bowuyaun 1985:Nos. 378, 379). Although similar to pieces from earlier dynasties, the connecting arch lacks the curvilinear tension of previous eras and is flat, straight, and rigid. Another marked Qianlong lacquer is a large platter with foliated rim and six concentric rows of *guri* patterns clearly divided into quarters (Fig. 10; Tokyo National Museum 1978:No. 228). This division contrasts with similar works of the Yuan period that emphasize the harmonious integration of all components (Fig. 11; Tokyo National Museum 1977:No. 466). Yet another lacquer box is marked Qianlong. It is eight-lobed and teems with deeply carved *guri* scrolls in four concentric rows on its cover (Fig. 12; Gugong bowuyuan 1985:No. 381). The depth of carving and the tightly packed scrolls create the effect of unrestrained embellishment typical of Qianlong.

Figure 12. Quianlong lacquer eight-lobed box, with deeply carved guri *scrolls in four concentric rows on the cover; Qianlong mark and period; Palace Museum, Beijing.*

Like so many aspects of Chinese civilization, it was in Japan that lacquers with *guri* designs continued to be created, from the Muromachi (1392-1568) through the Edo period (1600–1868). Inspired by the aesthetics of Zen and the tea ceremony, *kamakurabori* lacquered works were produced during the Muromachi era. Japanese tradition maintains that true carved lacquers were made as early as the Muromachi period, citing the works of Tsuishu Yosei in the late fifteenth century. Monnyu, another notable craftsman contemporary with Yosei, also created carved lacquer works. Tsuishu lacquers were patronized by the Ashikaga shoguns and subsequently by the ruling Tokugawa clan. Legend is confirmed by documentation in the late eighteenth century when tsuishu carvers are officially appointed as carved lacquer makers to the shogun (Noma and Tani 1952). Alas, it is only the lacquerware made in the eighteenth and nineteenth centuries that can be dated with assurance. The dating of carved Japanese lacquers is a complex issue and a few works now bearing Chinese attribution could be Japanese.

Bibliography

Chen, Jing

1979 "Ji Jiangsu Wujin xin chutu de Nansong zhengui gigi" (Important Southern Song lacquers newly excavated at Wujin, Jiangsu). *Wenwu* 3:46–48.

CPAM Sichuan Province

1985 "Nan Song Yu Gongzhu fufu hezang mu" (The tomb of Yu Gongzhu and his wife of the Southern Song dynasty). *Kaogu Xuebao* 3:383–402, Pl. 22, No. 6.1.

David, Sir Percival

1971 *Chinese Connoisseurship*. New York: Praeger Publications.

Garner, Sir Harry

1979 *Chinese Lacquer*. London: Faber and Faber.

Gugong bowuyuan

1985 "Gugong bowuyuan zang diaoqi" (Carved lacquer in the collection of the Palace Museum). Beijing: *Wenwu Chuban She* No. 269.

Hebei sheng bowuguan, Hebei sheng wenguanzhu

1974 "Hebei Gaocheng Xian Taixicun Shangdai yidu 1973 nian de zhongyao" (The Shang dynasty site at Taixicun, Gaocheng county, Hebei province). *Wenwu* No. 8:42–49, Pl. 1.

Huang, Xuanpei

1962 "Shanghai Song mu" (Song tombs in Shanghai). *Kaogu* 8:418–419, Pl. 10, Fig. 4.

Jiang, Zuanchu

1957 "Dan Hangzhou Laoheshan Song mu chutu de qiqi" (A discussion of the lacquer vessels unearthed from the Song tombs near Laoheshan, Hangzhou). *Wenwu cankao ciliao* 7:30, Fig. 9.

Kuwayama, George

1982 *Far Eastern Lacquer*. Los Angeles: Los Angeles County Museum of Art.

Lee, Yu-kuan

1972 *Oriental Lacquer Art*. Tokyo: Weatherhill.

Lu, Qi

1978 "Beijing Mentouguo qu Longchuanwu faxian Liaodai ciyao" (Liao dynasty porcelain kiln remains found at Longchuanwu in Mentouguo, Beijing). *Wenwu* 5:Figs. 3, No. 10; 7.

Mainichi Newspapers

1976 "Kokuho" (National treasures). *Mainichi Shimbun* 3:123–124.

Medley, Margaret

1974 *Yuan Porcelain and Stoneware*. London: Pitman Publishing.

Nanjing shi bowuguan

1973 "Jiangpu Huangyueling Nan Song Zhang Tongzhi fugui mu" (The Southern Song tomb of Zhang Tongzhi and his consort in Huangyueling, Jiangpu). *Wenwu* 4:65, Figs. 15,17,18.

Noma, Seiroku, and Tani, Shinichi

1952 Nihon Bijutsu Techo, Tokyodo: 430 and 600.

Shandong sheng bowuguan

1972 "Fajue Ming Zhu Tan mu jiyao" (A report on the excavation of Ming Zhu Tan's tomb). *Wenwu* 5:36, Fig. 24.

Shazhou County Cultural Office

1981 "Jiangsu Shazhou chutu baoyin zhutai qiwan" (A lacquer bowl with a silver lining and bamboo body excavated from Shazhou, Jiangsu). *Wenwu* 8:39.

Shen, Zhongchang

1961 "Sichuan Deyang chutu de Songdai yinqi jianjian" (Brief report on the silver objects of the Song dynasty excavated at Deyang, Sichuan). *Wenwu* 11:48–52, Fig. 6.

Tokugawa Art Museum and Nezu Institute of Fine Arts

1984 *Carved Lacquer*, No. 1.

Tokyo National Museum

1977 *Toyo no shikkikogei*, No. 431.

1978 *Toyo no Shikki Kogei* (Oriental lacquer arts commemorative catalog), No. 228.

Whitfield, Roderick

1985 *The Art of Central Asia*. Vol. 3. Tokyo: Kodansha International.

Wuxi shi bowuguan

1964 "Jiangsu Wuxi shi Yuan mu zhong chutu yipi wenwu" (A hoard of cultural objects unearthed from a Yuan tomb near Wuxi, Jiangsu). *Wenwu* 12:Pl. 6, Fig. 6.

Zhejiang City Museum, Jintan Xian Cultural Bureau

1977 "Jiangsu Jintan Nan Song Zhou Yu mu fajue jianbao" (Brief report of the discovery of the Southern Song tomb of Zhou Yu at Jintan, Jiangsu). *Wenwu* 7:18–27.

Zhejiang Provincial Museum

1973 "Ruian Bei Song Huiguang ta chutu wenwu" (Northern Song cultural objects excavated from the Huiguang Pagoda at Ruian, Zhejiang). *Wenwu* 1:48–58, Pl. 8.

Inro Research: Some Proposed Future Steps

Beatrix von Ragué
Curator, Museum of Far Eastern Art, Berlin

The aesthetic beauty of *inro*—a small, compartmentalized container originally for seals, but also used for medicines and other small objects—delighted some Japanese collectors as early as the end of the eighteenth century and many more Western collectors from the late nineteenth century on. But in spite of some good old catalogs like the Tomkinson catalog of 1898, and the profound article by the late U.A. Casal published in 1941, the history of *inro* and their place in the many schools of Japanese lacquer art has come into focus only recently.

Due to the combined efforts of collectors and art historians in many countries, *inro* research has reached a new level during the last fifteen years. An astonishing number of books and catalogs on *inro* have been published in Japan, as well as in Western countries (see bibliography).

Hirokazu Arakawa, evaluating early Edo period publications, has demonstrated that according to written sources the use of *inro* was known during the Kan'ei era (1624–1644), and *inro* were in general use in the second half of the seventeenth century. Text and drawings in the *Hinagata itoshigusa* (1705) and *Bankin sangyo-bukuro* (1732) show that in the early eighteenth century many *inro* shapes were popular. *Soken kisho* (1781), published by Arakawa in 1982, has an annotated list of thirty-seven *inro* artists. In addition to origins, shapes, and artists, the new publications describe the many lacquer techniques used for *inro* decoration, most of which were well established in the early Edo period.

All the modern publications agree on the almost limitless number of subjects for *inro* decoration, ranging from a single animal or flower to complete landscapes; from the visible world to the world of ghosts; from geographic maps to motifs with literary associations; as well as abstract ornaments. As early as 1723, Koami Nagasuku, in an essay for his descendants, mentioned that "everything in the universe is depicted in *makie*" (Hutt 1984:109).

However, the authors of most recent *inro* publications conspicuously refrain from dating. It is generally agreed that ninty-nine percent of existing *inro* were made during the eighteenth and nineteenth centuries, but there are very few reliable guidelines for exact dating—new research is necessary.

It is known that Ritsuo, who died in 1747, liked to inlay other materials into his lacquer bases. We know that beautiful *irotogidashi,* a *togidashi* technique employing a colored design, was in vogue at the end of the eighteenth century and that ivory *inro* or *inro* whose lacquer decoration imitated other materials were made during the last decades of the Edo period. We also have some knowledge of the impact of *Ukiyo-e* woodblock prints on lacquer decoration. But to date an *inro* to some specific time in the eighteenth or nineteenth century is at best just an educated guess and far from reliable.

The reasons for this are well known: the abundance of *inro* techniques, shapes, and motifs is so overwhelming, and the quantity of *inro* scattered over the whole world so enormous, that up until now we have been discouraged from solid art historical research.

Some writers believe in a stylistic development that would provide a chance of *inro* dating. But that seems hardly plausible for two reasons. First, in many lacquer schools the style of the original master was perpetuated by his followers while other schools were simultaneously working in different styles. Second, as we know from Japanese painting, one artist frequently mastered several styles.

Just as different styles are hardly helpful for *inro* dating, a similar difficulty occurs with *inro* shapes (although with shapes, certain trends can be detected). External and internal cord-runners (*himotoshi*), for instance, both occur as early as 1705. And it is misleading to suggest that "the longer, more cylindrical shapes are generally to be found in earlier *inro* while a flatter, larger size was in greater abundance during the late Meiji era" (Jahss n.d.:186). In fact, flat *inro* shapes occur throughout the nineteenth century. Such generalized statements tend to confuse the reader. However, I would like to suggest in this paper a few possibilities for dating *inro*.

Dating

Some *inro* are datable because their decoration is based on dated woodblock prints or book illustrations. Lacquer collectors are not always aware of the fact that many prints, especially those related to the world of theater and actors, are dated or datable. Theater prints were normally made immediately after the theatrical event they depicted.

Julia Hutt (1977) was the first to publish such a connection between a woodblock and an *inro*. The print, made by Katsukawa Shunsho, represents the actor Ichikawa Danjuro IV performing the Shibaraku role (which was performed exclusively by Ichikawa artists). The *inro* is signed "Koma Kyuhaku *saku*." The print is datable between 1768 and 1770, i.e., the year when Danjuro IV actually performed the Shibaraku role and the year when he changed his name and was henceforth not permitted to wear costumes decorated, as on the print, with the big squares that are the family livery of the Ichikawa artists. Most experts attribute the print to 1768.

The decoration of the *inro* obviously imitates the print and cannot have been made before 1768. Since at that time *inro* were objects of high fashion, the *inro*, too, was probably made before 1770; certainly it would not have been compatible with the taste of a man of fashion to carry an *inro* depicting a famous actor in a role he was no longer performing.

This short period of 1768–1770 makes it possible to identify the lacquer master Koma Kyuhaku, who signed the *inro*. There were several Koma artists who used the name Kyuhaku, but this *inro* must have been made by Kyuhaku III, who was active during the second half of the eighteenth century.

In cases like this, a datable print used as a model for an *inro* decoration can serve both to date the *inro* quite reliably and to identify the *inro* master. However, when an *inro* decoration is based on a woodblock illustration in a book, we often get only a date *ante quem non*. For instance, the decoration of a Kajikawa *inro* from the Casal Collection in the Osaka Municipal Museum of Art (Doi 1984:27, No. 43) is based on both sides on a Tachibana Morikuni print in the book *Ehon shaho-bukuro*, published in 1720. The *inro* cannot therefore have been made before that year. But since the subject of the print is not an actual event like the theatrical performance of Danjuro IV but a story from the old *Heike-monogatari*, one cannot be sure whether the *inro* was made very soon after the book was published or at a much later time. But at least the year of publication of the book provides the date "not before 1720" for the *inro*.

Both examples show that a close comparison of *inro* decoration with datable prints or book illustrations will offer some help for *inro* dating. The same method is currently used quite successfully by Margaret Medley in her research on the origin of blue-and-white porcelain decorations in printed book illustrations.

Signatures

One of the great difficulties in *inro* dating is, paradoxically, presented by their signatures. Although hundreds of thousands of *inro* are signed, the signatures very often give artists' names that were carried on through many generations. How many artists from the Kajikawa family (to name just one family) have made *inro*? Is there a way to tell whether a specific *inro* signed "Kajikawa" is a seventeenth, eighteenth, or nineteenth century piece? Is there a way to distinguish between several artists who use the same name? And if so, what can we learn about their chronological sequence? Are there means to decide that a signed *inro* cannot have been made before or after a specific year? If we find answers to some of these questions, dating of signed *inro*, and possibly even unsigned ones, might become much easier.

The most informative *inro* signature known to me was published in 1925 by Edward Strange. In his *Catalogue of Japanese Lacquer* in the Victoria and Albert Museum in London, he mentions an *inro* with the following signature: "Kobayashi Yasuaki of Yanagawa, age 65, Bunka 8th, year of Sheep" (No. 1175). There is no question, if the inscription is genuine, that the *inro* is dated 1811, a date confirmed by its execution in *sumi-e togidashi* (a style of design, imitating an ink painting), a technique often used at that time. The *inro* decoration shows a Chinese man with performing monkeys and a peasant family. Close to the man, we find the name of the painter Yasunobu. There were several painters with that name; the closest would be a Kano painter who died in 1798, thirteen years before the *inro* was made.

Often *inro* decoration was based on designs by famous painters. In such cases we find inscriptions like "made by Toyo after Hakugyoku Hogen." The painter Hakugyoku, also known as Kano Sukenobu, was born in 1730 and lived contemporaneously with the lacquer master Toyo. Hakugyoku received the title of *hogen* (eye of the law), a Buddhist title awarded to artists and craftsmen, in 1762. No *inro* explicitly made after a design of Hakugyoku Hogen can therefore have been made before 1762. Nineteen years later, in 1781, Hakugyoku was given the higher title of *hoin* (seal of the law), the highest Buddhist title awarded to artists and craftsmen, and we know of other *inro* carrying the inscription "made by Toyo after Hakugyoku Hoin."

It is unlikely that a lacquer artist working during the lifetime of Hakugyoku would have been so impolite as to mention the lower *hogen* rank after Hakugyoku received the higher title *hoin*. Furthermore, a lacquer artist producing fashionable *inro* objects would probably have preferred to use the latest design of a still-active painter, not an older one. If that assumption is correct, we can be sure that eighteenth century *inro* with the inscription "Hakugyoku Hogen" were made between 1762 and 1781, a much more specific date than "second half of the eighteenth century".

There are other painters, too, who received titles, for instance the seventeenth century painter Kano Tanyu. His drawings were used and reused as models for lacquer decoration by many generations of lacquer artists after his death. As far as I know, inscriptions on such lacquer objects only say "made by...after Tanyu," without mentioning the *hogen* or *hoin* title. It would be worth investigating whether eighteenth and nineteenth century lacquer masters ever mentioned Tanyu's titles on their *inro*. Could it be that titles were used only during the painter's lifetime? To know that would be a great help in other cases, too.

Sometimes not the painter's title, but his *go*, or pseudonym, can be a help in dating. The Greenfield Collection now in the Metropolitan Museum, New York, includes an *inro* made and signed by Jokasai; at the side of the decoration the name of the painter Hanabusa Itcho is added (Pekarik 1980:No. 85). Hanabusa (1652-1724) used the *go* "Itcho" in combination with his name only after his return from exile in 1709 (Akiyama 1977:182), and therefore 1709 would be the earliest possible date for the *inro*.

There were several lacquer masters with the name Jokasai; the first one, active from about 1681 to 1704, had many pupils, all using his name. If the first Jokasai did not work after 1704, it must have been one of his followers who, after 1709, made the *inro* with the name of Hanabusa Itcho.

Knowing that there have been several *inro* masters using the name of Jokasai, and that drawings by Itcho, as by other painters, have been handed down through many generations of lacquer masters, is there a possibility of finding out more about the date of the *inro* and the identity of the lacquer master? I believe there is. One way to deal with signed *inro*, so far rather neglected, is the study of their seals and *kakihan* (cursive monogram).

Seals and kakihan

Kajikawa artists made lacquerware for the Edo shogunate from the seventeenth century on. Until now, it has seemed impossible to differentiate the work of one Kajikawa master from those of earlier or later generations, especially because Kajikawa *inro* are very consistent in technique and style and therefore very difficult to date. Most of them carry a gold lacquer signature "Kajikawa *saku*" followed by a red lacquer vase-shaped seal (Fig. 1), but there are several other types of Kajikawa signatures:

1. Kajikawa (without "*saku*") and red vase-shaped seal (Doi 1984:No. 14).

2. Kajikawa *saku* (without seal; Figs. 2 and 3).

3. Kajikawa *saku* and gold lacquer vase-shaped seal (Schneeberger 1984:No. 115).

4. Kajikawa *saku* with red vase-shaped seal, both enclosed by a cartouche (private collection, Berlin, No. 11).

5. Kajikawa with square red seal "Kajikawa" (Schneeberger 1984:No. 56).

In many of these different types we also find different writing styles (compare Figs. 1 and 3) and different vase shapes. While many of the vases are roundish (Fig. 1), others are rather slim or have more elongated handles (Fig. 4), some show a very different arrangement of the characters inside the vase (Fig. 5), and at least one red lacquer vase seal has three little loops at the bottom (private collection, Berlin,

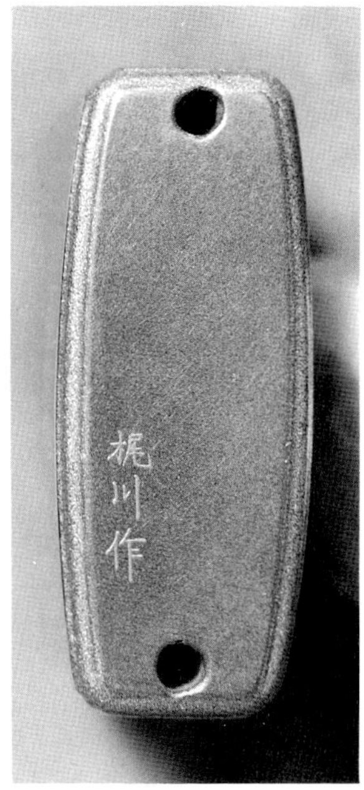

Figure 1. Signature Kajikawa saku *with seal.* Inro *Collection Baur, Geneva, No. 390.*
Figure 2. Signature Kajikawa saku *(Doi 1984: No. 45).*
Figure 3. Signature Kajikawa saku *(Doi 1984: No. 55).*

Figure 4. Signature Kajikawa saku *with seal (Doi 1984: No. 21).*
Figure 5. Signature Kajikawa saku *with seal (Doi 1984: No. 44).*

No. 35). To make things even more diversified, the nineteenth century master Kajikawa Hidetaka signed his full name, using a round seal (formerly in the collection of Kurt Herberts, Wuppertal, No. J 1817), while Kajikawa Bunryusai did not use a seal but a *kakihan* (Schneeberger 1984:No. 114).

We do not know how many Kajikawa artists made *inro*, but there must have been many whom we cannot so far separate and identify. Even when we have to deal with only three artists using the same name, for instance Koma Kansai, our present methods are not efficient enough to identify and date their work. Laurance P. Roberts in his *Dictionary of Japanese Artists* (n.d.) writes about Kansai:

> Kansai (?-1792).... Lacquerer. Born in Edo. Adopted into the Koma family; was the first to bear the name of Kansai: the artist who brought the work of the Koma school back to its original standard.... His son, Koma Kansai II (1766-1835), is said to have been the finest artist of the family.... In 1824 retired as a priest and Koma Kansai III succeeded to the name. It is, so far, quite impossible to say which one of these three artists made which of the many pieces, usually *inro*, signed Koma Kansai.

But a look at Kansai *inro* shows that there are at least three different types of signatures: (1) Kansai with *kakihan* (Fig. 6), (2) Koma Kansai with *kakihan* (Fig. 7), (3) Koma Kansai *saku* (Fig. 8).

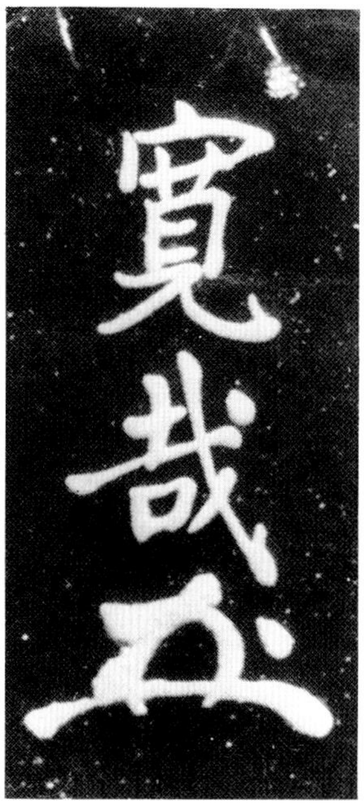

Figure 6. Signature Kansai with kakihan. Inro *Collection Baur, Geneva No. 176)*
Figure 7. Signature Koma Kansai with kakihan *(Doi 1984: No. 21).*
Figure 8. Signature Koma Kansai saku *(Doi 1984: No. 119).*

Could each of these signature types have belonged to one artist exclusively? Certainly it would be useful to arrange the existing Kansai and Kajikawa *inro* (and those from other schools as well) according to their types of signature, seal, and/or *kakihan*. By this means we should establish coherent groups. Other characteristic features such as technique and style could then be used to arrange these groups into a chronological order and, perhaps, connect each with one particular artist.

Conclusion

In 1964, I researched the lacquer master Toyo and his school (von Ragué 1964) using this method. At that time I found twenty-six different *kakihan* forms belonging to six major types (Fig. 9), all occurring with signatures from the Toyo school. The lacquer objects belonging to the different *kakihan* types (Fig. 9, A through F) quite obviously formed coherent groups, which by art-historical means could be arranged in chronological order. While general quality, style, techniques, sense of color, and other characteristics were not sufficient by themselves to give a closer date to the Toyo lacquer objects, all these features fit astonishingly well into the groups established by *kakihan* types. It became possible not only to recognize individual masters but even to identify a fake signature.

Several recent *inro* catalogs reproduce not only the *inro* but also the complete signature with seal or *kakihan*; evidence of the fact that a study of signatures, seals, and *kakihan* may open new and necessary avenues of *inro* research.

To sum up, I suggest further research on *inro* dating in three ways: by a close survey of dated prints and book illustrations used as models for *inro* decoration; by a more careful study of the dates provided by titles and artists' names, or *go*, mentioned on the *inro*; and by establishing groups of *inro* (and other lacquer objects) unified by the same type of signature, seal, or *kakihan*.

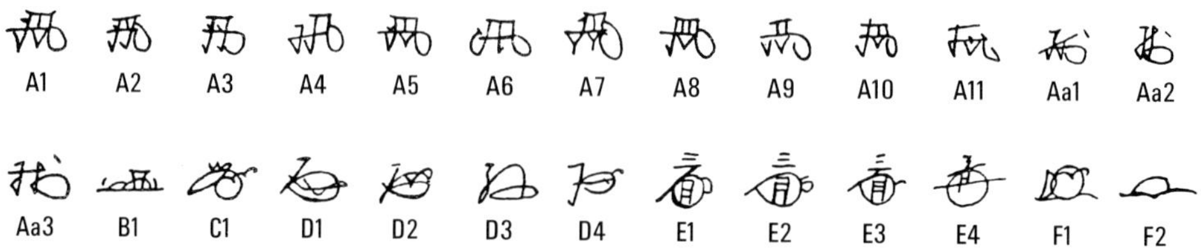

Figure 9. Kakihan *forms of types A–F from Toyo school objects.*

Bibliography

Akiyama, T.

1977 *Japanese Painting*. Geneva.

Arakawa, Hirokazu

1982 *Inro to netsuke*. Tokyo.

1984 "Inro to netsuke no hassei ni tai-suru ichi kosatsu." *Museum* 395. Tokyo National Museum.

Boyer, Martha

1970 *Catalogue of Japanese Lacquer*. Baltimore: The Walters Art Gallery.

Casal, U.A.

1941 "The inro." *Transactions and Proceedings of the Japan Society of London* XXXVII:1-53.

Champoud Collection

1984 *Inro Catalogue of the Champoud Collection in Bremblems, Switzerland*. Japan.

Doi, Kuniko

1984 *Soshingu. Osaka Shiritsu Bijutsukan zohin mokuroku* XIV. Osaka.

Hutt, Julia

1977 "Inro decoration: its individual character." In *Artistic Personality and Decorative Style in Japanese Art*. Edited by W. Watson. London.

1984 *Lacquer: An International History and Collector's Guide*. (Japanese section). Marlborough, Wiltshire: The Crowood Press.

Jahss, Melvin and Betty

n.d. *Inro and other Miniature Forms of Japanese Lacquer Art*. (c. 1970). London.

Pekarik, Andrew

1980 *Japanese Lacquer...from the Charles A. Greenfield Collection*. New York.

Roberts, Laurance P.

n.d. *Dictionary of Japanese Artists*.

Schneeberger, Pierre F.

1984 *The Baur Collection Geneva: Japanese Lacquer*. Geneva.

Strange, Edward

1925 *Catalogue of Japanese Lacquer*. Vol. II. London: Victoria and Albert Museum.

von Ragué, Beatrix

1964 "Materialien zu Iizuka Toyo, seinem Werk und seiner Schule." *Oriens Extremus* (Wiesbaden), Jg. 11 (the article appears in issue No. 2 and the reproduction of the Toyo *kakihan* list in the following issue.)

Wrangham, E.A.

1972 *Exhibition catalogue*. Oxford: Ashmolean Museum.

Sano Chokan, the Urushi Master, Studied through His Work

Akiro Haino
Kyoto National Museum

On 3 April 1925, the Kyoto National Museum opened its first lacquerware exhibition, *The Works of Chokan, the Urushi Master*. Although the exhibition lasted only five days, two months later a collection of illustrations of the major works exhibited was published by the Chokan Memorial Association. The foreword of this book introduces Chokan (1794–1856), born the second son of Nagahamaya Jihei, an urushi artisan of Kyoto. Chokan was a self-trained man who wandered Japan for eight years in order to learn the best works and techniques and who never accepted orders that did not meet his policy. Fame was not Chokan's goal; his works were collected and treasured for their high quality.

Along with this brief summary of the master's life, the book illustrates forty-nine items that were in the possession of major collectors of the Meiji and Taisho periods, including Hayashi Shinsuke, Nishimura Motokichi, Ohara Magozaburo, Nomura Tokushichi, and Imamura Teijiro. These forty-nine items can be divided into two groups according to their purposes. The first group consists of tea ceremony items, such as *cha-ire* (tea caddy), *natsume* (tea-powder container), *kogo* (incense container), *mizusashi* (water bowl), and *kashi-bon* (pastry tray). The second group consists of tableware, such as *wan* (soup and rice bowl), *jikiro* (food container), *jubako* (nest of lunch boxes), and *hai* (cup). When considered from the aspect of the designs employed in *makie* ("sprinkled picture") and *urushi-e* (true lacquer painting), the same forty-nine items can be roughly divided into two groups: those with traditionally Japanese designs taken from nature, plants, and flowers; and those that are modeled on Chinese designs.

In terms of design, the *jikiro* discussed in this paper to explain the personality and characteristics of Chokan's work, belongs to the latter group. This *jikiro* is not among the collection of illustrations published in 1925. However, according to the writing on the paper that covered the box containing the *jikiro*, it was given to the Kyoto National Museum on that occasion. Thus it can be said that the *jikiro* was among the works that commemorated the seventieth year after Chokan's death, and contributed to a reevaluation of the urushi master.

Figure 1. The jikiro.

The jikiro

The overall shape of the *jikiro* is round, with the edge of the container slightly curving outward and the side gently rounded (Fig. 1). It has a lid in the *otoshibuta* style (the lid dropping into the receptacle) and a black urushi ground coating. On the lid, a pair of phoenix and clouds are placed in gold, silver, and *aokin, hiramakie* and partly drawn in urushi colored with a mixture of red ocher and vermilion. At the center of the *kodai* (raised circular ridge) on the top of the lid, the characters for "made in the era of Kaei" (1848–1853) are placed in gold *makie* within a double circle (Fig. 2). Inside the lid, the elaborately drawn Chinese character for "wealth" is placed in gold *hiramakie* over a vermilion circle (Fig. 3). On the exterior of the receptacle are clouds and a pair of dragons. The receptacle has a raised base, on the inside of which the character for "noble" is placed in the same manner as the character on the lid (Fig. 4). On the outer surface of the base of the receptacle, there is a stamp reading "made by Chokan" in gold *hiramakie* inside a double circle (Fig. 5). From the base to the lower part of the side of the container, stylized flower petals are drawn next to each other. The *jikiro* is contained in a black urushi-coated box, at the bottom of which the following characters are found: "colored drawing"; "a sack of gold dust"; "lid attached"; "phoenix and dragon design in the Manreki style"; and "made by Chokan, the urushi master."

The *jikiro* is called "a sack of gold dust." This name derives from its round shape. The two characters signifying "noble" and "wealth" also suggest its name. As the writing on the bottom of the box indicates, the cloud, phoenix, and dragon designs follow the style of the Manreki (Wanli) period of the Ming dynasty (1368–1644). The stylized flower petals, seen together with the phoenix, dragon, and clouds, become symbols of the sacred mountains, typical of Chinese good-fortune patterns. Also, the inscription "made in the year of Kaei" imitates the wording "made in the Manreki period of the Ming dynasty," which was seen on all work made in the royal

Figure 2. The lid, showing the characters "made in the year of Kaei."
Figure 3. The character for "wealth" inside the lid.
Figure 4. The character for "noble" in the base of the receptacle.

Chinese studios of the Ming dynasty. Overall, then, the design follows the typical good-fortune pattern that is seen so often in Ming ceramics, textiles, and lacquerware.

Being a studious artisan, Chokan studied the basic techniques and patterns of Chinese lacquerware, especially the *zonsei* technique of the Manreki period. *Zonsei* is a technique of drawing patterns with colored urushi in vermilion, red ocher, green, and yellow, and then edging them in *chinkin* (incised gold decoration). Although Chokan learned from these Chinese traditions, he restructured the patterns and transformed them into "Japanized" Chinese design. In the original Chinese *zonsei,* the ground is completely covered with detailed patterns, but in Chokan's work the ground is left black. The colored *urushi-e* and *hiramakie* give the same effect as that of *zonsei.* But the stylized petals and the separation of the two characters in the lid and the receptacle suggest an attempt to create an impression of space through simplification. By disregarding the minute ground patterns and by using other original devices, Chokan achieved a truly Japanese artistic use of space.

Therefore, although this object was made under the influence of the chinoiserie popular at that time, it contains Japanese elements that make it an original piece of work. For example, unlike traditional Chinese dragons, which have five claws, Chokan's dragons have only four. Since Chokan was fully exposed to Chinese culture, and from the conspicuous way in which the dragons stretch out their claws, the difference in the number of claws seems intentional, not a result of ignorance.

Figure 5. The stamp "made by Chokan."

This *jikiro* was not an isolated piece of work, but was created as one of a pair or part of a series. The existence of a Chinese flower *urushi-e* and *makie jikiro* reinforces this assumption. It is made in the same technique and size, and with the same two characters, "noble" and "wealth." Moreover, the following are done in the same manner: the inscription "made in the era of Kaei"; a stamp on the bottom saying it was made by Chokan; and the description of the *jikiro*—"colored drawing"; "a sack of gold dust"; "lid attached"; "arabesque design in the Manreki style"; and "made by Chokan, the urushi master." From these facts, it seems obvious that, apart from the peony and arabesque design, the two works are identical and were clearly made during the same period.

Sano Chokan, the man

The life of Chokan can be traced mainly from books published in the Meiji and Taisho periods. From these records, the outline of Chokan's life emerges. Sano Chokan, often called Nagahamaya Jisuke, was the second son of Nagahamaya Jihei, born in Kansei 6 (1794) in Kyoto. He was a bright boy, who studied reading and writing from an early age and enjoyed making poems. On the death of his father, he succeeded to his father's name. Then he wandered Japan visiting many studios, masters, and collectors to see and learn from their best works. Among the places he visited were Negoro, Yoshino, and Edo (present-day Tokyo). In Edo he visited the shogunate's *makie* studio to try to study *murasaki-urushi* (a technique of making purple-colored urushi), but he was rejected. He found this insulting and later discovered the method on his own. In Bunsei 8 (1825) he returned to Kyoto and isolated himself in his studio. There, dressing in humble clothes and leaving his hair uncombed, he worked hard to create a new type of lacquerware. It was then that he changed his name to Chokan after Chokan, the great Korean master of Kourai and called himself a descendant of the Korean Chokan. Although he led a simple life, he enjoyed exchanging poems with his wife. All his creations were so rare and precious that collectors treasured them, but he never worked on orders that did not suit him, no matter how much money was offered. He died in Ansei 3 (2 March 1856).

Figure 6. A Portrait of Chokan by his son, Shosabura.

Since such an outline still gives only a vague image, I wish to add some more incidents to clarify some of the aspects of Chokan, the man. The first is Chokan's association with Oshio Heihachiro (1793–1837). In 1751, the divine armor of Genji was partially burned during a great earthquake. At that time, the way of the samurai had already withered and there was no one to mourn the disaster that had befallen the divine armor, which symbolized the righteous way of the samurai. Oshio, who was only a low-ranking bureaucrat, was concerned with the decline of the samurai spirit as well as with the social conditions in which the samurai class no longer considered itself a protector and leader of the suffering people. He felt that restoration of the urushi chest for the divine armor was equivalent to revitalizing the samurai spirit. Oshio donated his personal fortune and asked Chokan and his elder son to make the chest, which was completed in 1834. A long inscription on the bottom of the chest indicates that it was made by Chokan when he was forty years old. The association with Oshio suggests that Chokan, too, was in favor of the anti-establishment movement.

From a rough outline of his life, we might also conclude that he was not a sociable person. However, he was not completely cut off from social and cultural activities in Kyoto, for he was one of the important members of a salon led by Taiko (1770–1860), a Zen priest of Daitoku-ji. Daitoku-ji originally had strong ties with the Sen family (of the tea ceremony) and this made Taiko's salon a gathering place for people associated with the tea ceremony and sophisticated learning. The Zen influence that Chokan received could also be the reason for his life style, careless as he was of appearances. At the salon, Chokan found friends with whom he could not only enjoy poetry but also share opinions and knowledge on works of art.

One of Chokan's acquaintances at Taiko's salon was Eiraku (Nishimura) Hozen (1795–1854), a ceramicist who is remembered today for his profound talent for making Chinese-style pottery. The friendship of Chokan and Hozen was so close that Hozen adopted Shosaburo, the second son of Chokan.

Shosaburo, also known as Zenjiro or Kaizen, assisted the Nishimura family by making chinaware. Like his father, Shosaburo was not only artistically gifted but also knew how to enjoy the elegant scholastic pursuits of writing poetry and playing musical instruments. Shosaburo left a seated portrait of his father wearing a chrysanthemum-patterned kimono (Fig. 6). The eyes of the old man are wilful and leave a strong impression on anyone who sees the portrait. At the top there is a description of Chokan written by Taiko: "Chokan, the urushi master, a descendant of the Korean Chokan, died on March 2, Ansei 3, at the age of sixty-three." In this description Taiko also mentions Chokan's reputation.

Other tales that indicate Chokan's skill have been passed down to us. According to one, the lids of some soup bowls that Chokan made fitted so well that they could not be opened. It is said that upon hearing this, Chokan laughed, made a small hole in the bottom of each bowl to let air in, and opened them. Even after a whole night, the soup in the bowl was still warm. Another tale has it that a certain man, who did not know Chokan personally, criticized Chokan's work in front of him. Then Chokan put his soup bowl in boiling-hot water, stirred, and took it out. The ignorant man was embarrassed to see that the bowl remained unharmed in spite of such harsh treatment. Of course, there seems to be much exaggeration in these stories, as in many tall tales of famous men, but they help in assessing the reputation of Chokan's skill.

Chokan's personality, not only as an artist, but also as a virtuoso, is clearly reflected in the fact that he acknowledged in writing the value of the work of other artists. On a box of *mitsuda-e* bowls with plant and flower designs, Chokan wrote in Kaei 7 (1854) that they were made in the Keicho period (1596–1614). On another occasion he identified an ink box made by Koetsu in Tempo 12 (1841). Also, in Chokan's *Memoir* (private collection), he left an invoice and directions with diagrams for storing tableware he had made. From these accounts, we have a picture of a man who was devoted to his profession and studious in research.

Another good way of grasping a picture of the man in relation to his historical background is to compare Chokan to another urushi master, Tamakaji Zokoku (1807–1869). Zokoku was born thirteen years after Chokan in Takamatsu, Shikoku. Like Chokan, he was born in the house of a scabbard-painting master, learned techniques such as *tsuishu, tsuikoku, kouka ryokuyo, zonsei*, and *kinma*, which were introduced to Japan from China and other Southeast Asian countries, and left works that reflect these foreign styles. The work of the two artists, and the environments in which they were brought up, were similar but their lifestyles and their places of activity were quite different. Zokoku was under the patronage of the Lord of Takamatsu and his reputation was widely promoted under the protection of established society. Chokan, on the other hand, lived and died as a master without the protection or endorsement of the establishment.

Conclusion

I wish to conclude this paper by summing up the characteristics of Chokan. He was an urushi master who adapted foreign influences to create an original Japanese style and who worked in his own way without relying on any authority. The renowned masters of the Edo period can be separated into two groups: those under the protection of the Tokugawa shogunate and those living outside bureaucratic control. Chokan is a typical example of the latter group. The objective attitude he showed in his study of work by others and in his own work gives him the outlook of a modern artist. We have only just begun to research the master who was a forerunner of the modern period, and I believe that many works will be rediscovered to enrich our understanding of Chokan and his age.

Heidatsu and *Hyomon* in the Nara Period

Norimitsu Kimura
Shosoin Treasure House

The opinions expressed here are based on both the treasures and the writings about *heidatsu* and *hyomon*, the ancient urushi art techniques of Japan, which are stored at the Shosoin. I have the honor and privilege to be entrusted with the Shosoin's treasures and documents, and I believe it is my duty to write on the subject of urushi.

In *heidatsu* and *hyomon* silver and gold foil is cut into designs and adhered to the *urushi-shitaji* (ground) or *nakanurimen* (intermediate layers). After application of additional urushi, the topmost urushi layer is either peeled off the silver or gold foil or burnished to display the overall design. Finally, designs in fine lines are engraved in *kebori* technique.

Heidatsu and *hyomon* appear basically the same and today it is generally thought that *heidatsu* is the Chinese term and *hyomon* is Japanese. However, before this general conclusion was reached, there was much debate among scholars on this question. I believe that the debate should be reopened and all opinions should be given attention, respect, and reconsideration. In my personal opinion, although recent debate has touched upon method, it has been primarily semantic. I believe that it is not as important to discuss differences in terminology between *heidatsu* and *hyomon*, as it is to discuss them as they relate directly to urushi art techniques.

Treasures of the Shosoin

Both terms, *heidatsu* and *hyomon*, appear in the Todaiji *Kemmotsucho*, a list of many treasures dedicated to Todaiji Rushanabutsu (Daibutsu) by the Empress Dowager Komyo on 21 June 756 (Tempyo shoho 8) for the soul of her deceased husband, the Emperor Shomu. The following are included in this list of treasures:

1. Document in twenty volumes placed in a box decorated with silver *heidatsu*. The box was placed in a bag of Korai embroidery.

2. North storehouse 25. Four *gosu* (round boxes) decorated with silver *heidatsu* (Fig. 1 a–d). Each contains *go* (board game) pieces.

3. A harp decorated with silver *hyomon*; legs decorated with ivory. Inside the harp it is written that it was made by the Hyoi family; attached is a purple silk bag, lined in red silk.

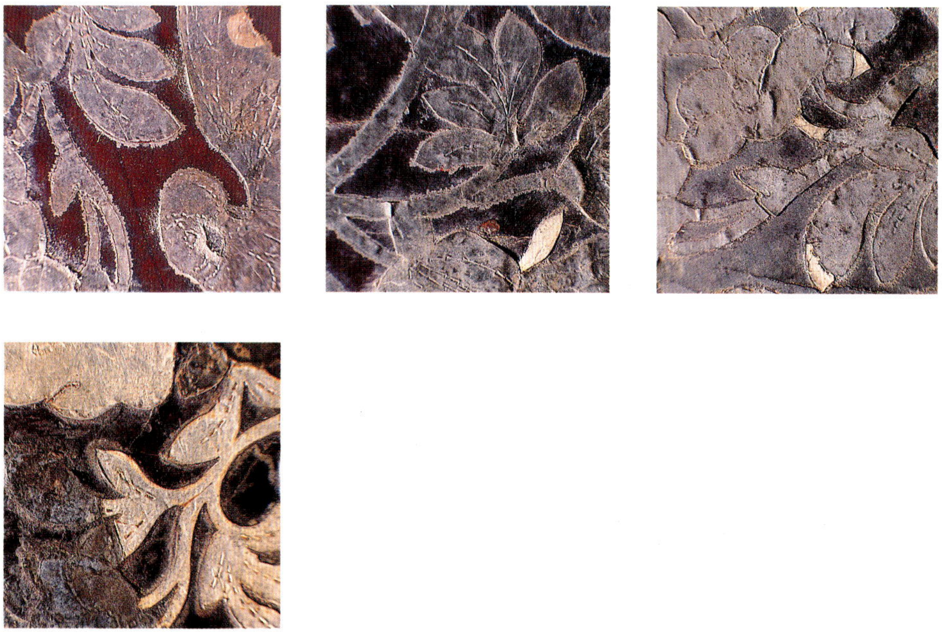

Figure 1 a–d. Gosu *(round boxes) decorated with silver* heidatsu *(N25): A C E I.*

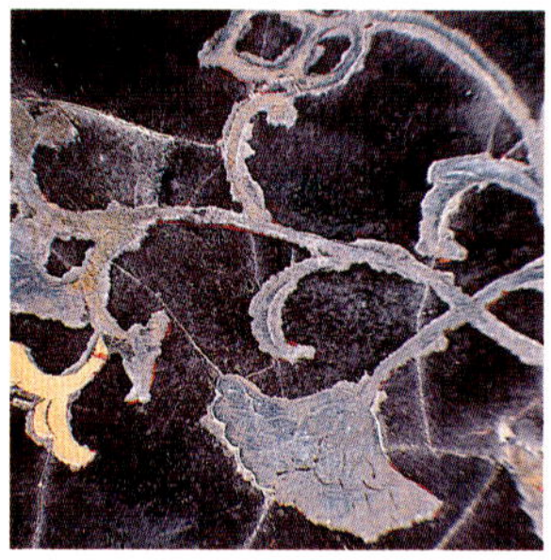

Figure 2. Fragments of a round mirror (N42-6): A C J.
Figure 3. An octagonal mirror (N42-12): A C E H.
Figure 4. Shikkohei *(ewer; N43): A C E H.*

Figure 5. A comb case decorated with silver heidatsu *(N154): A C E I.*
Figure 6. A harp decorated with gold and silver hyomon *(N26): B D F H.*

Legend to the captions

A. Nuritate (lustrous finish).
B. Roiro-shiage (mirrorlike finish).
C. Hagiokoshi (urushi over the design is peeled off).
D. Togidashi (urushi over the design is burnished).
E. Objects that have a rough urushi surface due to the peeling off of the urushi layer applied over the gold and silver foil.
F. Objects that have a distinct boundary between the gold and silver foil and the urushi, due to urushi left behind after kebori.
G. The gold and silver foil are lower than the urushi surface.
H. Gold and silver foil are level with the urushi surface.
I. Gold and silver foil are higher than the urushi surface.
J. Not clear, due to the peeling off of gold and silver foil.

4. A double-edged Chinese-style sword mounted with gold and silver, 89 cm in length. A dragon on a cloud is carved in silver. The grip is made of sharkskin. An ornament shaped like a mountain is in silver. The sheath has silver *hyomon* with ivory and dragon designs. A baldric in white leather is attached; the buckle is made of leather, too. There is a blackish-purple body-belt and a Korai green silk bag, lined in red.

5. A Chinese sword mounted with silver, decorated with the shape of a mountain in silver. Ivory and animal *hyomon* design.

6. A Chinese sword mounted with gold and silver, decorated with the shape of a mountain in silver gilt. Ivory and animal *hyomon* design.

7. A Chinese sword mounted with gold and silver and a Chinese sword mounted with silver gilt. Both are decorated with the shape of mountains in gold and silver. Dragon scale and ivory *hyomon* design.

8. A Korean-style sword mounted with silver, decorated with the shape of a mountain in silver. Ivory *hyomon* design.

9. North storehouse 42-6. Fragments of a round mirror, diameter 38 cm. Urushi back with gold and silver *heidatsu* (Fig. 2). Red silk belt. Urushi wooden box.

10. North storehouse 42-12. An octagonal mirror, diameter 29 cm. Urushi back with gold and silver *heidatsu* (Fig. 3). Red silk belt. Urushi leather box with red silk inner lining.

11. North storehouse 43. *Shikkohei* (ewer) decorated with flower and bird shapes in silver *heidatsu* (Fig. 4). Silver chain. A cap shaped like a bird's head. Capacity 2.7 liters.

Besides the above, on 26 July of the same year, an additional dedication was recorded in a different *Kemmotsucho*. It mentions "a comb-case decorated with silver *heidatsu*" (north storehouse 154; Fig. 5).

Thus, a total of sixteen treasures with *heidatsu* and *hyomon* decorations is reported. Out of these, there are nine examples of *heidatsu* and seven examples of *hyomon*. Eight of these items (Nos. 2,9,10,11) exist today. It is also clear from the list that they are *heidatsu*. However, the comb-case decorated with silver *heidatsu* is thought to have been deposited on a later occasion by a different route.

Next, in A.D. 814 (Konin 5), about sixty years after the above treasures were offered, a previously listed harp decorated with silver *hyomon*[1] was removed from the storehouse and was later replaced in A.D. 817 by the present harp decorated with gold and silver *hyomon* (Fig. 6; north storehouse 26).[2]

1. "...this harp is made by the Shiheii family..." is written on the harp and therefore it is thought to have been of the Zui period.
2. The harp is inscribed "...in the year of Otsuga Kishun..." and is therefore thought to have been made in Kaigen 23 (A.D. 735).

These are the treasures with *heidatsu* or *hyomon* recorded in either the *Kemmotsucho* or the *Dashi-irecho* (A.D. 817). However, there are other groups of treasures in the Shosoin that have come from other sources. In A.D. 950 (Tenryaku 4) the Kensakuin storehouse that belonged to Todaiji was destroyed and the many Buddhist altar fittings and utensils were moved to the south storehouse of the Shosoin. Later, some of these were again removed to the central storehouse. The quality and variety of these treasures are much greater than the items in the *Kemmotsucho*. Most of them indicate the year in which they were made and are datable to the Nara period. Those without any indication of date closely resemble the previously listed treasures in both quality and shape, suggesting that they are also of the Nara period.

Among these items are treasures which utilized techniques similar to *heidatsu* and *hyomon*. A list of these items was not drawn up when they were moved, however. Someone in the Meiji period studied them with reference to the *Kemmotsucho* and gave them official names. I do not know how the terms *heidatsu* and *hyomon*, which appear in the *Kemmotsucho* were interpreted but it is thought that the objects were identified by referring them to the items already recorded in the *Kemmotsucho*. With the exception of two or three of these items, all of them are *heidatsu*. The first two

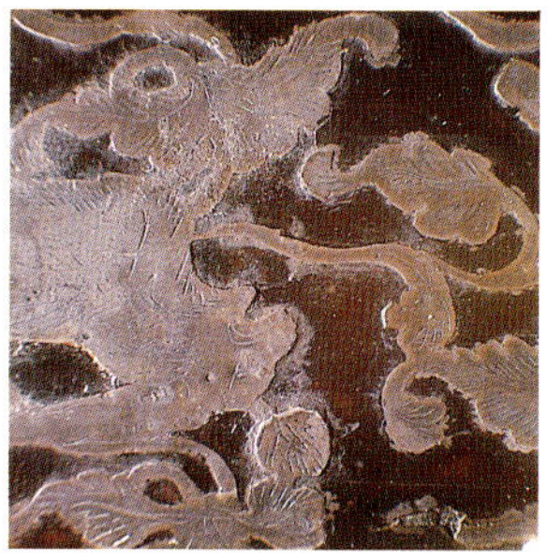

Figure 7 a,b. Chinese swords mounted with gold and silver (C8-2,3): B? D? H.
Figure 8. A sword mounted with gold and silver (C8-4):B? D? I.

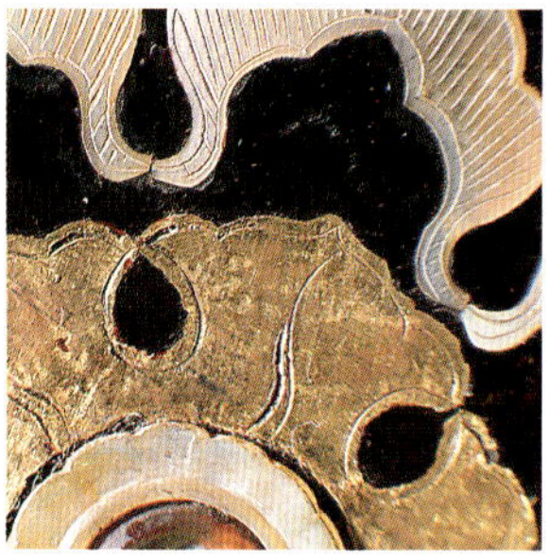

Figure 9. A mother-of-pearl inlaid box (C88): A C E H.
Figure 10. Kawabako *(leather box) decorated with gold and silver* heidatsu *(C138): A C E I.*
Figure 11. An inner tray decorated with silver heidatsu *(C164): A C E H.*

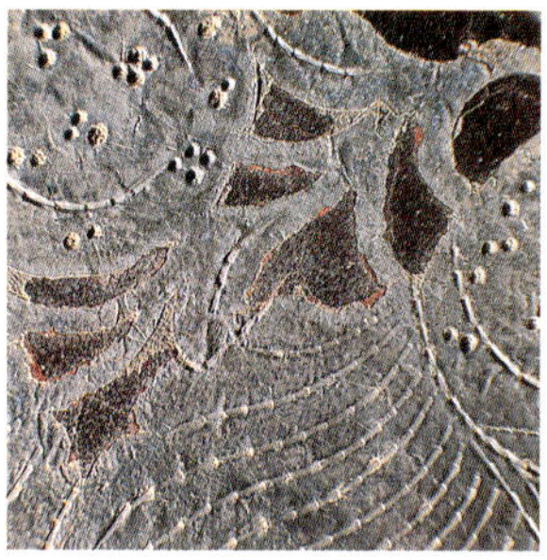

Figure 12. A box decorated with silver heidatsu *(S70-5): A C E H.*
Figure 13. An octagonal mirror case decorated with silver heidatsu *(S71): A C E G.*

Figure 14 a,b. A mirror case decorated with silver heidatsu *(S71): A C E G.*
Figure 15. Heidatsu tsubo *(pipe-holder for* wu*; S108): A C E G.*

Figure 16. Heidatsu tsubo *(pipe-holder for* wu*; S108): A C E G I.*
Figure 17. A sho *(musical instrument; S109): A C E G I.*
Figure 18. A phoenix head decorated with heidatsu *(S174-2): A J.*

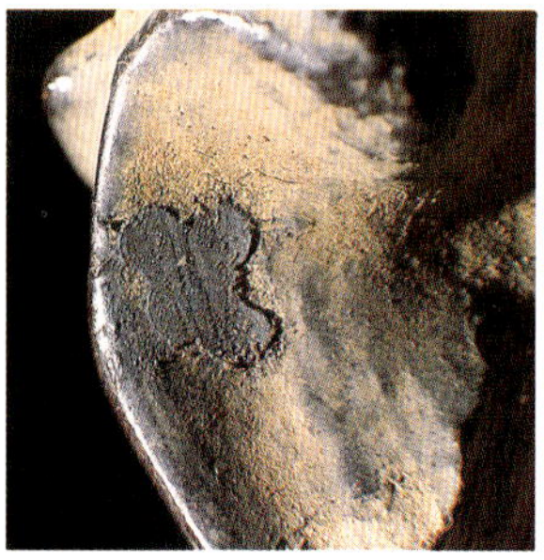

Figure 19. A dragon-shaped ink-box with silver heidatsu *(S174-3): A I J.*

Chinese-style swords mounted with silver gilt were reported in the special survey of swords (made between 1963 and 1966) to be *heidatsu*. The mother-of-pearl inlaid box and the phoenix head (originally called turtle head) decorated with *heidatsu* were also reported in a special survey of urushi art (made between 1968 and 1973) to be *heidatsu* or a combination of *heidatsu* and *raden*. However, neither survey was a study of *heidatsu* and *hyomon* alone.

1. Central storehouse 8-2,3. Two Chinese swords mounted with gold and silver (Fig. 7 a,b).

2. Central storehouse 8-4. A sword mounted with gold and silver. Gold and silver *heidatsu* in ivory and animal design (Fig. 8).

3. Central storehouse 88. A mother-of-pearl inlaid box (Fig. 9).

4. Central storehouse 138. Two *kawabako* (leather boxes) decorated with gold and silver *heidatsu* (Fig. 10).

5. Central storehouse 164. An inner tray decorated with silver *heidatsu* (Fig. 11).

6. South storehouse 70-5. A box decorated with silver *heidatsu* (Fig. 12).

7. South storehouse 71. An octagonal mirror case decorated with silver *heidatsu* (Fig. 13).

8. South storehouse 71. A mirror case decorated with silver *heidatsu* (Fig. 14 a,b).

9. South storehouse 108. Two *heidatsu tsubo* (pipe holders for *wu*) (Figs. 15,16).

10. South storehouse 109. A *sho* (musical instrument; Fig. 17).

11. South storehouse 174-2. A phoenix head (originally described as a turtle head) decorated with *heidatsu* (Fig. 18).

12. South storehouse 174-3. A dragon-shaped ink-box with phoenix head decorated with silver *heidatsu* (Fig. 19).

These fifteen treasures are in *heidatsu* and *hyomon*. When the previously listed nine items are added, there is a total of twenty-four items. They still exist at the Shosoin.

A personal perspective on the history of research

The following is a list of various opinions about *heidatsu* and *hyomon*, mainly about their terminology and technique, in order of publication.

In Mayori Kurokawa's *Nihon Shikki Shurui* (1901), *hyomon* is thought to be a simplified form of *heidatsumon* and, therefore, *heidatsu* and *hyomon* do not vary as techniques but are part of the same *togidashi* technique in which urushi is applied over a design and burnished with charcoal. However, I cannot completely agree with this explanation because among the treasures there are several items using the peeling off technique instead, and there are no references to *heidatsumon* being the origin of *hyomon*.

In Zentaro Ono's *Guide to the Shosoin* (1920), *heidatsu* is the art of raising the gold foil above the urushi surface so that it can be felt by the hand, while *hyomon*, like *raden*, has the design either on the same level as the urushi surface or a little below it. In both, urushi is applied over the design and later burnished, but they differ in appearance. However, some of the *heidatsu* pieces mentioned in the *Kemmotsucho* have the gold foil protruding above the urushi surface while in others it is lower than the urushi. Moreover, there is one obvious example of a design that is level with the urushi surface, thus bringing this opinion into question.

Shoka Tsujimura's *Tokyo Bijutsu Gakko Koyukai Geppo* (1927) mentions four points: (1) *kuro-urushi nuritate,* (2) *kuro-urushi nuri-togidashi,* (3) smoothness and unevenness of *kirigane* (gold flakes used in *makie*), (4) the existence of urushi in *kebori* (hairline engraving). Moreover, it introduces the concept that *heidatsu* and *hyomon* are different. When *heidatsu* is used, the urushi painted over the design is immediately peeled off. *Hyomon* uses *togidashi* and the design is always level with the urushi surface. However, as to the existence of urushi in *kebori,* although Tsujimura only mentions urushi remaining in *heidatsu* and *hyomon,* there are very few examples of *kebori* that have urushi. The condition of the urushi over the gold and silver foil in those objects that are thought to be *heidatsu* does not appear to have anything to do with *kebori;* it appears to have been simply peeled off. Meanwhile, in the harp, which alone is entitled to be called *hyomon,* the urushi remains only in the *kebori* and the gold and silver foil is burnished to the same height (of course, some parts have fallen off).

Shisui Rokkaku in his *Shikko-shi* (1928) and in his *Tokyo Shikko-shi* (1933) relates that although it may be convenient to think of *hyomon* as a simplified form of *heidatsumon,* if there is a difference it lies in the fact that in *heidatsu* either a thin layer of urushi is applied over the gold and silver foil and later removed or urushi application is avoided on the design, thus allowing the design to protrude above the urushi surface. Rokkaku's opinion is therefore similar to Tsujimura's. However, I do not completely agree with him, since there has been no report of applying urushi avoiding the gold and silver foil parts. Moreover, one cannot say that all designs with peeled off urushi are in relief.

Tomio Yoshino in his *Studies of the Shosoin* (1929) says that *heidatsu* consists in lightly burnishing and peeling the urushi off thin gold and silver foil designs. Therefore there are two types of gold and silver foil design, one being the relief kind and the other being the recessed kind. *Hyomon* uses a thick *kanagai* (gold or silver leaf) and urushi is applied several times until it becomes a beautiful piece of art when burnished. The separate categorization of *heidatsu* and *hyomon* is a more detailed observation than before. However, I feel that there is still room for more research.

In Tosen Hirose's *Studies of the Shosoin* (1929) *heidatsu* and *hyomon* are thought to be completely different techniques. *Heidatsu* uses the burnishing technique that created exquisite, elaborate items in the Tang (Nara) period. *Hyomon,* which flourished after the Heian period, is the simplified technical form, so to speak, of *heidatsu* design, but began in the Tang period. Thus it was also seen during the Nara period. Hirose also says that all the surviving items are *heidatsu,* which uses the burnishing technique. Thus, his interpretation is quite contrary to those previously held, for he states that *heidatsu* has an even surface and *hyomon* has a rough surface. I do not have any objections to calling *haritsuke-hyomon* simply *hyomon,* a simplified technique found in later generations. However, I cannot agree with the statement that all items of the Nara period were *heidatsu* using the burnishing technique, because it simply goes against the facts.

It was Tomio Yoshino's *Nihon Shikko-shi* (1934) that suggested for the first time that both techniques were practically the same but that *heidatsu* is the Chinese name and *hyomon* the Japanese name. This idea is based on the fact that there is no use of the word *hyomon* in Chinese documents but in Japan during the Nara period there are at least six examples (nine items) of *hyomon* plus numerous examples in later ages. He concludes that this is similar to calling the Chinese *"kyocho," "soku"* in Japanese and that this interpretation is the solution to the long argument over *heidatsu* and *hyomon.* This interpretation has since been widely acknowledged in the world of urushi art history and it is the most accepted view up to this day. However, I believe that *hyomon* after the Heian period (794–1185) refers to a different technique.

Shosoin Kyushitsu-hin Chosa Hokoku (1959) by Yuzuru Okada, Gonroku Matsuda et al. and *Shosoin-no Shikko* (1975) by Okada, Matsuda, Arakawa et al. support Yoshino's interpretation but also stress the fact that these items fall in the end either into the "peeled-off" group or into the "burnished" group. They maintain that there is only one example of a burnished object, the *hyomon* harp, and hint that Shoka Tsujimura's (1927) and Tomio Yoshino's interpretations (1929) are quite reasonable. However, this single example of a burnished object, the *hyomon* harp, is not reported in the *Kemmotsucho*. All the objects listed there are examples of *heidatsu*; not one example of *hyomon* is found. Thus they avoid separating *heidatsu* and *hyomon* in terms of technique after realizing the difficulty of matching the actual objects with the documents.

From this, it can be said that there is one interpretation which says that *heidatsu* and *hyomon* are different techniques and another interpretation that says that they are the same. Today, there is so much support for the idea that *heidatsu* is the Chinese name and *hyomon* is the Japanese name, that this is becoming accepted as fact. Yet during the compilation of the *Kemmotsucho* in the Nara period, there existed an explanation that the Chinese-style swords imported from China had *hyomon* designs. I cannot erase this from my mind, nor can I consider the peeling-off technique and the burnishing technique to be the same when obviously they are completely different. Therefore, on this point I support Shoka Tsujimura's view (1927) and the *Shosoin Kyushitsu-hin Chosa Hokoku* (1959), which suggest that *heidatsu* is the peeling-off technique and *hyomon* is the burnishing technique.

Conclusion

Looking again at the Todaiji *Kemmotsucho*, the items are listed in it as either *heidatsu* or *hyomon*. I feel there was a reason for this. There must have been several people engaged in compiling this list, which covers in great detail the quality, shape, and measurements of more than six hundred treasures. It is almost impossible to think that only one person did all this in such a short time. Most probably, different people dealt with different kinds of objects and, when the records were combined, different names had been used to describe the same technique. It can be seen from the way the *Kemmotsucho* is written that it was probably drafted by one learned man from many notes handed on to him by several people, in order to unify the mode and content of the list. The amazing orderliness of the text and that person's familiarity with the field and his extremely deep knowledge of every kind of art are evident.

Next, we cannot simply say that what was then called in China *heidatsu* was known as *hyomon* in Japan, as if we were calling the Chinese "*kyocho*," "*soku*" in Japanese. There surely must have been an accepted distinction in technique to differentiate the two. As in later times, *togidashi makie* was called *hiramakie* and *choshitsu* was called *kamakurabori*. Although *heidatsu* and *hyomon* look alike at first glance, the techniques involved in their creation differ immensely. I think the person who compiled or drafted the list must have realized the differences involved and given them their names. That person did not want to differentiate objects according to whether they were originals or copies or whether they were made in China or Japan. He wanted to differentiate between the technique in which gold and silver foil designs are cut out and placed on urushi, urushi is once again applied and later peeled off, and the second technique, in which the design is revealed by burnishing the urushi instead of peeling it off.

As already mentioned, in the last stages of *kyushitsu* (coating) it is necessary to burnish the surface with repeated applications of *dozuri* or *suri-urushi* because there are minute lumps or a frosted surface resulting from the *sumitogi*. This is called *roiro-shiage* and should be considered completely different from *nuri-hanashi*. In other words, *roiro* takes more time and effort. It can be clearly seen from other records that the art of *roiro-shiage* already flourished in the Nara period.

In this paper, I have studied the *heidatsu* and *hyomon* reported in the *Kemmotsucho* and have seen that among the surviving objects most are called *heiatsu* and only one is described as *hyomon*, a harp that was recorded sixty years after the rest. To say that this harp is everything there is to know about *hyomon* in the Nara period is extremely dangerous and we must be very careful when speaking about it. However, since there is a clear difference between *heidatsu* and *hyomon*, my interpretation given above should be considered.

Acknowledgments

I should like to thank the General Director of the Shosoin, Mr. Yoshihiko Hashimoto, for his sympathetic understanding; Mr. Hiromu Abe, Chief of Conservation Section in the Shosoin; Mr. Daitsu Kitamura, Professor Isao Fuyuki, Mr. Kenichi Kitamura, and others for their advice on urushi art techniques; and Professor Masaki Nakano and Ms. Toyoko Takahashi for helping with the collection of data.

Bibliography

Hirose, Tosen

1929 *Studies of the Shosoin.*

Kurokawa, Mayori

1901 *Nihon Shikki Shurui.*

Okada, Yuzuru, Gonroku Matsuda et al.

1959 *Shosoin Kyushitsu-hin Chosa Hokoku.*

Okada, Yuzuru, Gonroku Matsuda, Hirokazu Arakawa et al.

1975 *Shosoin-no Shikko.*

Ono, Zentaro

1920 *Guide to the Shosoin.*

Rokkaku, Shisui

1928 *Shikko-shi, Kokogaku Koza.* Vol. 3.

1933 *Tokyo Shikko-shi.*

Shinryu, Seharpe

1979 *Annual Report of Office of the Shosoin Treasure House.* Vol. 1.

Tsujimura, Shoka

1927 *Tokyo Bijutsu Gakko Koyukai Geppo.* Vol. 27, No. 2.

Yoshino, Tomio

1929 *Studies of the Shosoin.*

1934 *Nihon Shikko-shi.*

Origins of the Use of Urushi in Japan and Its Development

Yoshiyuki Kuraku
Archaeologist

The use of urushi in the prehistoric period

It was not until the 1930s, when urushi objects were unearthed from historic sites, that it became clear that urushi was used in Japan from the prehistoric era. This realization first came with the discovery of urushi bows, goblets, combs, and sword-shaped wooden items of the final Jomon period (c. 1000–400 B.C.), during the excavation of Korekawa, Hachinohe, Aomori prefecture.

Since then, various urushi objects have been found on many sites, mainly in eastern Japan. Recent national land development, even on low-lying marshy ground, has yielded numerous urushi objects from over five hundred sites in east and west Japan that date to the prehistoric era (the Jomon, Yayoi, and Kofun periods). Although the lacquerware objects were buried for a long time, they still maintain their original shapes and colors. The inner wood may have rotted away, but the urushi itself is extremely strong and free from erosion.

The oldest urushi found in Japan to date is from the Torihama shell mound in Obama-shi, Fukui prefecture. It includes arrows with red urushi coating, parts of shallow wooden bowls with red and black urushi coating, and a comb with red urushi. Because these objects were found with pottery of the early Jomon period (c. 4000–3000 B.C.), they are also dated to this period. Furthermore, in the same shell mound, other objects from the middle of the early Jomon period were also found such as earthenware with red urushi partially applied on the surface, with black urushi arcs drawn on red urushi ground, and with pictorial designs formed by narrow lines of red urushi on black urushi ground. These objects prove that during this period there existed quite advanced techniques of mixing pigments and of applying different urushi colors. Such examples from the early Jomon period were also excavated from the site of Ondashi, Yamagata prefecture, in the summer of 1985, proving that urushi was produced in various regions (Fig. 1). Moreover, the fact that the urushi technique in this period was highly evolved means that the first use of urushi must have been much earlier.

Figure 1. Pottery decorated with urushi, early Jomon, Ondashi site, Yamagata prefecture.

So far, urushi objects from the middle Jomon period (c. 3000–2000 B.C.) have been found only in eastern Japan. Besides earthenware with urushi coating, there are bows and containers such as pots, bowls, spoons, and trays made of wood with red and black urushi. Though not many urushi remains have yet been discovered, more are expected to be found in the future.

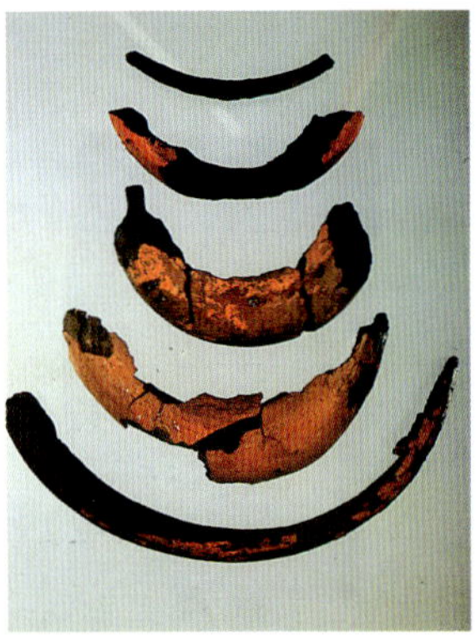

Figure 2. Pottery bowl with red urushi decoration on a black urushi ground, final Jomon, Kamegaoka site, Aomori prefecture.
Figure 3. Rantai *(woven bamboo) lacquerware, final Jomon, Korekawa site, Aomori prefecture.*
Figure 4. Wooden bracelets, final Jomon, Funagatani site, Ehime prefecture.

During the middle to late Jomon period (c. 2000–1000 B.C.), not only wooden containers, but also accessories decorated with urushi, such as combs, necklaces made of bone, and clay earrings and earplugs, were made. In addition, many bows decorated with red urushi belonging to this period have been discovered.

Towards the final Jomon period, extraordinary strides were made in urushi art in eastern Japan and it became amazingly rich and varied. In western Japan, too, the use of urushi seems to have spread, and numerous sites that contain such objects are gradually being found. Among these objects are pottery jars covered completely with red urushi and jars and plates with distinctive curved designs in red urushi on black urushi ground (Fig. 2). The brilliant contrasts created by the use of red and black demonstrate the richness of artistic expression.

The remains of *rantai-shikki* (lacquerware with a woven bamboo substrate), especially popular in the final Jomon period, are found mainly in the Tohoku district, but also in other regions from Hokkaido to the Kinki district. This advanced technique is found on containers, baskets, and pots covered completely with red urushi (Fig. 3) and on plates with designs in red urushi on black urushi, similar to those painted on pottery.

In various locations in western Japan, a number of bracelet-shaped objects made of wood with red urushi have been excavated (Fig. 4). These objects were produced continuously until the Yayoi period.

In addition, in the Jomon period, urushi was used not only as a paint, but also as an adhesive and fixing agent; for example, it was used to repair damaged pottery and to fix arrowheads to their shafts.

During the Yayoi period (c. 400 B.C.–A.D. 250), the availability of iron made possible the use of a lathe, which added a new dimension to the production of substrates. Urushi technique, however, continued to follow the traditions of the Jomon period, whereas, compared to that period, the use of urushi decreased. It was rarely used to decorate pottery and was mostly limited to application on wood. Also, urushi on containers for daily use is difficult to find. It seems likely that the use of urushi became restricted to such objects as decorative bows and accessories, headpieces and combs. In contrast to the Jomon period, usually only one color—black or red—was used and the drawing of designs was rare. Exceptions to this are some goblets and armor colored with red ocher (Fe_2O_3) or mercuric sulfide (HgS) and outlined with black urushi. The red urushi used in the Jomon period became neglected and disappeared after the late Yayoi period (c. A.D. 100–300). Moreover, in the Kanto and Tohoku districts where lacquerware was popularly produced during the Jomon period, little evidence has been found so far of the use of urushi dating from the Yayoi period. The only known example is the use of lacquer for repairing damaged pottery found in Saitama and Niigata prefectures.

The Nabatake site in Karatsu-shi, Saga prefecture, excavated in 1981, is important in that it exhibits the beginning of Japan's agriculture, which was influenced by Chinese and Korean cultures. On this site, about ten *shikki* (lacquerware vessels) were found, dating from 500 B.C. to 400 B.C., the latter part of the final Jomon period to the early Yayoi period. With one exception—a jar covered with red urushi on both inner and outer surfaces—the vessels had geometrical designs in red urushi over black urushi. Such designs were not found in western Japan during the Jomon period. This technique was probably imported from the Korean peninsula along with agricultural techniques; it was never widely handed down nor developed much, and later it was overtaken by the traditional techniques dating from the Jomon period.

The use of urushi spread almost throughout Japan in the Kofun period (c. A.D. 300–A.D. 600). The advanced state of the technique can be seen in the various accessories—weapons and armor—that were interred along with bodies in ancient burial mounds. New substrates, such as bamboo, leather, and metals, were introduced and, especially in the late seventh century, the technique of pasting layers of cloth with urushi began to be employed in making caskets.

The origin and technique of urushi

The origin of the use of urushi in Japan was thought formerly to have been influenced by China. However, the excavation of the Torihama shell mound in 1975 revealed that it was used in Japan from the beginning of the early Jomon period. On the other hand, the discovery of a red urushi bowl (height 5.7 cm, diameter 10 cm, diameter of base 7 cm) belonging to the beginning of the Neolithic age, during the second excavation (1977–1978) of the Hemudu site in Zhejiang province, shed further light on the relationship between Japan and the mainland. An Zhimin, the assistant director of the Institute of Archaeology CASS, People's Republic of China, has recently and emphatically stated that the origin of urushi techniques in Japan was strongly influenced by the Neolithic culture of the lower Yangzi River region; they were introduced to Japan and put to use along with such techniques as split stone earrings and elevated floor construction. Yet only one piece of urushi has been excavated from the Hemudu site and two lacquerware vessels from the remains of Yü-tun belonging to the Majiapin culture, which followed the Hemudu culture. However, it is not possible to make a definite statement from this evidence that the urushi technique of the Neolithic age in China is directly related to the objects excavated from the Torihama site. In fact, it may be said that there is a very subtle difference in the archaeological concept of time between Japanese and Chinese specialists. To discuss this matter we must wait for further scientific examinations and analyses. For example, carbon-14 analysis of the stratum of the Hemudu site from which the urushi was excavated dates it to around the latter half of 4000 B.C., which is also the date given to the remains of the Torihama shell mound.

Conclusion

The urushi technique of the Jomon period was very advanced from the outset, for example, in the use of both red and black urushi for painting. In general, red urushi was red ocher (Fe_2O_3) and cinnabar (red mercuric sulfide, HgS). It is considered, so far, that cinnabar was first used in the middle of the late Jomon period. After that, cinnabar and red ocher were both used, on different objects or on the same object, to produce different tones and gain effectiveness, a very complicated technique (Fig. 5). Sometimes the thickness was composed of five or six layers, about 150–200 m in total. Since each layer is almost completely smooth, it can be supposed that each layer was well-polished before the next was applied. This advanced technique is indeed fascinating. Occasionally even alternate applications of red and black were carried out. Already in this period, plaster materials such as clay, *kokuso-urushi*, or granite-sand *urushi-shitaji* were employed for making objects with irregular surfaces (for example, *rantai-shikki* or combs). Therefore it is possible to say that the fundamental urushi technique had been established by the final Jomon period. Through the Yayoi and Kofun periods, urushi applied art techniques developed from the tradition of the early Jomon period. This became the basis for the great advance in urushi technique after the seventh century, under the influence of the Tang culture in China.

Figure 5. Photomicrographs of cross sections of urushi layers. (a) Wooden bowl, late Jomon, Juno site, Saitama prefecture. (b) Lacquered woven bamboo bowl, final Jomon, Morigafuchi site, Ishikawa prefecture. (c) Wooden ring (see also Fig. 4), final Jomon, Funagatani site, Ehime prefecture.

Bibliography

Hemudu Archaeological Team

1980 "Excavation of 2nd season at Hemudu site in Zhejiang province." *Wenwu* [Cultural Relics] 5.

Hopei Museum and CPAM, Hopei Province

1974 "The Shang dynasty site at T'ai-hsi-ts'un, Kao-ch'eng county, Hopei province." *Wenwu* [Cultural Relics] 8.

Kaneko, Hiroyuki

1981 "Wooden lacquerware of special shapes in Funagatani sites." *Gekkan Bunkazai* 218.

Kenjo, Toshiko

1984 "Analysis on ingredient of urushi-like film in *shikki*." In *Report on the Excavation of Juno Bog Site*. Saitama Prefecture: Board of Education and Prefectural Museum.

Kida, S. and S. Sugiyama

1932/1980 *Plates Showing the Vegetal Relics of the Japanese Stone Age.*

Kobayashi, Yukio

1962 "Kyushitsu." In *Techniques of Ancient Times*. Hanawa Shobo.

Kobayashi, Y. and N. Mino

n.d. "On the structure of Jomon combs in lacquer excavated at Misawagawa site, Hokkaido." In *Annual Report of the Historical Museum of Hokkaido* No. 7. Historical Museum of Hokkaido.

Kuraku, Yoshiyuki et al.

1984 *A List of the Sites where Lacquer (Urushi) Implements were Recovered. East Japan.* Nara National Cultural Properties Research Institute.

1985 *A List of the Sites where Lacquer (Urushi) Implements were Recovered. West Japan.* Nara National Cultural Properties Research Institute.

1986 "Lacquerware techniques of the Yoyoi period." *A Study of Yayoi Culture* 6. Yuzankaku.

Morikawa, Masakazu et al.

1979 *Torihama Kaizuka—An Examination of Bog Sites of the Early Jomon Period.* Vol. 1. Fukui Prefecture: Board of Education.

1981 *Torihama Kaizuka—An Examination of Bog Sites of the Early Jomon Period.* Vol. 2. Fukui Prefecture: Board of Education.

1983 *Torihama Kaizuka—An Examination of Bog Sites of the Early Jomon Period.* Vol. 3. Fukui Prefecture: Board of Education.

Nagashima, Masaharu

1984 "Urushi techniques of the Jomon period seen in ornamental bow excavated from Juno Bog site." In *Report on the Excavation of Juno Bog Site*. Saitama Prefecture: Board of Education and Prefectural Museum.

1985 "Lacquerware techniques of the Jomon period—a study of the *rantai* lacquerware of Tohoku prefecture." *Bulletin of the National Museum of Japanese History*. Vol.6.

Nakajima, Naoyuki et al.

1982 *Nabatake*. Karatsushi City: Board of Education.

Nakasato, T., Y. Emoto, and R. Ishikawa

1971 "Note on the structure and technique of a Neolithic comb of lacquer work unearthed at Sanno site, Miyagi prefecture, and preservative treatment on it." *Science for Conservation* 7(3).

Naora, Nobou, and Esaka Teruya

1941 *On Objects Excavated from Kamegaoka Bog Site*. Kodai-Bunka. Vol. 12–3.

Naruse, Masakazu

1984 "Red-pigmented pottery, urushi-painted pottery and urushi wares: chemical analysis of their red pigments." In *Report on the Excavation of Juno Bog Site*. Saitama Prefecture: Board of Education and Prefectural Museum.

Suzuki, Kimio

1984 "Meaning of the excavation of Juno Bog Site—with special emphasis on wooden products and urushi objects." In *Report on the Excavation of Juno Bog Site*. Saitama Prefecture: Board of Education and Prefectural Museum.

Takahashi, Tadahiko

1985 "Remains of urushi workshop of the Jomon period." *Ethnos* 26. Shin-nihon Kyoiku Tosho.

Takahashi, Tadahiko et al.

1984 *Report on the Excavation of Nakayama Site*. Gojome Town, Board of Education, Akita Prefecture.

Yamada, Masahisa and Masae Yamaura

1984 "Types of *shikki* and selection of trees: on techniques of production." In *Report on the Excavation of Juno Bog Site*. Saitama Prefecture: Board of Education and Prefectural Museum.

Wu Su

1978 "Excavation of the Neolithic site at Yü-tun in Ch'ang-chou, Kiangsu province." *Kaogu* (Archaeology) 4.

Zhimin, An

1985 "Effect of the prehistoric cultures in the lower reaches of Changjiang River on ancient Japan." *Journal of the Archaeological Society of Japan* 70(2).

Lacquerware in the Unified Silla Period, with Special Reference to the Finds at Anap-chi

Jongseok Lee
Cultural Properties Research Specialist

When the historical development of lacquerware in Korea is considered, the works from the Unified Silla period are rarely discussed. The Unified Silla period dates from the late seventh century to the early tenth century. It spans the 250 years between the time when Silla unified the Three Kingdoms (Koguryo, Paekche, and Silla) and the fall of Silla at the hands of the new conqueror, the Koryo dynasty. Although it was in this period of Korean history that the most brilliant craftworks flourished, the study of lacquerware during these years has been surprisingly neglected. I surmise that this is because there is a lack of research materials from this period.

However, ten years ago an excavation was carried out at a lake named Anap-chi, located in the Kyongju district, where a Unified Silla palace once existed. Numerous objects were discovered at the bottom of the lake, including lacquerware, metalware, porcelain objects, tiles, stoneware, wooden objects, and various other remains.

Judging from the state of the collapsed stone walls of the lake and the characteristics of the excavated remains, there is no evidence of extensive reconstruction or dredging work since the lake was built in A.D. 674. It has probably been in ruins since the fall of Silla in A.D. 935. This suggests that the remains at Anap-chi consist mainly of objects from the Unified Silla period, with a very few from the Koryo period (918–1392).

The lacquerware found at Anap-chi consists of a number of broken pieces, but more than forty of these have recognizable shapes. There are twenty-five broken multilayered boxes, several pots and bowls, small juglike jars for oil, wooden inkslabs that show distinct signs of use, two pieces of *heidatsu* (applied metal decoration) material that were presumably used to decorate special objects such as Buddhist altars, and colored picture boards whose use is unknown (Cultural Treasures Conservation Department 1978). These remains differ from those excavated from the tombs of the preceding Three Kingdoms period (37 B.C.–A.D. 668), displaying a phase of luxurious life in the Silla palace. They seem to have remained at the bottom of the lake since the day of the destruction of the kingdom and thus exhibit consistent characteristics.

According to the *Samguk Sagi* (History of the Three Kingdoms), the government of Silla had a special department for lacquer. While there were other departments that handled general wooden products, the lacquer department was not only engaged in the production of lacquerware but also managed the cultivation of lacquer trees.

Korea has had a long tradition of regulating the use of lacquer objects. The government used its strict authority to determine the use of lacquerware according to rank. The *Koryo-sa* (History of Koryo) records that the government encouraged the development of lacquer tree plantations throughout the country, and, especially in the Yi dynasty, each local official was obliged by law to supervise and manage the lacquer tree plantations in his territory. Such laws were intended to prevent lacquer being wasted; royalty and public officials had priority in the use of lacquer objects, items that have been consistently regarded as valuable.

The lacquerware of Anap-chi should be considered from this standpoint. The lacquer items, as well as objects of gold, silver, and bronze, were found only on the banks of the lake where there were old buildings. In other words, compared with the Silla stoneware, which was found scattered randomly, the lacquer objects were found in a specific location. Therefore, these lacquer objects must have been used in the daily life of the palace, unlike the earthenware of the Three Kingdoms period. All known lacquerware from the Three Kingdoms period has been excavated from tombs and includes such utensils as pots, bowls, cups of various shapes, trays, wooden coffins, etc.

There are also reports of discoveries of lacquerware fragments from sites dating to the end of the Bronze Age and the beginning of the Iron Age, which preceded the Three Kingdoms period. These are prehistoric remains from before the period when the Naknang culture flourished, and they are regarded as the oldest evidence of lacquer objects in the Korean peninsula. While prehistoric remains of lacquer objects are very scarce, in the Three Kingdoms period there is evidence that lacquer was used to decorate wooden coffins and their accessories. Lacquer coffins were used in the kingdoms of Koguryo, Paekche, and Silla, and throughout the Koryo period and Yi dynasty, and are still used at luxurious funerals and as a sign of particular respect.

There probably were similar funerary customs in the Unified Silla period; unfortunately, however, no remains have been discovered from that period to prove it. Possibly because of a change in funeral regulations, discoveries from the Kyongju district consist predominantly of stone tomb chambers and containers of bones, etc. Additionally, there is no research material available because there has not been any excavation of tombs of this period. For these reasons, research in the area of craftwork is quite poor. Thus the discovery of the remains of Anap-chi has given us not only important research material but also an opportunity for tremendous progress.

Characteristics of objects at Anap-chi

The majority of the tableware from Anap-chi has a very thin body and an interior finish of red lacquer. In the Three Kingdoms period, the main tablewares were pots (Lee n.d.). Wooden wares were made by hollowing out wood; thus the remains from the tombs of that period are characterized by thick centers. (Alternatively, dry lacquer was used.) Although it is rare to find signatures in the lacquerware from the Three Kingdoms period, a number of lacquer objects of the Unified Silla period are signed. Such signatures are either written in red or done by the technique of hairline carving. These signatures seem to be related to the official signatures of Silla.

A precise manufacturing technique for the wooden substrate developed during this period. There are two different methods of making wooden vessels: one is to form the side by bending a thin piece of wood, as seen in multilayered boxes; the other is to shave willow (*Salix kriyanagi*) or, less often, fir (*Abies holophylla* Max.) twigs into a diamond shape and weave them to form the body of a bowl or a pitcher.

The shallow trays of multilayered boxes are found in great numbers among the remains of Anap-chi. They have high bases that are larger than the openings at the top, so that they can be stacked. They are lacquered in red inside and black outside. The material is fir, with a beautiful grain. After careful investigation of a broken part of a vessel, it was found to be formed from two or three layers of thin veneer about 1 mm thick. This shows that there was quite a precise manufacturing process. These round-brimmed vessels are reminiscent of Japanese multilayered boxes or tea ceremony utensils, not seen in present-day Korea. Only the technique of using thin boards of stripped fir to make the brims of sieves or containers and the lids of wicker trunks has been carried down through the ages.

The same technique of making curved objects using fir is seen in a wooden inkslab. The inkslab itself is made of solid wood and lacquered, but the base is made of thin wood that is overlapped and bent to form the four corners. This is one of the techniques used in manufacturing curved objects and is a truly astonishing method.

The method of making willow twigs into a diamond shape and weaving them into a desired form is a unique and precise manufacturing process, characteristic of the techniques of this period. After fixing the round bottom area by attaching some pieces of wood where needed, small twigs that have been shaved to form an angular shape are woven to construct the side of the vessel. This not only prevents a thin vessel from becoming cracked or deformed but is also a very practical way of making wooden bowls with narrow brims, or jars with small necks. I am familiar with the examination of lacquered jars from the Shosoin (1975) using x-radiography, etc., and assume that they too were made by this method of using willow twigs.

There was a "willow department" in the Silla government, which supervised manufacturers who used wood from trees of this family. Wicker trunks have been traditional furniture in Southeast Asia since ancient times; in Korea, where bamboo grows only in limited areas, willow was naturally used more frequently. Until the time of the Yi dynasty in Korea, demand for willow products was considerable and there were special ranks of people who were appointed to attend to this matter as an occupation. It is no coincidence that there was a willow department in Silla, and in the following years the technical level was unrivaled.

Heidatsu

Next, the development of the *heidatsu* technique in Silla lacquerware should be mentioned. It is recorded in the "Miscellaneous" section of the *Samguk Sagi* that "it is prohibited for those whose rank is lower than the fourth rank to use gold, silver, brass and [certain] lacquer ware." The lacquerware mentioned here is red with black outside, and gold and silver *heidatsu* lacquer. In other words, this regulation says that such vessels are allowed only to the upper three ranks, i.e., *shinkotsu*, sixth, and fifth. Those who belonged to the fourth rank were lower government officials in the palace and village headmen. This seems to indicate that at the height of the Unified Silla period's prosperity, the use of precious novelties such as lacquerware spread among ordinary citizens, which presumably led to the creation and enforcement of this new prohibitive law.

Heidatsu lacquerwares from Anap-chi are not vessels in common use and there is no way of knowing what their original purpose would have been. However, there are two wooden objects that might have been used as parts of a Buddhist altar or ceremonial utensils. One consists of eight multilayered flower petals (17.5 cm long, approximately 1 cm thick), on the outer surface of which are butterflies and animal faces made of thin silver plates. These petal-shaped decorative pieces are similar to the half-opened water lily decoration on the lid of the copper jar (a container for *sari*, a relic of the Buddha) in the iron tower of the Nara Seidaiji temple. It is difficult to speculate in what way the wooden pieces from Anap-chi, which resemble enlarged versions of these pieces from Japan, were used.

Another type of *heidatsu* object consists of several half-tubular pieces shaped like bamboo split in half. They seem too delicate to decorate poles or walls. However, from the religious symbol at the bottom end, the band of beads, and the silver *heidatsu* flower patterns over the whole surface, they must have formed part of an object that had some sacred meaning, and were made with the utmost care and precision. There are several reasons for assuming a religious association with the *heidatsu* wooden objects. More than twenty Buddhist images were found among the remains from Anap-chi, the practice of Buddhism was popular among the royalty and nobility in Silla and Koryo, and there is an historical record that describes frequent Buddhist ceremonies in the prince's palace, at the same time as festivals for kings and nature gods.

Heidatsu is a lacquer technique that gained popularity in the seventh century during the Tang dynasty in China, and is thought to have spread to the Korean peninsula by the time Silla culture was at its peak in the eighth century. In tracing this route, attention should be paid to the wooden pillow and wooden footrest (Cultural Treasures Conservation Department 1973) in the tomb of King Munyong of Paekche (sixth century), because the features of these objects (thin sheets of gold attached to the lacquered surface to achieve a turtle shell design) suggest the possibility of the *heidatsu* technique. Other unusual examples from old tombs are a silver *heidatsu* hexagonal tray with grapevine design (Kokura collection), which is thought to have been unearthed in the province of Kyongsang Namdo, and the gold and silver *heidatsu* copper mirror in the possession of the National Museum of Korea. It is doubtful, however, whether this was manufactured in Silla, because of a great similarity in design to those known to be from the Tang dynasty.

In any event, the *heidatsu* designs on the wooden decorative objects from Anap-chi are not delicately made; they are not symmetrical and the lines are unclear. At the same time they depict realistic flowers and plants in an artistic way. This is probably a natural result of the *heidatsu* technique being applied to a wooden surface, or it could have been a trend in that period. Until now it was thought that *heidatsu* was a technique for applying designs on metal surfaces, such as the Tang mirror. The Anap-chi remains, however, are remarkable new evidence that this technique was used to decorate lacquered wood. In other words, the discovery of the wooden *heidatsu* suggests that a new technique was developed to fit the characteristics of a new locality. This is related to the subject of how thin-shelled mother-of-pearl lacquerware appeared in Korea.

The technique that naturally accompanied the development of *heidatsu* was mother-of-pearl lacquerware. This is commonly regarded as having enjoyed popularity in the Tang dynasty; however, objects from the Tang dynasty decorated with mother-of-pearl are scarce. Unexpectedly few such remains have been discovered in either Korea or Japan.

There is a great similarity between the technique of gold and silver *heidatsu* and that of mother-of-pearl applied to wooden objects. Considering the fact that at the beginning of the tenth century mother-of-pearl products were regarded as a specialty in Koryo, it is reasonable to assume that the technique had existed since the end of Silla. The lacquer department of Silla was renamed the "decorative vessels department" in the eighth century during the reign of King Kyongdok, which seems to show that lacquerwares were no longer traditional simple products and the types of decorative manufacturing processes had become a lot more versatile and numerous. In A.D. 930, which corresponds to the period between the end of Silla and the beginning of Koryo in Korea, there is a record that Koryo's royal government sent mother-of-pearl objects as a gift to the rulers of the late Tang dynasty. Around A.D. 1000 a system of manufacturing mother-of-pearl boxes seems to have been set up, indicating that the demand for mother-of-pearl had increased so rapidly that it was necessary to build government-controlled factories. The Koryo remains, such as the mother-of-pearl sutra casket (reputedly one of the best in the world), are the products of this historical background. Lacquerware craftsmen of Korea concentrated their efforts on mother-of-pearl, unlike those of China or Japan, who had more colorfully diversified products.

The oldest mother-of-pearl object is the copper mirror with a design of flowers and animals, which was discovered in another excavation. The remains of the Three Kingdoms period tombs in Kyongju include the black lacquer saddle that was discovered in Hwangnam, to which semicircular shells are attached with metal nails and bronze objects with shell decoration. These shell decorations were not done by the mother-of-pearl lacquer technique, but they are interesting examples of designs that used shells. Although mother-of-pearl products superior to those of the Unified Silla have not been discovered, judging from the historical evidence it is reasonable to assume that certain techniques had been tried and established since the early days.

Lacquer pictures and colored paintings

Finally, I would like to discuss lacquer pictures and colored painting. In the tomb in Kyongju from the Three Kingdoms period, an unexpectedly large quantity of lacquer pictures and lacquerware was found. Most of these are objects such as trays, square goblets, etc., decorated with designs of flowers, plants, animals, and so forth, rendered with red lacquer. In contrast, no lacquer painting was discovered at Anap-chi. Instead, there were pieces that seem to be broken parts of black trays, which have delicate drawings in red lacquer and gold. Red lacquer was used for the rough rendering of plant designs, and objects were drawn with gold using the Chinese method. Details were done with thin silver lines. There are five pieces, such as those described above, that are very important material for the study of decorative techniques and require more thorough examination.

Buddhist pictures rendered with red and yellow pigments on black lacquerware were also found, but the pieces are so small that it is difficult to determine what type of vessel they formed part of, or whether the design was of flowers or clouds. One of the red-painted pieces seems to be part of a disc-shaped object, and it has a Chinese phoenix in the center that is surrounded by eight very precise and complex flower petals. Another object of interest is the piece shaped like a mortuary tablet that has a plant design at the edge and obvious traces of red and yellow coloring.

It is difficult to carry out a comparative study of these lacquer paintings. However, parts of some pieces are of equal quality to those of the Shosoin. There are objects other than lacquerware that seem to be particularly related to the Shosoin treasures, which should be the subject of further investigation.

With regard to the Buddhist pictures, another subject to be discussed is yellow lacquer. The yellow lacquer mentioned here is not a mixture of yellow pigment and lacquer, but the clear sap of semitropical evergreen trees *Dendropanax morbifera* Lev., which grow only on the southern coast of the Korean peninsula and nearby islands. This natural sap is mentioned in historical records in both Korea and China, some from the Tang dynasty, which call it Paekche or Silla lacquer. A record of the Song dynasty equates lacquer with yellow lacquer, which illustrates how famous Korean yellow lacquer was both inside and outside Korea (Lee n.d.). No objects decorated with this yellow lacquer have yet been discovered. However, the wooden pillow for a queen that was found in the tomb of King Munyong from sixth century Paekche may have had yellow lacquer as a background. Also, yellow lacquer may have been used on the paintings on the bark of white birch that were found in the Ch'onmach'ong mound, or on others from Anap-chi. Although the collection of sap from this kind of tree has been limited, it is a subject for study that should not be ignored, because of its historical importance, being equated with lacquer until the end of the Yi dynasty.

Bibliography

Cultural Treasures Conservation Department

1973 *King Munyong Excavation and Examination Report.* Seoul.

1978 *Anap-chi Lake Excavation and Examination Report.* Seoul.

Kim Pu-Sik (1075–1151)

1928 *Samguk Sagi* (History of the Three Kingdoms). Choson sahakhoe edition. Seoul: Kyongsong.

Lee, Jongseok

n.d. *Study of Ancient Korean Lacquer Ware.* Seoul: Tankuk University Press.

1975 *Lacquer Ware in the Shosoin.* Tokyo: Heibon Publisher.

Urushi Coating and Color Painting Applied to Japanese Architectural Cultural Monuments

Yoshio Hasegawa
Agency for Cultural Affairs

Figure 1. Ise-jingu: the shaden *(building complex) of the shrine.*

As the *shaden* (building complex) of the Ise shrine illustrates, Japanese buildings were originally of plain wood (Fig. 1). When Buddhist architecture was introduced from the Asian mainland in the sixth century, however, the technique of using vermilion or other red colors on buildings was also imported. There are traces of vermilion paint on wood excavated from the corridor of Yamadaji, an architectural monument of the Asuka period (mid-seventh century), much discussed in the press at the time of its excavation. The application of vermilion to angle rafters and common rafters has also been confirmed by the excavation of Shitennoji *kodo* (lecture hall; eighth century).

The interior and exterior of Horyuji (Fig. 2), which is the oldest existing wooden building, dating to the late seventh century, were painted in red pigment dissolved in liquid glue. Lotus flowers and *hosoge* (arabesque) designs were drawn in colors on the ceilings of the inner and outer sanctuaries (Fig. 3), and Buddhist pictures were drawn on the walls of the inner sanctuary. Red pigment was also used for painting the interior and the exterior of Yakushiji *toto* (eastern pagoda; 730, Tempyo 2) and Toshodaiji *kondo* (central sanctuary; 770–780, around Hoki), and lotus flowers and arabesque designs were drawn in colors inside the rooms (Fig. 4). All these show

Figure 2. Horyuji kondo *(central sanctuary building).*

Figure 3. The ceiling of the outer sanctuary of Horyuji kondo.

Figure 4. Painted designs inside the rooms of Toshodaiji kondo.

that Buddhist architecture in the eighth century was very colorful.

As early examples of urushi used on some parts of buildings, Byodoin Hoodo (1053, first year of Tenki; Fig. 5) and Chusonji Konjikido (1124, first year of Tenji) should be mentioned. At Byodoin Hoodo urushi was used for *raden* (mother-of-pearl inlay) works on the *shumidan* (altar) canopies of the Honzon Amidanyorai. The ceiling and the *kumimono* (bracket complexes) inside the Hoodo were painted in rich colors, and the outside of the building was painted with a red pigment. The original painting of Chusonji Konjikido (Fig. 6), which was partly modeled on Byodoin, has been well maintained as the building was protected by the *sayado*, a protective structure built in the medieval period. The temple exhibits the best of the arts and crafts of that age, such as *raden, ikakeji, makie,* and *heijin* (similar to *makie*), decorating the inner sanctuary; *urushi-haku* was executed on the other visible parts of the whole building, the roof structure excepted (Fig. 7).

The Tamamushi *zushi* (shrine) and the Tachibana Fujin Nenjibutsu *zushi* of Horyuji are the oldest in existence, made between the seventh and the eighth centuries, and they were coated with urushi. The *zushi* of Taimadera *hondo* (main hall), however, is the first example made as part of an entire building. This *zushi*, recognized as having been made in the early Heian period (around 794), is approximately 5 m wide, 1 m deep and 5 m high. It was originally coated with black urushi and decorated with pictures drawn with gold and silver powder. The undersurface of the roof and the eaves was decorated with *heidatsu* (applied metal decoration). Graining in black urushi and red oxide was applied to the balustrade. The *zushi* of Taimadera was repaired in 1243, the first year of Kangen, and a door decorated with *makie* was added at this time. Much of the *zushi* and the *shumidan* was repainted in 1453 (Kyotoku 2).

Figure 5. Byodoin Hoodo.

Figure 6. Chusonji Konjikido: (a) the shumidan *(altar), (b) outside, showing the* kumimono *(bracket complexes).*

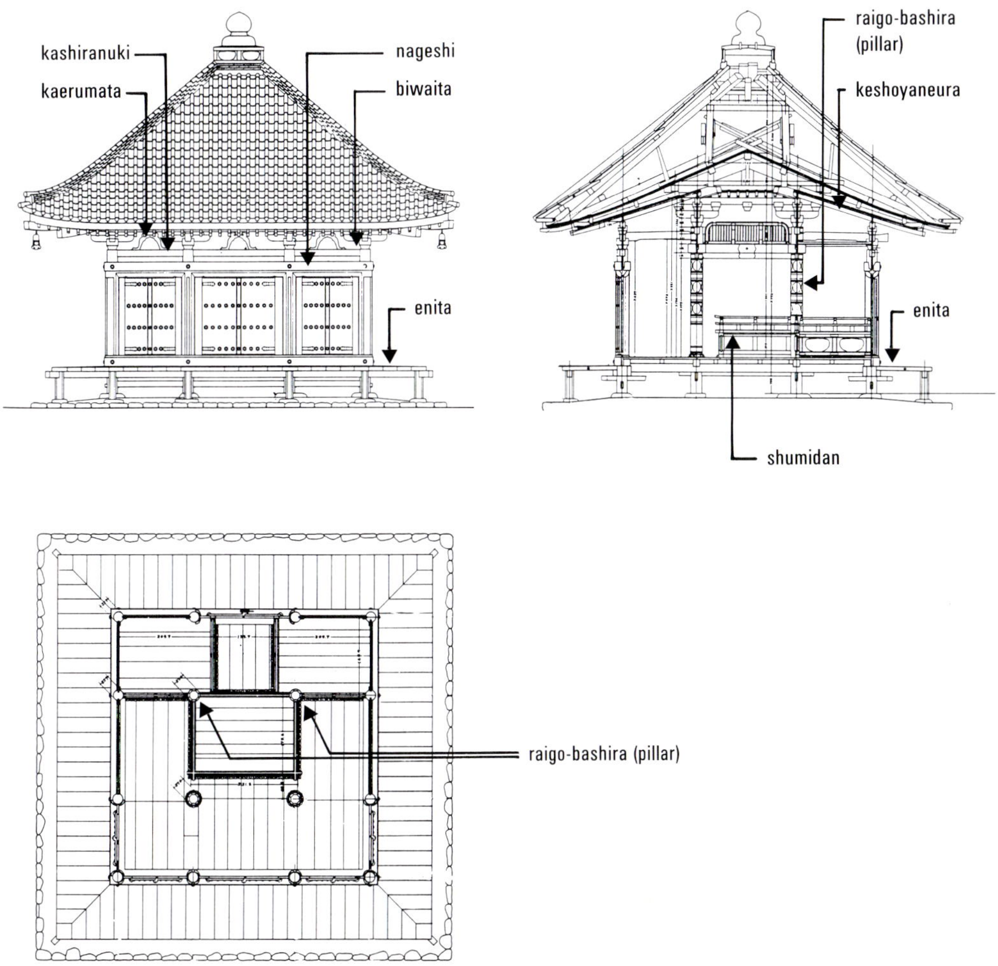

Figure 7. Chusonji Konjikido, showing typical architectural features.

Figure 8. Taimadera hondo.

The Taimadera *hondo* was constructed in the early Heian period (794–929) and was largely remodeled in 1161 (Eiriyaku 2). The building (Fig. 8) is 21.1 m wide and 18.0 m deep and is composed of an inner and an outer sanctuary. The framework of the inner sanctuary, including the pillars and the roof structure, was left as it was originally, and the outer sanctuary was remodeled. Repairs of various kinds were made on nine subsequent occasions. The *raigo-bashira* (columns; see Fig. 7) of the inner sanctuary, which had been plain wooden pillars, were coated with vermilion urushi around 1386 (Shitoku 3). Urushi-coated columns and altars, like those in the main building of Taimadera, are commonly found in medieval architecture.

From the Momoyama period (late sixteenth century) urushi began to be used for coating the exterior as well as the interior of buildings. Moreover, the *kaerumata* (frog-leg struts), the *biwaita* (wooden plates placed between bracket complexes) and other structural elements (see Fig. 7) were decorated more and more with carvings of flowers and birds, and the carvings were painted in rich colors.

As has been made clear, until the Middle Ages urushi was used only for *raden*, *makie*, or with gold leaf for buildings. Moreover, its application was limited to small areas inside the rooms such as altars and shrines. The technique of urushi work at that time was, therefore, like a craftwork. The extension of urushi to the interior of large buildings, like Chusonji Konjikido and Rokuonji Kinkaku in Kyoto (1398, Oei 5), was made possible because the responsible official at that time had great financial power.

The application of urushi to the exterior of buildings came as a result of the realization that, as a protection against weathering, urushi is superior to pigment dissolved in liquid glue. Although urushi was, and still is, much more expensive than glue, the use of urushi was made possible by the financial strength and influence of those in power during this period.

The following are some representative urushi-coated buildings from the Momoyama period that exist today.

1. The *honden* (main hall) of the Tsukubusuma shrine (Chikubujima, Shiga prefecture; Fig. 9). This shrine consists of a building erected in 1567 (Eiroku 10) and a building removed from the Fushimijo castle and reconstructed in 1602 (Keicho 7). The main framework of the shrine is believed to have been brought from Fushimijo. *Hiramakie* was applied on black urushi for the framework (Fig. 10), including the pillars, and flowers of the four seasons were drawn in color on the ceiling.

2. Hogonji *karamon* (gateway; Chikubujima, Shiga prefecture). The gate was removed from the Hokokubyo *karamon* and reconstructed in 1603 (Keicho 7). Its framework, the pillars, the *daiwa* (architrave; Fig. 11), and the doors were coated with black urushi. Carvings fitted into the doors and the walls, as well as the bracket complexes on the top of the pillars, were painted in rich colors.

3. Osaki Hachiman shrine *honden, ishinoma, haiden*; Sendai, Miyagi prefecture; Fig. 12). The shrine was constructed in 1607 (Keicho 12) and is an early example of a building in which urushi was applied on both the interior and the exterior. On the outside the gableboards were coated with black urushi; the eaves above the *koryo* (rainbow-shaped beam) and the *uchinori-nageshi* (horizontal wooden plank) of the *kohai* (roof over steps) were painted with colors. A dragon was carved on the *kohai-bashira* (pillars); celestial nymphs were carved on the *biwaita* (Fig. 13) between the bracket complexes. These carvings were also painted in rich colors. Parts of the *hafuita* (decorative wooden piece on the gable), the *taruki-koguchi* (rafter ends) and the *koran* (balustrade) were decorated with gold-plated metal fittings.

Figure 9. Tsukubusuma shrine honden.
Figure 10. Hiramakie *on black urushi used in the* honden *of Tsukubusuma shrine.*

Figure 11. Hogonji karamon *(gateway).*

Figure 12. Osaki Hachiman shrine honden.
Figure 13. Carved and painted decoration on the biwaita *of Osaki Hachiman shrine.*

Figure 14. Toshogu honsha *and* honden.
Figure 15. Toshogu honden *showing the use of black urushi and* makie.
Figure 16. Urushi-haku *used in Toshogu* honden.

The painting technique using urushi for the *hafuita* and lower parts of the building, and attractive colors for the *nokimawari* (eaves) above the *nageshi* (Fig. 13) in order to make the building more colorful, was often used until the end of the Edo period. It makes sense, as the *hafuita* and the lower parts of the building are subjected to wind and rain.

The *shaden* of Nikko Toshogu (Fig. 14) serves to represent the architecture of the early Edo period (the first half of the seventeenth century). The present Toshogu *shaden* was built between 1635 and 1636 (Kanei 12–13). The carvings, urushi coating, color painting, and metal ornaments used on the inside and the outside of the main building represent the highest level of architectural decoration at the time.

The building was richly decorated with *jimonbori* (carving done directly onto an entire surface; Fig. 15) and *ukibori* (embossed carving; Fig. 16) added to the pillars, *nageshi*, and *kashiranuki* (uppermost beam), which were finished with *gofun*; the bracket complexes of the *nokimawari*, *taruki* (rafters), *enmawari-koshigumi* (balcony brackets; Fig. 15) and *koran* were coated with black urushi. *Urushi-haku* was used abundantly in many places such as the *kesho-uraita* (visible wooden ceiling board; Fig. 16), *nokimawari-kumimono*, *madomawari* (windows and frames) and so on. *Sankarado*

Figure 17. Richly painted kashiranuki *(uppermost beam) at Shizuoka Sengen shrine.*
Figure 18. The haiden *of Shizuoka Sengen shrine.*
Figure 19. Details of the decoration of the haiden *shown in Figure 18.*

(frame-and-panel doors) and *itado* (doors of slab construction; Fig. 15) were decorated with *makie*. Carvings on the round beams and the walls were painted in rich colors and the *chuto* (column tops) and the *nageshi* were decorated with gold-plated metal fittings. The door with *makie* decoration follows the example of the urushi-coated buildings of the Momoyama period.

There is a view that the *gofun*-painted parts that are seen today, including the pillars, the *nageshi*, and the *kashiranuki* on the exterior, were not originally painted; instead the beauty of the grain of the zelkova tree was used to decorative effect.

From the seventeenth century onward, many buildings, modeled after the example of Toshogu and painted with urushi and colors, were built in various parts of the country. The painting technique for these buildings, however, was not as fine as that employed in lacquer vessels, since the purpose of using urushi on buildings was for protection from wind and rain. Fine techniques like *makie* and *raden*, therefore, were not found in the buildings of the Edo period.

As an example of urushi-coated and color-painted buildings of the late Edo period (nineteenth century), the *shaden* of the Shizuoka Sengen shrine may be mentioned. The present building complex was rebuilt between 1813 and 1851 (Bunka 10-Kaei 4). While black urushi was used for coating the pillars, walls, and doors of the *kanbe* of the Sengen Ryosha *honden*, which forms the center of the shrine, *shu-urushi* and *bengara-urushi* (urushi colored with cinnabar and red ocher, respectively) were used for coating the *kohai-bashira*; the carvings and bracket complexes placed over the *kashiranuki* (Fig. 17) were painted in rich colors. Furthermore, urushi leaf was used to decorate the *kesho-uraita*, the boards of the *shitomido* (shutters), and the *nokisaki-gawara* (eave end tiles); gold-plated fittings and fittings finished with urushi leaf or *nikurome* (metal with black patina) were put on doors, *chuto,* and *hafuita* for decoration. Colors in the adjoining buildings such as the *haiden* (Fig. 18), *kairo* (roofed corridor), and *romon* (two-storied gate) were comparatively modest. They were painted red with *bengara-urushi*, except for fittings like doors and *shitomi,* which were coated with black urushi. *Urushi-haku* was used only for a few areas, including the *nokisaki-gawara*, and colors used for the carvings and the *kumimono* (Fig. 19) above the *kashiranuki* were also plain. Such decorative techniques are typical of urushi coating and color painting techniques employed in buildings in the Edo period.

Method of urushi coating

There is a detailed description of the old method of urushi coating for buildings in the report on repairs to Chusonji Konjikido, issued in 1968. The report shows the original techniques used in the early twelfth century.

For outer eaves and the ceiling of the outer sanctuary, *suki-urushi* (raw urushi) was applied once or twice directly over *kokusogai* made of a mixture of flour and linen fibers, without any *kijigatame;* gold leaf was applied over the urushi. Urushi was used on areas under the beams both inside and outside the temple following these steps: *kokusogai, kiji-jinokozuke, nunokise,* and *jinokozuke.* Gold leaf was then applied over the urushi. The four to seven steps used for priming are a relatively simple process.

The report says that the upper surface of the roof of the Taimadera *hondo zushi* originally had *nunokise* (cloth reinforcement) for urushi-coated parts, but the priming was relatively thin.

The *urushi-shitaji* (ground) of the Osaki Hachiman shrine was also thin (see the report on repairs, 1968). The gold leaf was adhered with glue directly on the *biwaita* and the carvings of the *haiden* inside the *ishinoma. Suri-urushi* (thinned raw urushi) was directly applied to the *biwaita* outside the *ishinoma* and finished with gold leaf. According to the report, therefore, the grain of the wood showed clearly through the gold leaf.

Figure 20. Danzan shrine honden.

Although the *Zoeicho* (a record of how the building was made) of the present building complex of Nikko Toshogu *honden* issued in 1636 makes comments on *nunokise* and *roiro-sinnuri,* the process of *shitaji re*mains unknown. According to the report on the repairs issued in 1966, there were three layers of priming for the final coating, which was considered to be the original urushi coating of the temple; the bottom layer was a double layer of thinly applied coarse and fine *shitaji* followed by a coat of raw urushi. It is also reported that the sequence for the process of *nunokise* was: *kiurushi-gatame, nunokise, nunomedome, shitaji, kurourushi-nuri.* The *shitaji* was a little thicker in the first recoating with urushi and a thick *shitaji* was used in the second recoating, according to the report.

The *Hondocho* (a similar record) issued in 1797 and owned by Nikko Toshogu may be mentioned as an early record of the urushi coating process. Since the document states the cost of urushi involved in each step of *urushi-nuri,* the sequence of priming can be identified: "*kijigatame, kokusogai, nunokise, mefusagi, ji, kiriko, jigatame, sabi, nakanuri....*" Thus the three layers of *shitaji—ji, kiriko,* and *sabi*—were all coarse *shitaji.*

Figure 21. Urushi-coated sample boards prepared for the rebuilding of Danzan shrine in 1850.

Eight urushi-coated sample boards for Danzan shrine (Fig. 20) in Nara prefecture give detailed examples of the method in the late Edo period. These boards were made as samples of urushi coating when the present main building of the shrine was rebuilt in 1850 (Kaei 3). Urushi was coated on the boards in layers from the *shitaji* to the final coating to show the order (Fig. 21). There was also a detailed explanation of the process in a document accompanying the boards. For example, the process of *shitate-roiro*, according to this document, was done in the sequence: *kokuso, honji-gatame, sarashinuno, urushizuke, mefusagi, honjizuke* (first *jizuke*), *nidome-jizuke* (second *jizuke*), *jitogi, kirikozuke, sabitsuke, sabitogi, sabi-suri-urushi, nakanuri, muranaoshi, suri-urushi, uwanuri, roiro-togi.*

This process is basically the same as the process mentioned in the *Hondocho* of Toshogu. It should be noted, however, that *jizuke* was done twice in this process. This is similar to the repair process employed for the *shaden* of Toshogu in the 1870s and to standard methods of repairing architectural monuments in current use. In other words, the existing method of repairing these buildings has been handed down from the late Edo period.

As described above, the method of urushi coating for buildings developed notably at the beginning of the modern age. Urushi was first applied to the exterior of buildings and the *shitaji* gradually became thicker with every periodical recoating, since the substrate had to be smoothed out.

Like the *Zoeicho* of Nikko Toshogu, the *Kanjocho* of Shizuoka Sengen shrine issued in the same period says "*nunokise-kuro-nuri, shu-nunokise, haku-shitaji-nuri, shin-kakiawase-nuri.*" *Gushikenki* (Vol. 9, *Sho Tsumori*), which was famous as a carpenter's manual in the Edo period, also mentions "*nunokise-roiro, roiro, nunokise-joshin, joshin, tome-nuri, shinno-kakiawase, kakiawase-nuri.*" A postscript reading "*Kanbun* 7–8 [1167–1168] *kogi shufuku nyusatsu* [official bid for restoration]" is added.

Although the technical terms mentioned in documents and manuals up to that time had not been standardized, the Nikko Onmiya Narabi Owakidosha Kekkosho (official records) issued in 1753 (Horeki 3) used terms that are still in use today, such as *nuritate-roiro, shitate-roiro, shin-kakiawase-nuri, shippaku, tame-nuri.* These terms were also used for the urushi coating of Shizuoka Sengen shrine and Danzan shrine. This is an indication that around 1753 the shogunate standardized both the urushi coating method and the terminology.

Conclusion

Pigments dissolved in liquid glue were used for the interior and exterior of buildings until the Middle Ages, and urushi was only applied to limited areas inside the building, such as altars and pillars. Urushi started to be used for the exterior of buildings from the beginning of the sixteenth century, while the carvings increasingly made to decorate both the inner and outer walls and the *kumimono* (bracket complexes) of the *nokimawari* (eaves) were painted in colors. Moreover, metal fittings with urushi leaf, *nikurome* finish, and gold plate were used to decorate the *hafuita* (gables), *nageshi* (horizontal wooden planks), and *chuto* (column tops). Decorations both inside and outside the buildings, therefore, became very bright.

On the other hand, unlike lacquer vessels, the exteriors of buildings required periodic recoating with urushi because of exposure to rain, wind, and direct sunlight. Besides, the coated area is much larger than that of lacquer vessels. Therefore the urushi coating technique is very different from those used on lacquerware, which have developed in a more delicate way.

Bibliography

Committee on the Preservation of Cultural Property of Nikko Nisha Ichiji

1967 National Treasure Toshogu Honden, Ishinoma, Haiden: Report on Repairs.

Committee on the Preservation of Horyuji National Treasure

1956 National Treasure Horyuji Kondo: Report on Repairs.

Committee on Repairs for Preservation of the National Treasure Chusonji Konjikido

1968 National Treasure Chusonji Konjikido: Report on Repairs.

Committee on the Restoration of the National Treasure Osaki Hachiman Shaden Shrine

1968 National Treasure Osaki Hachiman Shrine Shaden: Report on Repairs.

Department of Preservation of Cultural Properties, Board of Education, Nara

1960 National Treasure Taimadera Hondo: Report on Repairs.

Office for the Repair of the National Treasure Tsukubusuma Shrine

1937 National Treasure Tsukubusuma Shrine Honden: Report on Repairs. Shiga.

Daiichi Hoki Shuppan

1976 Lecture on Cultural Properties, Japanese Architecture 4, Modern Period 1 (Architecture and Urushi).

Architectural glossary

biwaita	a wooden plate placed in the space between bracket complexes
chuto	a column top; a capital
daiwa	an architrave; lowest division of entablature resting immediately on the capital of the column
enmawari	a balcony around the building
hafuita	a decorative wooden piece at the gable of a roof
haiden	a free-standing building for public worship, in front of a sanctuary
honden	the main hall of a Shinto shrine
hondo	the main hall of a Buddhist temple
ishinoma	a room or hall connecting the main shrine and a hall for worship
itado	a door of slab construction
jikido	a refectory
jimonbori	carving done directly onto an entire surface, usually of uniform pattern
kaerumata	straddle-legged shaped wooden piece in the entablature, usually a decorative piece; frog-leg strut
kairo	a roofed corridor or peristyle
kashiranuki	the uppermost beam that runs through the columns
karamon	a Chinese-style gateway
kesho-uraita	a visible wooden ceiling board behind the visible rafter (*kesho-taruki*, q.v.)
kesho-yaneura	a roof-structure visible from the interior
kesho-taruki	a visible rafter
kodo	a lecture hall
kohai	a roof built over the steps leading up to a building
kohai-bashira	pillars of the *kohai*
kondo	the central sanctuary building of a Buddhist temple, usually houses the most sacred images
koran	a balustrade
koryo	a rainbow-shaped beam that connects two columns
koshigumi	bracket complexes under balconies
kumimono	bracket complexes

madomawari	a window and frame
nageshi	a horizontal wood plank, usually near the top of a room, resembling a beam between posts
nikurome	a black metal ornament, patinated with potassium sulfide and smoke
nokimawari	eaves
nokisaki-gawara	a tile at the end of the outermost part of the roof
raigo-bashira	a pair of columns standing at the corners of the altar (*shumidan*)
romon	a two-storied gate
sankarado	a door of frame-and-panel construction
shaden	the building complex of a Shinto shrine
shitomido	wooden latticed shutters
taruki	a rafter
taruki-koguchi	a rafter end
toto	the eastern pagoda
ukibori	deep carving done directly onto an entire surface, usually of uniform pattern, creating a relief; embossed carving
zushi	a small shrine placed in the inner sanctuary

Conservation and Techniques

Preparation of gold makie *work.*

The Protection of Ancient Chinese Lacquerware

Zhou Bao Zhong
Museum of Chinese History, Beijing

Figure 1. Fragments of lacquerware from the Shang dynasty site in Taixicun, Gaochen, Hebei province. Hebei Provincial Museum. Wengu No. *8:42–9 (1974).*
Figure 2. Tomb guardian from the Zhan Guo tombs.
Figure 3. Fragments of a lacquer painting of dancers and hunters from the Zhan Guo tombs.

The craft of lacquerware has an age-old history in China. Records in ancient books and documents and findings from archaeological excavations confirm that raw lacquer has been used since the beginning of the Neolithic age.

In 1978, at excavation sites in Yuyao and Hemudu (Zhejiang province), lacquer-coated wooden bowls from the beginning of the Neolithic age were discovered among other remains. These bowls, which were covered inside and out with a red lacquer coating, are the earliest findings of lacquerware so far, dating back 6000 years.

In 1973, at excavation sites in Gaocheng and Taixicun (Hebei province), containing remains from the Shang dynasty, fragments of lacquerware were unearthed (Fig. 1). Although the objects were broken, the fragments were still bright and colorful. The designs were elegant and it was obvious from a study of the fragments that the original objects were plates and large bowls with black flowers on a red background and designs of *taotai* (a ferocious, mythical animal). Some of the dishes were inlaid with turquoise, which was cut in circular, rectangular, and triangular shapes. At that time, the techniques of carving, lacquer painting, inlay, etc., were already in use. This shows that the art of lacquerware in China had already reached advanced levels from the sixteenth to the eleventh century B.C. (Shang dynasty).

By 1100–221 B.C. (Xi Zhou and Dong Zhou dynasties), lacquerware was widely used and the art, which was fully developed, had become a genuine profession. In Xinyang, Henan, Changsha (Hunan province) and in Jiangling and Suixian (Hubei province) lacquer-coated furniture, household items, musical instruments, weapons, coffins, etc., from the Spring and Autumn period and the Warring States period are continually being discovered.

In 1957, in Jiangtaiguan and Xinyang (Henan province), lacquer-coated coffins were excavated at the Zhan Guo (Warring States) tombs. Among the items found was a colorfully painted wooden animal, 1.4 m high, which was supposed to protect the tombs (Fig. 2), an equally colorful *se* (a stringed musical instrument similar to the zither), and the remaining third of a lacquer painting of dancers and hunters (Fig. 3).

In 1980, in Linli (Hunan province), six mythical animals with heads like tigers and bodies like dragons were discovered at the Jiuliyihao tomb. These figures, 97 cm high, were carved from one piece of wood and painted with a dark brown lacquer.

Figure 4. Carved screen from Wangshan, Jiangling (Hubei province). Hubei Provincial Museum.

In 1965, in Wangshan, Jiangling (Hubei province), a colorful screen carved with animals, from the Chu state of the Warring States period, was excavated (Fig. 4). The screen was in a rectangular frame. Fifty-one different animals, including deer, phoenix, sparrows, frogs, snakes, and boas, were engraved in the middle of the screen. It was done in red, gold, silver, gray, green, etc., on a black background. The animals were done in a lifelike manner and it is truly a masterpiece of lacquer art.

In 1978, in Leigutun, Suixian (Hubei province), the tomb of Zenghouyi from the early part of the Warring States period was disinterred. Twenty-two colorfully painted coffins were discovered. The main coffin was done in black on the outside with red and gold designs on the top and red on the inside walls. One hundred and twenty-four musical instruments, including *bianzhong* (similar to xylophones), were found near the tomb. Among other instruments were colorfully lacquered *zhen* (harps), *paixiao* and *shen* (flutes), dating back 2400 years.

The development of lacquerware flourished between 206 and 24 B.C. (Western Han period). Several areas became established as centers for lacquerware. The technique was perfected and the products were sold throughout the country. Lacquerware became one of the most widely used household items. From 1972 to 1973, the first and third tombs of Mawangdui, Changsha (Hunan province) were excavated. Over 500 pieces of lacquerware from the early Western Han period, dating back over 2100 years, were found in good condition. They were shiny and like new. Inside the outer coffin of the first tomb of Mawangdui were four smaller coffins, one inside the other. The next largest coffin inside the outer coffin was colorfully painted on a black background with wisps of cloud, animals, and gods (Fig. 5). The next largest coffin inside the second coffin was red inside and out, with dragons, tigers, sparrows, and deer (all considered animals that bring good luck) painted in bluish-green, brown, yellow, and white (Fig. 6). The elegantly formed Western Han lacquerware from Mawangdui had magnificent designs of all kinds, such as vessels with cloudlike designs, boxes with phoenix, and *zhunxinshi* dishes (Figs. 7–10).

As time went on, the technique of lacquerware took on new dimensions. Lacquerware inlaid with gold, silver, and jewels from the Tang dynasty (A.D. 618–907) and carved lacquerware from the Song dynasty (A.D. 960–1279), the Yuan dynasty (A.D. 1279–1368), the Ming dynasty (A.D. 1368–1644) and the Qing dynasty (A.D. 1644–1911) greatly enriched the collection of this art form. Lacquerware, one of the more outstanding cultural remains, advanced for several thousand years. Modern lacquerware takes a variety of forms and has developed into an exquisite art form.

Lacquerware from each dynasty is collected in museums in order to preserve forever this magnificent achievement of mankind. Scientists dealing with the preservation of cultural relics do a great deal of research and experimentation on ways to protect lacquerware.

Figure 5. Painted inner coffin from the first tomb of Mawangdui, Changsha (Hunan province). Hunan Provincial Museum.

Figure 6. The third in the series of coffins from the first tomb of Mawangdui. Hunan Provincial Museum.

Figures 7–10. Western Han lacquerware from the first tomb of Mawangdui. Hunan Provincial Musuem.

Lacquerware consists of a combination of a substrate and an outer coating. Raw lacquer is an excellent natural paint. If raw lacquer is mixed with pigment, it forms a paint that can be applied on a substrate and becomes a skinlike layer. Only after the stages of polishing, painting, molding, inlaying, carving, and burnishing with gold and silver, does it become "lacquerware." The coating serves as both protection and decoration.

Natural raw lacquer comes from trees of the angiosperm class. Originally, lacquer trees grew only in the wild. Starting with the Han dynasty, however, they began to be cultivated. More and more lacquer trees were planted in the Song and Yuan dynasties and they began to flourish in the Ming and Qing dynasties. The production of raw lacquer spread over fifteen provinces. The trees grow mainly in high mountainous areas and hilly regions. The lacquer sap is stored in channels within the tree. If the tree is damaged, the channels are broken and the sap flows from the tree. If a cut is made in the bark of a living tree, the milklike substance that flows out is raw lacquer. Fresh lacquer is milky-white or grayish-yellow and changes color when exposed to air.

The raw lacquer is heated, some of the water is removed, and it is mixed with oil (probably tung-oil). Then it is mixed with pigments to obtain bright and colorful lacquer. Raw lacquer does not peel or chip and is a protective, noncorroding, acid-, scratch-, heat- and waterproof, bright and shiny, long-lasting coating.

The substrate forms the physical structure of the lacquerware. Various materials can be used as substrate: wood, bamboo, cane, silk, ramie, leather, porcelain, gold, silver, copper, tin, lead, etc. Since wood is the most commonly used, the following discussion will concentrate on wooden substrates.

Wood is an organic substance that will decay to different degrees according to whether it is buried underground or exposed to air. The decay process can involve the wood rotting, decomposing, growing mold, changing color, becoming infested with termites, drying and cracking, deforming, and/or becoming saturated with water. The natural characteristics of the wood and its environment determine the degree of decay. If wood is immersed in water for a long time or buried in wet soil, it can become waterlogged, sometimes to the point of resembling *tofu*. Wooden substrates that have been excavated contain 100–400% absorbed water and sometimes up to 700%. This particular decay process of lacquerware has therefore become an important area of research. The dewatering of lacquerware must be strictly controlled to prevent or minimize deformation.

Some of the lacquerware buried several thousand years ago is well preserved. Some of it has decayed because of its environment. Temperature, moisture content, the acidity of the moisture, the chemical makeup of the soil, and the quality of the air in the tomb can affect the lifespan of lacquerware. In the first tomb of Mawangdui (Western Han dynasty), the coffins, burial articles, and even the body were well preserved. There were 180 pieces of lacquerware among the burial articles. Inside the lacquerware containers, various fruits, beef, and lotus-root remained. The major factor in the preservation of these items was the depth of the tomb, about sixteen meters. The coffins were surrounded by charcoal, about 30–40 cm thick. The charcoal was covered by about 60–130 cm of thick white soil. The white soil was covered with a thick red clay which does not easily permit seepage. Using this technique, the tomb was sealed securely. The temperature of the coffin chamber was between 15–20°C. When the tomb was excavated, the temperature inside the outer coffin was 18°C. This environment, characterized by low temperature, constant moisture content, lack of oxygen and light, and no bacterial growth, preserved the lacquerware and other cultural relics. When the tomb was uncovered and the chamber opened, the sudden exposure of the lacquerware to the outside air disrupted the relatively stable environment. As a result, the lacquerware began to crack shortly after excavation. Many of the coffins of the Warring States period and the Western Han dynasty, although buried in tomb chambers, were found to be decayed at the time of excavation. Some, which had not decayed, were found immersed in water. The destruction of lacquerware by saturation is a big problem and the environment of lacquerware is therefore very important.

Lacquerware preserved in museums must be kept in a special environment and well cared for so as to prevent the process of decay and maintain the original condition. While environment has an effect on lacquerware, the process of decay is slow and difficult to observe. It becomes apparent, however, after much time has passed. Change of color, brittleness, drying and shrinking, distortion and loss of luster are caused by exposure to oxygen, light, and air pollution. Lacquerware is prone to shrinkage and expansion of the substrate. The shrinkage is caused by dryness and the expansion by moisture. Therefore environment is a major concern. The temperature must be kept between 18–22°C and the relative humidity between 60–65%. The lacquerware must be shielded from high light-levels and pieces that have been dewatered must be kept in boxes. If the lacquerware coating begins to crack, buckle, or chip, repair must take place immediately so that the substrate is not harmed.

The key to preservation of ancient lacquerware that has been saturated is to reinforce it so that it keeps its form. The excavation of a great deal of lacquerware has demonstrated that raw lacquer is an excellent coating. It is easily applied, durable, and resistant to decay. Some of the substrates had been destroyed by saturation and only the raw lacquer coating remained. If lacquerware is left to dry naturally after it is excavated, it will shrink, crack, buckle, and chip, and finally be destroyed. A great deal of ancient Chinese lacquerware has been found. At times numerous pieces have been excavated from a single tomb. This was the case with the Chu tomb in Xingyang, Henan province, the Warring States tomb in Suixian, Hubei province, and the Western Han tombs in Jiangling, Hubei province, and Mawangdui, Changsha, Hunan province. In order to preserve these rare remains of ancient civilizations, a good deal of research and experimentation has been done by scientists. This falls into two areas.

First, the dewatering of the wood and bamboo substrates of lacquerware, while at the same time maintaining the original form of the object, has been investigated. Second, attention has been given to the choice of the proper types of material with which to fill and reinforce the lacquerware to increase its durability. These two areas of concern are often considered together, though the first is of greater importance. If the original form of ancient lacquerware is to be preserved, the addition of any new material is not desirable. As long as the objects are periodically dewatered, it is not necessary to add any new materials to reinforce them. If, however, the objects are already seriously decayed, they cannot be preserved without reinforcing them with new materials. The purpose of using new materials is to protect the original material rather than to replace it. The new coating material must be of the same quality as the original. In cases where the original material is not available, the use of other materials is acceptable. In this way, lacquerware that has already begun to decay can be rescued before it decays further. When it is necessary to use different materials, it is best to choose those which are removable so that if more suitable material becomes available, the previously used material can be replaced. The various methods of reinforcing saturated lacquerware are presented in the following paragraphs.

Drying naturally

Lacquerware can be placed in a special environment and left to dry naturally. This environment must be very stable. The relative humidity must be above normal, yet not to the point of complete saturation. In this way, the water can evaporate slowly from the objects. The slower the process of evaporation, the better the results. The saturated lacquerware is in a state of equilibrium. In order to dewater the lacquerware, the state of equilibrium must be destroyed. Drying naturally is the best method of gradually changing the state of equilibrium and evaporating the water slowly until dewatering is finally achieved. The various possible environments are listed below:

1. Sealed drying boxes. The lacquerware is placed in a sealed glass box in order to dewater slowly. The condensation on the object inside the box and on the surface of the box must be removed frequently until the weight of the object becomes constant. Then the dewatering process is complete. This method is best suited to the dewatering of small objects.

2. Plastic wrapping. Larger objects that cannot be contained in drying boxes are wrapped in plastic and then placed in sealed containers. These containers are periodically opened so that the condensation can be removed. After a period of time, the object is completely dewatered.

3. High-humidity controlled chambers. The objects are placed in controlled chambers with high relative humidity, which is stabilized at about 95%. The objects are dewatered slowly. In basements with high relative humidity, the dewatering process can be completed in one to three years.

4. Burial. Large objects are buried in wet sand or sawdust. The dampness of the sand or sawdust must be measured periodically and it must be replaced by sand or sawdust containing less water. In this way, the relative humidity of the environment can be controlled and the objects dewatered.

5. Constant temperature. The objects are placed in boxes with constant temperature and relative humidity. The temperature is set at 32°C. The objects are taken out every four hours and immersed in distilled water for ten minutes. They are then replaced in the boxes. The process is continued for several months until the objects are dewatered.

6. Glycerol spraying. In order to keep the surface of the objects moist and evaporate the water, glycerol is sprayed over them. Since glycerol absorbs water, the surface temperature of the lacquerware can be controlled and the water evaporated.

7. Environment controlled by inorganic salts. Inorganic saline solutions can be used to control relative humidity so that lacquerware can be dewatered. The various salts: $CuSO_4 \bullet 5H_2O$, KNO_3, KCl, NaCl, $NaNO_2$, and $Mg(NO_3)_2$, are dissolved in water and put in different containers. Each saline solution creates a different level of relative humidity, ranging from 98% to 53%. The lacquerware is first placed in a container of saline solution, which creates the highest relative humidity. It is then transferred to a different saline solution with lower relative humidity, then to a solution with still lower relative humidity, etc. In this way the lacquerware can be dewatered slowly. The lacquerware, however, must not come into direct contact with the saline solution.

8. Silica gel. Silica gel is a common hygroscopic agent. Different quantities of silica gel are placed in different containers. Different degrees of relative humidity are achieved by varying the amount of silica gel. The lacquerware is placed in a drying box, along with a container of silica gel, and dewatering takes place.

These eight methods require different conditions. They are all, however, simple, economical, practical, and have no negative side-effects. In addition, the original form of the object is maintained while dewatering takes place. The drawback of these methods is that they are time-consuming. They usually take several years. In order to use these drying methods, the object must be completely intact, with a solid substrate and low saturation.

Vacuum drying

Saturated lacquerware can be kept under vacuum to be dewatered. These methods require the use of drying techniques. There are three specific methods that can be followed.

1. Vacuum drying. Saturated lacquerware is placed in a vacuum chamber. After the machine is turned on, water is evaporated from the objects. After the vacuum pressure reaches 700 mmHg (9.3×10^4Pa), the machine is turned off. The vacuum pressure decreases naturally to 500 mmHg (6.7×10^4Pa). The machine is turned on again and the processs is repeated until the object reaches constant weight.

2. Vacuum drying with heat. The lacquerware is first placed in a vacuum chamber at 70°C with the machine turned off. After some of the water has escaped, the machine is turned on for dewatering.

3. Vacuum drying with refrigeration. After the water in the lacquerware has been turned to ice in a low-temperature environment, the objects are transferred to a vacuum chamber where they are lyophilized. The key to using this method is to have a system that can provide low temperature and high vacuum pressure. In a low-temperature refrigerator or a container with dry ice that can reach -20°C, the water can be frozen quickly. This method is suitable for objects with rough coatings and thick wooden substrates or those that are already broken or decayed.

Solution displacement

This is the classic method of dewatering. When the lacquerware is immersed in water mixed with an organic solvent, the solvent can permeate the object by osmosis. Thus the water in the lacquerware is displaced. The advantage of this method is the speed with which dewatering occurs. It is best suited to small, thin objects with smooth textures.

1. Alcohol and ether. Alcohol is used to displace the water in the substrate and then the alcohol is displaced by diethyl ether. Afterwards, the object is placed in an environment at ambient temperature or under vacuum. The diethyl ether then evaporates automatically and the object is dewatered. While diethyl ether, propan-2-ol, propan-l-ol or 2-methylpropan-2-ol can be used, diethyl ether is the most common. This method is best suited to objects that are already decayed or are made of soft wood.

2. Acetonitrile. Acetonitrile is the most effective for dewatering lacquerware and does not harm the original coating. The water is displaced by acetonitrile, which is then displaced by diethyl ether. Afterwards, the lacquerware is placed in an environment at ambient temperature or under vacuum and the diethyl ether evaporates naturally. This method is best suited to delicate objects.

Permeation reinforcement

Organic or inorganic high polymer material can be used to permeate and fill the structure and cells of the wood. In this way, the cells of the wood are supported and reinforced and shrinkage is prevented. This is one of the methods for reinforcing saturated lacquerware.

1. $AlK(SO_4)_2 \cdot 12H_2O$. Aluminum potassium sulfate dodecahydrate is very easily dissolved in hot water. At room temperature, it can barely be dissolved. The state of condensation of $AlK(SO_4)_2 \cdot 12H_2O$ changes with temperature. At high temperatures, the $AlK(SO_4)_2 \cdot 12H_2O$ and water solution can displace the water in the lacquerware. When the temperature is lowered, $AlK(SO_4)_2 \cdot 12H_2O$ crystallizes and remains in a solid state in the wooden structure and the object is reinforced.

2. High polymer material.

(i) Monomer permeation reinforcement. The monomer of high polymer material permeates the wood and polymerizes. Thus the structure of the lacquerware substrate is reinforced.

(ii) Water-insoluble high polymer reinforcement. First, an organic solvent is used to displace the water in the lacquerware. Then the wooden structure is permeated with a water-insoluble high polymer material combined with diethyl ether. When the diethyl ether evaporates, the water-insoluble high polymer material remains and reinforces the structure. This method is also called the alcohol diethyl ether resin immersion method.

(iii) Water-soluble high polymer reinforcement. A water-soluble high polymer material permeates the substrate of the lacquerware and displaces the water, then solidifies. Among the many kinds of high polymer material are polyethylene glycol, phenolic resins and others. Polyethylene glycol is easy to use, cheap, nontoxic, nonodorous, nonpolluting, and not easily flammable. It also has the reputation of not being harmful to workers.

The steps taken to preserve cultural relics must follow the principle of not changing the original state of the object. The original state includes all the materials of which the object is made. Therefore filling ancient lacquerware with modern high polymer materials or inorganic materials is never the best method. It is only acceptable for those pieces which are already decayed.

Each piece of lacquerware has its own characteristics. Therefore the methods of dewatering and reinforcement must be applied with consideration for the individual piece. These methods should be used only after experimentation and then with great care. New methods should continually be sought.

The Preservation and Handing Down of Traditional Urushi Art Techniques in Japan

Shin Yagihashi
Agency for Cultural Affairs, Tokyo

The traditional techniques of the applied arts and the performing arts are regarded in Japan as "intangible cultural properties." Active measures have been taken for their preservation and transmission since the enactment of the Cultural Properties Preservation Law in 1950. This law was passed because after World War II there was a great revival in the applied arts, including urushi, but there was no system to deal with the rapid changes in lifestyle that threatened the handing down of these traditional arts.

The new law enabled the government to take measures to preserve those intangible cultural properties that were selected as being threatened with extinction; an amendment to the law in 1954 revised the original selection and introduced the following measures:

1. Designate important intangible cultural properties; acknowledge the holders of intangible cultural properties (or holding groups); and offer financial assistance to them. To date, eighty-one persons are recognized as holders, ten in the field of the urushi arts.

2. Select and record intangible cultural properties, other than *important* intangible cultural properties that need special attention.

3. Provide financial support to establish training institutions and for training activities organized by local government and other bodies.

4. Select traditional preservation techniques that are indispensable for the preservation of cultural property and assist with necessary expenses. (This provision was added after a further amendment of the law in 1975.)

The following is a complete chronological list of all those who have ever been appointed and selected in the field of urushi art.

Holders of important intangible cultural properties in the field of urushi art

Yusai Akaji (1906–1984). **Kyushitsu (Fig. 1)**

All the holders of *kyushitsu* (sequential lacquer coating technique) are skillful not only in applying urushi but also in molding shapes. Akaji was born in Kanazawa, Ishikawa prefecture, to a family that made *magemono* (bent-wood substrates) for lacquerware (Himono-ya) and was trained in *kyushitsu*. Later he went to Tokyo and became a pupil of the eldest son of the renowned master of urushi art, Kisaburo Watanabe. At first he produced lacquer objects for the tea ceremony, but after World War II he started making creative works and became famous for his technique of *magewa-zukuri*, in which he utilized the *magemono* technique. Usually the entire *magemono-kiji* (wooden substrate) is covered with linen, but Akaji covered each

Figure 1. Yusai Akaji, colored urushi dish in magewa *technique, 1961, height 10.2 cm, diameter 42 cm. Notice the simple and forceful shape of the* magewa.

Figure 2. Mashiki Masumura, tray with handle in kanshitsu *technique, 1963, 26.6 cm x 46.6 cm x 31.3 cm. Notice the animated and dynamic shape characteristic of* kanshitsu *technique, in which an object is made by pasting together layers of hemp cloth.*

magewa (thin strip of wood used in making *magemono*) and then joined them to form the shape. The slits between each *magewa* correct the distortion caused by the drying of the wooden base so that the *magemono* becomes firm and free from deformity. Furthermore, the *magewa* itself becomes an element of the design and expresses lucid and modern beauty.

Mashiki Masumura (1910–). Kyushitsu (Fig. 2)

Masumura was born in Kumamoto prefecture where he studied the basics of the art of urushi. Later he became a pupil of Yusai Akaji in Tokyo. Masumura excels in the *kanshitsu* (dry lacquer) technique, in which the artist first shapes a prototype in clay, then makes a plaster mold and pastes on linen to make the substrate. In the technique of *kyushitsu* he is skillful in *roiro-nuri*, in which the surface of the top coating is ground down and polished to mirrorlike brilliance, thus adding to the beauty of the form.

Kodo Otomaru (1898–). Choshitsu (Fig. 3)

Choshitsu is a technique of applying urushi thousands of times in order to make a thick layer, and then engraving a design on this surface. Otomaru is from Takamatsu, Kagawa prefecture, famous for *Kagawa-shikki*. From his early teens he engaged in wood carving. Later he went to Tokyo in order to devote himself to the art of *choshitsu*. Originally only a few colors were used for urushi: black, vermilion, yellow, and green, but today a variety of colors is possible. Using these richly colored urushi, Otomaru demonstrated the complicated and delicate effect of urushi and perfected the modern *choshitsu* technique.

Figure 3. Kodo Otomaru, kogo *(incense container) in the shape of a guinea fowl in* tsuishu *technique, 1970, 4.6 cm x 5.7 cm x 7.7 cm. The shape of the guinea fowl is carved from a lump of urushi made by layers of* urumi *(red and black) and black urushi applied alternately.*

Figure 4. Taiho Mae, small box with cat design in chinkin *technique, 1959, 6 cm x 12.5 cm x 26 cm. Except for the whiskers and eyelashes of the cat, the rest is carved in pointillist technique in order to give dimension to the body.*

Taiho Mae (1890–1977). **Chinkin (Fig. 4)**

Chinkin is the typical decorative technique found in *Wajima-nuri* (lacquerware from Wajima), in which the urushi surface is carved with a special knife and the incisions are filled with gold leaf and gold filings, thus exhibiting sharp lines of gold color. In order for the gold leaf to be laid in precisely and give out an intense brilliance, the carving knife is shaped so that the base of the incision is smooth.

Taiho Mae was a leading *chinkin* artist in Wajima. *Chinkin* is basically a linear technique, and it is difficult to produce a dimensional expression; Mae, however, developed a new method of multidimensional expression with pointillism.

Figure 5. Joshin Isoi, eight-sided food container with designs of plants and flowers in kinma *technique, 1955, height 15.2 cm, diameter 46.5 cm. The substrate is* rantai, *bamboo woven according to the traditional* kinma *technique and the decoration employs classical designs of plants and flowers.*

Joshin Isoi (1883–1964). **Kinma (Fig. 5)**

Kinma is a technique that uses a special knife called a *ken* to engrave designs on the urushi surface. Colored urushi, such as vermilion, is embedded in the incision and burnished. The lines of vermilion often depict trees and flowers. *Chinkin* exhibits delicately sophisticated lines, whereas *kinma* tends to create accentuated and forceful lines. After Zokoku Tamakaji, who was a leading exponent of *kinma* technique in the late Edo period in Takamatsu, actively put this technique to use, it became the principle technique for *Kagawa-shikki.*

Joshin Isoi learned the art of urushi at Kagawa Prefectural High School of Applied Arts. Not only was Isoi skilled in the classic *kinma* technique, but he also experimented with variations in depth, the angle of the knife blade, and the degree of density of the engraving. Furthermore, he introduced the use of pointillist engraving to express perspective and dimension. It is interesting to note that Isoi devised and developed a new technique for both *chinkin* and *kinma,* which similarly use line engraving for decoration. Isoi taught at the high school where he himself had studied, and worked enthusiastically for the establishment of the Kagawa Prefectural Institute of Urushi Arts.

Figure 6. Masami Isoi, box with design of stone pavement in zonsei *technique, 1980, 10.5 cm x 17.4 cm x 27.5 cm.* Zonsei *technique is used on the band of stones. The speckled design of the surface of the earth is expressed in* kinma *technique.*

Masami Isoi (1926–). **Kinma (Fig. 6)**

Masami Isoi is the third son of Joshin Isoi. After army service in World War II, he was trained by his father and studied *kinma*. Masami succeeded to the creative techniques of his father, who was widely skilled in many techniques besides *kinma*, introducing new designs one after another. For example, among the traditional techniques of *Kagawa-shikki*, the *zonsei* technique is as famous as *kinma*. *Zonsei* is a technique in which the outline of the design drawn in colored urushi is line-engraved and then either left as it is or filled with colored urushi or gold. Masami Isoi created a new form of artistic expression by uniting the *zonsei* and *kinma* techniques.

Figure 7. Shozan Takano, small box with squirrel design in kijimakie *technique, 1940, 16.5 cm x 24.5 cm x 34.5 cm. The contrast of the smooth* makie *surface with the rough surface of the paulownia substrate, and the simplification of the design such as the loquat tree trunk, are particularly effective.*

Shozan Takano (1889–1976). **Makie (Fig. 7)**

Makie—sprinkling metallic powder on a wet lacquer surface to form a picture or design—is one of the principle decorative techniques in the art of urushi. Since there are numerous types of *makie* technique, there are a number of holders, each with his own characteristics. Takano was from Kumamoto prefecture and is senior to Mashiki Masumura. After studying basic urushi technique in Kumamoto, he attended the Department of Urushi Arts at the Kyoto Municipal Art College and went on to the Department of Urushi Art at Tokyo National Art College. He was especially successful at the finely detailed *makie* technique of Shosai Shirayama, under whom he studied in Tokyo. The characteristic of Takano's *makie* is that he raises hard urushi in order to make shadows along the sides, thereby creating a forceful effect. Takano's works in *kiji-makie* are modern in style, contrasting the smooth surface of *makie* with the rough surface of the paulownia wood base. He is one of the contributors to the development of the *makie* technique from the delicate and detailed art of the Edo period to one that is adaptable to modern art forms.

Figure 8. Gonroku Matsuda, shelf with heron makie, *1938, 69.5 cm x 90.5 cm x 39.3 cm. The heron is expressed in* shishi-ai togidashi makie *and its soft feathers by scratching technique.*

Gonroku Matsuda (1896–1986). **Makie (Fig. 8)**

Matsuda was born in Kanazawa, Ishikawa prefecture. From childhood he was engaged in *makie* and had a well-developed technique. He went on to graduate from the Department of Urushi Art at Tokyo National Art College. For many years he was a professor at his own college, now the Department of Urushi Arts at Tokyo National University of Fine Arts and Music. He was vigorous in the research and study of classical works, the preservation and restoration of cultural property, the training of successors, and the conduct of international cultural exchange, as well as in his own creative work. He was a leading figure in the art of urushi in Japan. He contributed to the establishment of the Wajima Training Center for Urushi Techniques, Ishikawa prefecture, and he was still teaching fervently up to his death. He was not only very skilled in the technique that involves the sprinkling of gold filings, but was also capable of using a variety of materials such as *raden* (shell), *hyomon* (metal) and *rankaku* (eggshell). These techniques were acquired through his profound study of urushi art objects from all over the Orient, including those of ancient China. Apart from the subject of decoration, he actively pursued the quest to produce a durable substrate for lacquer objects that would resist damage and distortion in any conceivable environment.

Figure 9. Shogyo Ohba, box with design of snake-gourd in hyomon *technique, 1981, 14.6 cm x 15.8 cm x 26.4 cm. The snake-gourd leaves and flying birds are depicted by* hyomon *technique. The snake-gourd fruit is represented by* kawari-nuri *technique, in which cracks are made deliberately in the surface of colored urushi.*

Figure 10. Naoji Terai, water container in kintai *technique, 1976, height 16.4 cm, diameter 18.8 cm. The red blossoms are made by eggshells pasted on red urushi, while white urushi is used for the white blossoms.*

Shogyo Ohba (1916–). **Makie (Fig. 9)**

Ohba, born in Kanazawa, Ishikawa prefecture, learned *kyushitsu* from his father and then went to Tokyo where he studied *makie* under Gonroku Matsuda. While he was engaged in the preservation and restoration of classical urushi with Matsuda, he deepened his knowledge of *hyomon* (metal set flush into urushi) and is now the leading exponent of this technique. Ohba's *hyomon* technique is so detailed that even the finest lines express the intensity of his stroke. He is now in Kanazawa, engaged in creative work. At the same time, he is enthusiastically assisting in the training of successors.

Naoji Terai (1912–). **Makie (Fig. 10)**

Terai was born in Kanazawa, Ishikawa prefecture, and studied under Gonroku Matsuda and others in the Department of Urushi Arts at Tokyo National University of Fine Arts and Music. After graduating, he studied materials for urushi, such as those used in *kintai* lacquerware, at the Institute of Chemical Research. Then he taught for a long time at Ishikawa Prefectural Technological High School, where he himself had studied, as had Ohba and Matsuda. He also served as director of the Wajima Training Center. Besides teaching, he worked on his original *makie* creations and has gained a high reputation, especially for his notable technique of *rankaku*. This technique uses selected parts of quail's eggshells, which are broken into minute fragments and then pasted over the design. The *rankaku* technique was formerly used merely to depict the white sections of the design; Terai, after several improvements, succeeded in expressing dimension and perspective solely by employing this technique. He also made it possible to depict colors other than white and to exhibit gradation. Terai is also very skillful at making *kintai* substrates. Usually, in the *kintai* technique, raw urushi is heated on metal, but Terai has introduced a new method of using an aluminum substrate. He has made use of the nature of the aluminum surface, which forms an oxidized film when electrolytically treated, to absorb urushi.

Each of the holders described here has not only contributed to the preservation of traditional techniques but has also striven to develop the rich expressionistic ability needed to create modern applied arts. They have pursued artistic superiority as well as technical perfection. Each holder has worked to improve his own skill but at the same time has been enthusiastically involved in the training of successors. Today, their contributions have brought about satisfactory results: many young talented artists have been brought into the field of the traditional arts.

Urushi art techniques selected as intangible cultural properties

The concept of "important intangible cultural properties" was designed to assure the training of successors and the handing down of techniques to future generations. This section is concerned with urushi techniques considered valuable enough to be recorded for future generations.

Hida-shunkei. ***Hida-shunkei Art Preservation Association***

Shunkei-nuri is a technique of applying transparent urushi over wood grain so that the beauty of the natural wood pattern shows through the urushi surface. The Hida region of Gifu prefecture has developed as a large *shunkei-nuri* producing district, and many varieties of *shunkei-nuri* are found in this area.

Noshiro-shunkei. ***Shojuro Ishioka (deceased)***

The Ishioka family, under the special care of the Satake *han* (feudal clan) of Akita prefecture, has handed down this technique from generation to generation, thereby preserving the pure form of the *shunkei* technique of the Tohoku district. This technique differs in minute details from *Hida-shunkei.*

Awano-shunkei. ***Giryo Inagawa (deceased) and Shozo Inagawa***

The technique of *shunkei* is found in various places throughout Japan. Of these, *Awano-shunkei* from Ibaragi prefecture is significant because it is the oldest form of the technique. The wooden substrate for lacquer objects made by the elder brother, Giryo Inagawa, and the urushi coating applied by the younger brother, Shozo Inagawa, were both completed in the same workshop. The use of a hardwood substrate and the application of transparent urushi without taking measures to prevent the penetration of the urushi into the wooden substrate are the significant characteristics of this technique.

Zonsei. ***Soseki Kagawa (deceased)***

Zonsei is the leading technique of *Kagawa-nuri* (Takamatsu, Kagawa prefecture). The outline of the design in colored urushi is carved with a knife to achieve a very distinctive appearance. Soseki Kagawa learned the *zonsei* technique from his father and became an outstanding practitioner. His devotion to the *zonsei* technique was highly valued.

Raden. ***Kako Kataoka (deceased)***

Raden is the technique of inlaying engraved shells such as *yakogai, awabi-gai, shirochogai,* and *kurochogai* in urushi and wooden substrates. In Japan, we sometimes make a distinction between *raden,* which uses *atsugai* (thick shell), and the *aogai* technique, which uses *usugai* (thin shell). Kataoka was especially skilled in using *atsugai.* While active in his own creative work, he was also enthusiastically engaged in the restoration and preservation of the Konjikido of Chusonji, a National Treasure.

Murakami-tsuishu. ***Shuko Suzuki (deceased) and Koichi Itagaki (deceased)***

Tsuishu usually means the engraving of several layers of urushi applied one on top of another. However, there is another technique—carving wooden substrates and then applying urushi—that resembles the *tsuishu* technique. The skill of Koichi Itagaki's application of urushi and the engraving by his younger brother Shuko Suzuki have been very highly praised.

Makie *tools.* *Matabei Komiya (deceased)*

There are various techniques for *makie,* and many special tools for these techniques have been devised. For example, there are numerous *makie* paintbrushes using all sorts of animal hair. Komiya was an outstanding maker of the most complex *makie* brushes, and his talent was highly appreciated by urushi artisans.

Training institutions

Kagawa Prefectural Institute of Urushi Arts, established in 1954, and Wajima Training Center, established in 1966, are the principal training institutions for professional and advanced urushi arts. Fundamental training in urushi techniques is carried out by each urushi-producing district, and urushi techniques are taught in some schools, such as the Tokyo National University of Fine Arts and Music.

Kagawa Prefectural Institute of Urushi Arts, which has a very long history, gives practical training in the traditional techniques of *Kagawa-shikki,* including *kinma, zonsei,* and *choshitsu.* The institute also teaches the skills necessary for creative work, such as painting and the molding of forms.

Wajima Training Center has been rebuilt recently and possesses good facilities. The institute offers courses in four different skills: *makie, chinkin, kyushitsu,* and *kiji* (preparing the wooden substrate). These courses are provided because *Wajima-nuri* is outstanding for its decorative techniques, such as *chinkin,* and for firm and beautiful *kyushitsu* technique. Trainees who are permitted to enter this institution have passed a national qualifying examination.

Both training institutions accept only a limited number of trainees. Because the courses are very advanced, graduates of these institutions now form the backbone of Japan's urushi art world. In addition, the instructors are holders of important intangible cultural properties in the field of urushi art, and they give their time enthusiastically to their students.

Besides the already mentioned individual holders of important intangible cultural properties, there are group designations. In the field of urushi art, the Association for the Preservation of Wajima-nuri (Wajima, Ishikawa prefecture) was approved in 1977 as "a holding group of important intangible cultural properties (*Wajima-nuri*)." This association consists of about twenty leading technicians from each field of wooden substrate making, such as *rokuro* (lathe-turning), *magemono*, *sashimono* (joinery), and *ho*-body (the production of parts such as legs); of coating, such as *kyushitsu*; and of decorating, such as *makie* and *chinkin*. Since there was traditionally a division of labor in *Wajima-nuri*, practitioners of each process were selected to form this important association.

Urushi art techniques selected for the preservation of cultural property

Techniques for the preservation of cultural property include both the materials and tools used in the process of preservation and restoration, and the actual techniques used to preserve and restore cultural property, such as art and craft objects or architectural monuments like temples.

Repair of urushi objects. Daitsu Kitamura (1910–)

Born in Nara, Kitamura studied urushi technique with his father and graduated from the Department of Urushi Arts, Tokyo National Art College. From an early age he worked on the repair of urushi objects in the Shosoin and in temples and shrines. He also researched the urushi technique of the Nara (646–794) and Heian (794–1185) periods. From his long years of experience, he has developed skillful repair techniques and his competence is highly valued.

Repair of urushi objects* (raden). *Kako Kataoka (deceased)

Kataoka's work has been described in the section on techniques selected as intangible cultural properties.

Manufacture of urushi brushes. Seikichi Izumi (1911–)

Among the urushi painting techniques, surface coating is particularly difficult. The surface should be free from irregularity, painted evenly, and clear of any dust. The making of the urushi brushes used for this process also requires careful work.

Human hair is usually used for an urushi brush. Izumi prefers hair cut from living Japanese women; to be suitable, it must be well dried, it must not have been damaged by chemicals, nor be very oily. Extensive experience is needed to differentiate brushes according to the characteristics of the painting methods employed in different urushi-producing districts. The fact that Izumi's brushes are superior is known throughout the urushi world.

Manufacture of urushi* koshigama (Yoshino-gami). *Kazuo Konbu

Any dust particles should be eliminated from urushi used for *uwanuri* (final coating) by filtering it through specially manufactured Japanese paper. *Yoshino-gami* (Yoshino-cho, Yoshino-gun, Nara prefecture) used as urushi filter paper is immediately dried on a drying board without first squeezing it to get rid of water; hence the paper is supple and suitable for filtering. At the time of writing, Konbu is the only manufacturer of traditional urushi filter paper.

Production and refining of Japanese urushi. Association of Japanese Urushi for Cultural Properties

Japanese urushi is translucent and lustrous. It also produces good colors and creates beautiful surfaces so that it is indispensable for the preservation and restoration of old cultural properties and the production of urushi artworks. This association is engaged in the planting of urushi trees (Iwate prefecture).

Conservation of Chinese Urushi: Methods and Difficulties

Toshie Nakajima
Tokyo National Research Institute of Cultural Properties

I have experience in the conservation of both Japanese and Chinese *shikki* (lacquerware objects). This paper, however, will discuss problems I have encountered in the conservation of Chinese *shikki*. Although the basic technique of producing lacquerware objects is the same everywhere, each region has its own modified technique and each urushi artist has his own technique. Conservation work, likewise, is conducted by conservators with their own peculiar techniques. Because of such diversity, there are many specialists who work in ways that differ from mine.

Classification of conservation techniques based on types of decoration

In conserving Chinese *shikki*, surface decoration determines the conservation policy. Here I should like to address problems found during the conservation of four typical kinds of decoration: *tsuishu* and *tsuikoku*, *guri*, *chinkin-zonsei*, and *raden*.

Tsuishu *(carved red lacquer) and* tsuikoku *(carved black lacquer)*

Many pieces of Chinese *tsuishu* and *tsuikoku* are found with cracks and missing parts because of changes over time and defects in the materials themselves, despite the fact that the objects originally had firm *shitaji* (ground), and surfaces coated with thick layers of urushi, and were clearly carved. Some damage is restricted to the *urushi-shitaji*, but much is rooted in the substrate. This is because, generally speaking, many Chinese *shikki* have roughly made substrates. Damage also differs according to whether the substrate is a joined work or a bent work. In joined works, poor construction at the edges often triggers damage.

First, damaged areas are fixed by injecting raw urushi. The operation is done in several steps. In the first step, raw urushi diluted with a solvent such as ligroin (petroleum benzine) is recommended. However, due consideration must be given to the fact that each object is in a different condition, each requiring steps and methods appropriate to it alone.

One way to determine whether the raw urushi has set is to rub it with a fingernail after it has dried. If set properly, the *urushi-shitaji* on the substrate does not peel off. This does not mean the harder the better. If too much urushi is applied, the area becomes extremely hard as a result of the low penetrability of urushi and this may have unfavorable effects on the surrounding undamaged area. Intuition and common sense are required for the conservator to make proper judgments at this stage.

Once raw urushi has been injected, the damaged area is filled with *kokuso*, a paste made of urushi mixed with various thickeners such as clay and wood powders.

This process consists of several repeated operations as *kokuso* is applied little by little. A knife is used to shape the *kokuso* filling after it has dried completely. *Sabi-urushi* (urushi mixed with pulverized ceramic) is applied over it, if necessary. A spatula can be used for this purpose unless the spreading of *sabi* may cause problems (as when the damaged area is adjacent to fine ground carving). In such cases, the *fude-sabi* technique, which utilizes a *makie* brush to apply a soft paste made from ordinary *sabi-urushi* and water, should be used. *Sabi-urushi*, like *kokuso*, is shaped with a knife after it has dried. This process is repeated if necessary.

The *sabitsuke* (*sabi* coating) is then coated with *hidori-urushi*, which is made by evaporating the moisture contained in raw urushi in sunlight. This urushi becomes transparent as time passes. After it dries it becomes mat, similar to old urushi surfaces. I use this *hidori-urushi* for all my conservation work. One or two layers of *hidori-urushi* are usually applied, followed by *suri-urushi* (raw urushi plus camphor), if necessary.

It is rather difficult to reproduce the same pattern and color tone on the restored area as on the original. There are arguments for and against color matching. The recent tendency is not to match colors, in order to differentiate the restored area from the original. On the other hand, some people believe in coordinating colors with adjacent areas. It is not easy to coordinate the tone of the new color with the original.

Guri *(multicolored carved lacquer)*

In essence, the conservation method used for *tsuishu* can be employed for *guri*. First of all, dust in the hollows of the *guri* and old conservation treatments (coloring) must be removed. The urushi surface will be kept free from scratches and other damage if a *kido* made from a sharpened willow chopstick is used for this operation. If any damage is found after the removal of dust, the area is filled with *kokuso* using a tool such as a bamboo spatula, followed by *sabitsuke* and coating with *hidori-urushi* and *shu-nuri* (urushi colored with cinnabar). Excess *kokuso* and *sabi-urushi* on the surrounding area can be removed when they dry to a certain degree. However, once they dry out completely, it becomes quite difficult to remove them.

Layers of *guri* often become detached from the *shitaji* (ground). In such a case, the *guri* should never be pressed down. Instead, the space should be filled with *mugi-urushi* (urushi with adhesive properties made of wheat flour, rice paste, water, and raw urushi) containing wood powder. This method seems to cause less damage in the future. Even if the restored area is still lifting after this operation, the object as a whole will be in a stable condition and the restored part will not be noticeable once the object has undergone all the necessary work.

Coating with *hidori-urushi* and *shu-nuri* are the next steps. *Suri-urushi* must then be applied over the entire surface once or twice, according to the condition. Finally, excess urushi is carefully removed with soft paper.

Chinkin-zonsei *(incised gold decoration)*

Basically the same conservation policy as for *tsuishu* can be employed. However, in this case, *kokuso* must be infilled using a bamboo spatula and *sabi* must be brushed on in the *fude-sabi* technique in order to keep the line carving free from urushi.

Raden ***(shell inlay)***

Most *raden* found on Chinese lacquerware is made of *usugai* (*aogai*), i.e., thin shell. A problem arises if urushi is used to fill the spaces between the inlay and the substrate since the urushi under the shell may cause it to change color. In the past, I have tried urushi for this purpose but have come to the conclusion that a resin adhesive is better. Due care must be taken in selecting such a conservation material, however, so that if urushi is coated over the restored area the drying process is not inhibited.

Further remarks

There are many more points that need care and attention. Some of them are listed here:

1. When *shikki* (lacquerware) is to be conserved, x-radiographs are sometimes taken in order to analyze the technique used for the substrate. Special attention must be paid to the thickness of the *shitaji* (ground) and urushi layers and to the material of the substrate in order to determine the power level needed to obtain a clear image. The conservator and the technician must cooperate fully.

2. Careful studies should be made of the type of material used and how it was applied.

3. In the case of jointed works in which many fine cracks are found, raw urushi diluted with ligroin is applied, wiped off with tissue paper and allowed to dry. This procedure is repeated until the cracks are filled. The greatest possible care should be taken if the cracks are so deep that urushi may leak through to the other side.

4. In many cases, it is hard to detect how the urushi was processed when Chinese *shikki* is coated with thick layers of *tsuishu*. In some extreme cases, the *urushi-shitaji* becomes "sandy." Conservators must bear in mind the possibility of very dilute urushi leaking out before it dries.

5. It often happens that damage on Chinese *shikki* is covered up instead of being repaired. *Shikki* that has undergone this type of repair sometimes suffers problems after being taken overseas.

Repairs using wax

Recently I had the opportunity to conserve some forty pieces of Chinese *shikki* at the request of the Linden Museum in the Federal Republic of Germany. All of them had been repaired with wax. This method of repair seems to have been developed in Europe. Careful observation of these repaired areas revealed that wax injected for the purpose of repair had in turn caused further cracks; both kinds of cracks had to be repaired. Since urushi coated over wax never dries, more time and labor were spent on the removal of the wax than on the conservation of the actual object. This experience shows that the selection of material for repairs should be made carefully.

Conclusion

Studies on the conservation of Chinese *shikki* have just started and most of the problems are left for future consideration. I have come to the conclusion, based on my personal experience, that minimal conservation work should be done and only on areas that are actually damaged. It is the role of conservators to bridge the past and the future by trying to preserve the object while leaving the original material as far as possible untouched.

When restorers are given their assignments, they work upon advice from persons responsible for conservation and display in museums. It is my sincere hope that those in authority will further deepen their knowledge of urushi and its techniques.

The *Kyushitsu* Technique Demonstrated on a *Natsume*

Shogyo Ohba
Lacquer Artist and Restorer

There is great variety in the technique of *kyushitsu* (coating with lacquer), reflecting both the individual urushi artist and the district where the *shikki* (lacquerware) is made. The model of the process given here shows four of the various *kyushitsu* techniques: *urushi-nuri, maki-ji, hon-ji,* and *honkata-nuri*. Until now most models have been made on sample boards or bowls, whether by lacquerware-producing districts, training or research institutions, or individual artists. However, after consulting Naodai Sakashita, a *kyushitsu* artist living in Kanazawa, we decided to make this model using a *natsume* (tea-powder container), an especially difficult *kyushitsu* technique. In September 1984, we started work. Ryozo Kawakita, a woodwork artist of Ishikawa prefecture, made the *kiji* (wooden substrate); Naodai Sakashita worked on the *kyushitsu*, and I supervised the work. The completed *natsume* was presented to Kanazawa Municipal Arts and Crafts College by Naodai Sakashita and is being kept there as study material.

This is a brief explanation of the *kyushitsu* process, based on the technique I learned from my father, from my master Gonroku Matsuda, and from colleagues, as well as the method of application suggested by Sakashita.

Steps in the kyushitsu *technique*

The model consists of four major stages: substrate, priming, intermediate coating, and top coating. These stages comprise seventy steps, the most significant of which are explained here.

I. Substrate.

A. First we cut the wood for the substrate from the trunk of a Japanese cypress. We cut a transverse section a little thicker than the total height of the body of the *natsume* and its lid (Fig. 1). This Japanese cypress had at least 318 wonderfully fine annual rings. Kawakita suggested that the tree was about four hundred years old.

B. *Kidori*. Several round shapes for the *natsume* are drawn on the wood (Fig. 2) and cut out with a belt saw (Fig. 3a).

C. *Arabiki*. The pieces are turned on a lathe until they are a little larger than the final size (Fig. 3b). Usually unseasoned wood is used, since it is easier to cut and shape.

D. Boiling the *kiji* (wooden substrate). The roughly shaped *natsume* is boiled in a large container for two hours, and the resin from the *kiji* flows out into

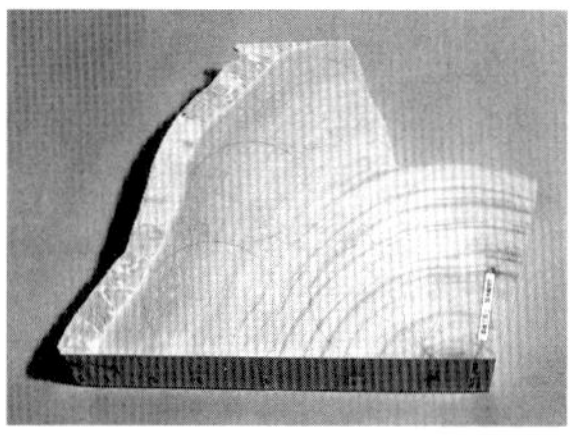

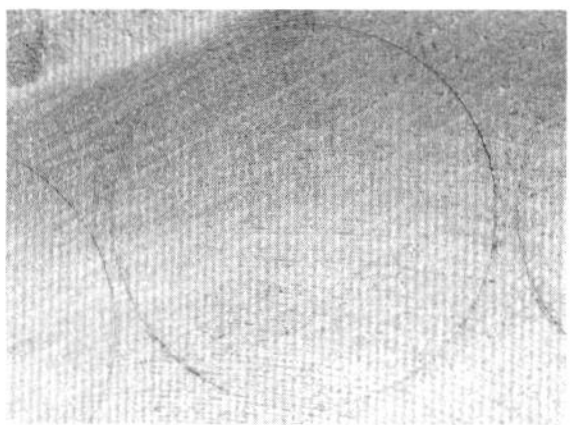

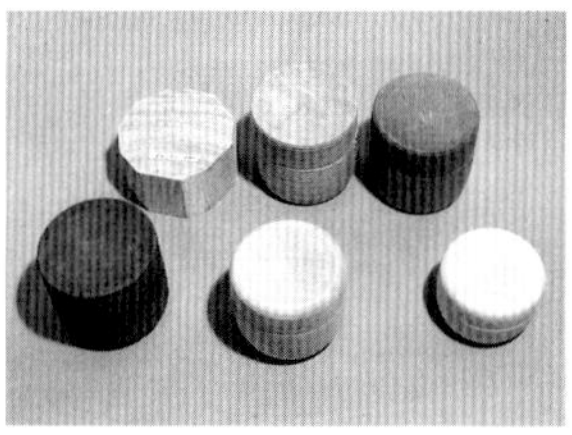

Figure 1. Transverse section of Japanese cypress.
Figure 2. The shape of the natsume *is marked out on the wood.*
Figure 3. Stages in the production of the wooden substrate.
Figure 4. The final shaping of the kiji *is completed.*

the hot water. This facilitates the drying of the wood and helps to prevent distortion (Fig. 3c).

E. Drying and heating (followed by at least two years' wait). Since boiling the *kiji* moistens the wood, this moisture must be completely evaporated, so the *natsume* is placed in a drying chamber for ten to fifteen days. Then the roughly shaped *natsume* is placed in an electric oven and heated for seven to eight hours at 80°C; if it is heated above 80°C , the wood will burn. This prevents distortion. The wood is charred to a depth of 2 mm from the surface (Fig. 3d). Then the *natsume* is allowed to stand for at least two years; if it is worked on the lathe immediately after heating, the speed of turning will cause cracks. Some *kiji* manufacturers carve the *natsume* after only six months but this will result in cracks and distortions later.

F. *Nakabiki*. After intermediate carving (Fig. 3e), the *natsume* is left to dry naturally for at least two months.

G. Final shaping. After final lathe-working, the shape of the *natsume* is completed (Figs. 3f,4).

II. Priming.

A. *Kokuso-bori*. When the *kiji* has knots, cracks, or gaps in the wood, that section is chiseled away with a small knife. Sometimes pieces of wood may be used to fill knotholes.

B. *Kijigatame*. This is performed twice. First, raw urushi diluted with ligroin (petroleum benzine) is applied to the *kiji*. Next, raw urushi alone is applied. Urushi penetrates horizontally cut surfaces more easily than vertically cut ones.

C. *Kokuso-kai*. This is performed twice. Small gaps resulting from *kokuso-bori* are filled with *kokuso* using a spatula: *shofu* paste (4 parts) + raw urushi (6 parts) + *kokuso-men* + wood dust. Since the urushi content is high, this *kokuso* is soft. The second *kokuso* consists of raw urushi (10 g) + rice paste (10 g) + *kokuso-men* (1 g) + wood dust (5 g; Sawaguchi 1966).

D. *Hikikomi-sabitsuke*. A thin layer of *sabi* is applied over the *kokuso*. *Sabi* consists of *tonoko* (10 parts) + water + raw urushi (6 parts).

E. *Kokuso-togi*. Grinding powder with water is used to rub down the *sabi*.

F. *Suri-urushi*. Raw urushi is applied.

G. *Kami-kise*. *Minogami* paste (10 parts) + raw urushi (10 parts) is applied. This mixture is called *handa-urushi*.

H. *Sabitsuke*. *Sabi* is applied to the edges of paper covering areas of *kokuso*.

I. Polishing. Sandpaper is used.

J. *Urushi-nuri*.

1. Coating followed by rubbing down. *Kurome-urushi* (dewatered urushi) plus some carbon black is used for the first coat. Subsequent coats are done with urushi alone. Each coat is followed by rubbing down.
2. Edge-making. Urushi has now been applied ten times and the *aikuchi* (where the sections join) has become blunt. *Sabi* is applied once on this part.
3. Heating. Donut-shaped rings are placed in the body and the lid of the *natsume*, to prevent distortion. The rings are made of plaster (7 parts) + cement (3 parts). After this mixture has dried completely, it is ground to shape and placed in the body and the lid (Fig. 5). Then the *natsume* is placed in an electric oven and the heat is increased slowly (about 10°C in two hours) until it reaches 70°C . After eight hours at this temperature, the *natsume* is removed and cooled.

Figure 5. Plaster rings placed in the two parts of the natsume *prevent distortion during heating.*

Figure 6. Hon-ji: *first* jitsuke.

4. Rubbing down with charcoal. The *urushi-uwanuri* (see II.J.1) and the *sabi* of the edge (see II.J.2) are rubbed down with charcoal.

K. *Maki-ji.*

1. First *maki-ji.* Raw urushi (10 parts) diluted with ligroin or volatile oil (10 parts or a little more) is applied with a rabbit's hair brush. The surface is immediately sprinkled with *jinoko,* 60-80 mesh (*Wajima nihen jinoko*).
2. *Fun-gatame.* Raw urushi is diluted with ligroin and applied with a rabbit's hair brush. A mixture that easily evaporates is preferred.
3. Second *maki-ji. Jinoko,* 80-100 mesh (something like *Wajima sanben jinoko*) is used.
4. *Fun-gatame,* as in II.K.2.
5. *Ji-togi.* The surface is smoothed with emery. The next step is separated into the application of urushi and the *honkata-ji* process; therefore *maki-ji* was performed twice.
6. Preparation for urushi coating.
    a. First coat and rubbing down.
    b. Second coat and rubbing down.
    c. Third coat.
    d. Edge-making and heating.
    e. Rubbing down.
7. Preparation for *honkata-ji.*
    a. *Uwa-sabitsuke.*
    b. Heating.
    c. *Sabi-togi.* The *sabi* is rubbed down with a grindstone.
    d. *Shitaji-gatame.* Excess urushi is wiped off.

L. *Hon-ji.*

1. First *jitsuke.* Raw urushi is placed on the work bench and mixed with some *jinoko. Heratsuke* is performed. Too much *jinoko* makes application difficult; not enough *jinoko* causes the *maki-ji* to shrink. The first application is 60-80 mesh (Fig. 6).
2. Second *jitsuke.* 80-100 mesh.
3. *Ji-togi.* The surface is rubbed down using emery with water, so that the *jinoko* covers it evenly in the next process.
4. Third *jitsuke.* >120 mesh. A fine powder is sprinkled on.
5. Heating.
6. *Shitaji-togi.*
7. *Shitaji-gatame,* as in II.J.7.d.

M. *Honkata-ji.*

1. First *jitsuke.* 80–100 mesh. *Tonoko* (10 parts) + water + *jinoko* (a little) + raw urushi (6.5 parts).
   Ji = *jinoko* (100 g) + water (40 cc) + raw urushi (50 g; Sawaguchi 1966).
   Kiriko = *jinoko* (50 g) + *tonoko* (50 g) + water (45 cc) + raw urushi (50 g; Sawaguchi 1966).
2. Second *jitsuke.* 100–120 mesh.
3. *Ji-togi.* This should be performed lightly.
4. *Uwa-sabitsuke.*
    a. First *sabitsuke.*
    b. Second *sabitsuke.*
    c. Application of *uwa-sabi* (final coat).

Figure 7. Stages in the intermediate coating procedure.

5. *Shitaji-gatame.*
   a. Heating.
   b. *Sabitogi.*
   c. *Shitaji-gatame.*

III. Intermediate coating.

A. Intermediate coating (Fig. 7a). Non-oily urushi such as *nakanuri-urushi* and *roiro-urushi* is used for the intermediate coating.

B. Rubbing down the intermediate coating, *uchi-konaka-nuri* (Fig. 7b). The surface is rubbed down with charcoal and *konaka-nuri* is applied to the inside of the *natsume*.

C. *Kiri-aikuchi-zukuri*

1. *Aikuchi go-zume* (Figs. 7c,8). *Tonoko* (10 parts) + water + cooked rice paste (2 parts).
2. *Aikuchi-sabitsuke* (Fig. 7d). *Sabi* is applied to the *aikuchi.*
3. *Aikuchi-sabitogi* (Fig. 7e). This area is rubbed down.
4. *Aikuchi-nakanuri* (Fig. 7f). *Kuro-urushi* is applied to the *aikuchi.*
5. *Aikuchi-nakanuri-togi* (Fig. 7g). The *kuro-urushi* is rubbed down.

D. *Konaka-nuri.* Urushi is applied.

E. *Konaka-nuri-togi.* The *konaka-nuri* is rubbed down.

F. *Aikuchi-wari*: *uchi-konaka-nuri-togi.* The *konaka-nuri* is ground down. The *natsume* is placed in water heated to about 50°C (hot to the touch). The expansion of air causes the lid to separate from the body of the *natsume.*

G. *Suri-urushi.* Water is kneaded into the raw urushi, making it easy to handle. The excess is carefully wiped off with paper.

Figure 8. Kiri-aikuchi-zukuri.

IV. Top coating.

Top coating is the final step. There are two types of top coating: *nuritate* and *roiro-nuri.* For *nuritate* the final coating completes the process. For *roiro-nuri,* after the last *uwanuri,* the processes of rubbing down (coarse and fine), *suri-urushi, dozuri, sutezuri, suri-urushi,* polishing and *kesho-zuri* are done in that order. The *roiro-nuri* process was not performed in this model of the process.

Acknowledgments

I should like to express my gratitude to the Kanazawa Municipal Arts and Crafts College and Assistant Professor Tsuneo Ueda for lending the model of the process and for their great cooperation. Furthermore, I wish to express my appreciation to Naodai Sakashita for making the model and Ryozo Kawakita for his valuable advice on *kiji*-making and the *kiji* process.

Bibliography

Sawaguchi, Goichi

1966 *Nippon shikko no kenkyu* (Studies on Japanese urushi techniques). Revised edition.

Lacquer Examination and Treatment at the Freer Gallery of Art: Some Case Histories

W.T. Chase

Smithsonian Institution, Washington, D.C.

The treatment of lacquer objects is still in its infancy. Many of our treatments are aimed at symptoms rather than attacking the causes of deterioration. Many of the treatments commonly employed do not adhere to modern conservation principles and may later cause damage that is difficult to undo. In some cases, better treatment procedures can be found by a synthesis of traditional and modern techniques. In this paper I shall illustrate our current thoughts on treatments for lacquer objects with a series of case histories of objects treated at the Technical Laboratory of the Freer Gallery of Art and the Arthur M. Sackler Gallery.

Similar ideas have been expressed by Anita Franke in articles on the restoration of a very large and important Chinese imperial throne with screen that is now in the Museum of East Asian Art in Berlin (Franke 1978, 1982) and recently in a pair of articles by Nicholas Umney (1987a, 1987b). Mrs. Franke used methods and materials very similar to ours described here, and she expresses very similar reasons for their adoption. For details of a conservation treatment of a complexity far beyond any here described, the reader is advised to consult Mrs. Franke's papers.

The current progression is away from making the object look like our conception of it and towards letting as much original material as possible show, and also towards treating the object in such a way that our treatment can be undone in the future. The two ethical principles of conservation that emerge are reversibility and minimum intervention.

Some procedures (cleaning, for instance) are by their very nature irreversible and must be undertaken with great care and with as full knowledge as possible of what their effects will be. Often a choice must be made between procedures; the more irreversible ones should be chosen only if they offer some benefit not available in reversible procedures.

One of the best formulations of the principle of minimum intervention was voiced by the late Per Guldbeck, while discussing the conservation of ethnographic objects: "Don't do anything more than you have to, consistent with not getting fired!" Perhaps the Bauhaus dictum "less is more" is another way to phrase it. Minimum intervention frees one from attempting to make something look like new, when all the object really needs for its continued preservation may be the setting down of some minor cleavage and a stable environment. The choice of treatments should be made not only with the end result in mind but also (and more importantly) the least change in the object.

In the specific case of the repair of Oriental lacquer objects, to apply these two principles would mean not using urushi, a material that is totally insoluble after it has set. In some cases (see cleaning of the inkstone box S87.0386a–pp, below) old urushi repairs can be removed mechanically. In others, where surface coats of urushi have been applied to even out the appearance of the restored object, removal is impossible. The color of the newly applied lacquer layers changes slowly with time; many old repairs have discolored and are now impossible to remove without damage. While discoloration can be dealt with, to some extent, by inpainting the discolored areas (see treatment of S87.0373a,b, below) it would be preferable to be able to remove the old, discolored repairs and to redo them in the correct color in a reversible material. Thin layers of old lacquer repairs also render it difficult (and sometimes impossible) to assess the present condition of a lacquer object and say how much of the original remains.

Perhaps a shining example of minimum intervention was demonstrated to the Urushi Study Group participants by the lute preserved at the Kasuga-taisha shrine at Nara, where the original lute had been stabilized and left in fragmentary condition and a replica had been made by a modern lacquer master to show what the lute looked like when new.

The application of the principles above, however, must be tempered with judgment, judgment seasoned with experience and common sense. In conservation we attempt to understand and apply not only the principles of material science and the methods of the craftsman, but also artistic and aesthetic judgment. Our results must succeed on many levels. Perhaps this paper should be viewed not as a plea to discard the traditional methods of lacquer repair practised so successfully by many of our colleagues in Japan and elsewhere, but as a demonstration that other methods of repair, using reversible materials and attempting to minimize intervention in the work of art, can be worthy of consideration.

Some of our Japanese colleagues are also considering and using modern reversible synthetic materials in lacquer repair; Nakasato (1986) used Paraloid B-72® in a 5% solution in xylene to consolidate the foundation layers of urushi on two decorated lacquer doors from Ryuzo-ji.

Effective housekeeping procedures

In the case of lacquer objects particularly, one must understand the causes of deterioration and attempt to halt the deterioration by attacking these causes. The primary cause of deterioration of lacquer objects is the movement of the substrate layers, coupled with loss of adhesion of the lacquer (and ground) layers. One of the most memorable things said during the Urushi Study Group's trip was Mr. Nakasato's pronouncement that deterioration of lacquer objects is always due to problems in the substrate or supporting layers.

For that reason, we attempt to keep our lacquer objects at a constant relative humidity. In Washington, D.C., we have standardized on 50% relative humidity, a level that is possible to maintain in our climate year round. Lacquer objects (and wooden objects) often can be seen to change when brought into this climate from the Orient. Any further cracks or cleavage from the change in climatic conditions are repaired, using the methods outlined below, and the objects then seem to be quite stable. As we proceed with the program of the new Arthur M. Sackler Gallery, we plan to exercise care in lending our lacquer objects only to borrowers who can maintain the correct level of relative humidity. Some of our lacquer objects are so fragile and vulnerable both to climatic change and to dangers in handling that they will not be lent at all. Conversely, it is planned to have the climatic conditions in the storage and exhibition areas adjustable so that special conditions required by lenders can be accommodated.

We are also careful to keep dust to a minimum, maintaining a good filtration system, and keeping the lacquers in closed cabinets, or in vitrines while on display. Gloves are used in handling the lacquer objects in the collection, to avoid fingerprints. The most effective, reversible treatments in terms of minimizing intervention are good housekeeping procedures.

Conservation treatments

For this paper, we reviewed the examination and treatment records of some eighty-three lacquer objects in the collection of the Freer Gallery of Art and selected specific case histories for presentation below. We have also treated a number of objects for private owners, usually because of the extremely challenging nature of the problems involved; we have been able to learn a great deal from these treatments of objects from outside the collection. Two are included in the case histories.

We have also recently completed the examination and treatment of approximately one thousand objects for the opening exhibition of the Smithsonian Institution's new Arthur M. Sackler Gallery. Included among these objects were forty-three Chinese lacquer pieces, many of which presented interesting problems in treatment. Five case histories of objects now on exhibition in the Sackler Gallery are included below.

Before we discuss specific case histories, let us see what generally happens in our laboratory when a lacquer object is treated. The object has usually been brought to the laboratory for correction of a particular problem, either in appearance or stability. Before any treatment is done, the object is given a thorough technical examination, including careful examination under a low-power stereomicroscope and examination of fluorescence in ultraviolet light. We use both a high-powered mercury arc lamp with Wood's glass filter (longwave ultraviolet) and a strong shortwave ultraviolet lamp. Operators always wear ultraviolet protective goggles for eye protection and to eliminate the fluorescence of the cornea, which adds flare. Visual acuity and perception of the colors of the fluorescence are much improved when wearing ultraviolet protective goggles.

The examination usually enables one to detect any repaired or re-lacquered areas. In many cases, x-radiography is used as well, both to detect old repairs and to ascertain the underlying structure of the object.

The technical examination has enabled us to formulate a treatment proposal, which is then discussed with the curators responsible for the object. After any necessary changes are made, work can proceed. Curatorial involvement is particularly necessary in the case of lacquer objects; in addition to the initial review of the treatment proposal, constant contact must be maintained with the curatorial staff as the work goes forward and options in treatment thoroughly discussed when the need for choice arises.

After a technical examination, objects are often cleaned. Cleaning is usually done with distilled water or a weak detergent mixture and cotton swabs; in some cases, organic solvents such as petroleum benzine (V.M.&P. naphtha, ligroin, petroleum ether), toluene or xylene are used to remove greasy surface accretions. Occasionally, the traditional Japanese polishing powder (*tsunoko*) is used; we have been using the modern version of this material, made from finely-divided anatase (titanium dioxide), with some success.

Consolidation of loose areas, or areas with cleavage and lifting of lacquer, is often done with wax or wax-resin mixtures, applied hot (60–70°C) with a small tacking iron. The heat used in application of the wax mixture softens the lacquer and allows it to be laid back into the correct position. In the case of small loose flakes without distortion, plastic resins such as the polyvinyl acetates or the methacrylates (Acryloid B-72®, an ethyl methacrylate-methyl acrylate copolymer, or Acryloid B-48N®, N-butyl methacrylate) are used, generally in solution in acetone or toluene, although we have also employed these resins in emulsion form. The same materials are used for fixing loose inlay, especially mother-of-pearl. In the case of extremely fragile lacquer pieces, or ones where the original surface is heavily deteriorated and loose, it seems safer to consolidate first and then to clean.

Small losses in the lacquer coat can be filled at the same time as consolidation, by employing some pigment of an appropriate color in the wax mixture. Larger losses, or losses at corners or rims, are usually filled with a harder wax mixture, containing some carnauba wax. This mixture can be carved to shape and then polished to the required sheen.

Rebuilding of larger losses can be done with a number of different materials; we have employed the harder wax mixtures, AJK (Alvar®-jute-kaolin) dough, and Japanese paper impregnated with polyvinyl acetate, all with success.

After the piece has been cleaned and the lacquer and inlay are in stable condition, inpainting is done with reversible paints. We have used colors ground in dammar resin (Maimeri® colors), methacrylate solution colors (Magna® and our own colors ground in Acryloid B-72®), and colors based on vinyl acetate emulsions (Liquitex® and Winsor & Newton). Inpainting must be done after polishing with *tsunoko*, if this step is included in the procedure.

In some cases, where the lacquer has lost its sheen or gloss, an overall surface coat may be applied. Usually this is a very thin application of the British Museum wax formula, commercially available as Renaissance Wax®. In some cases a thin coat of synthetic resin varnish may be applied to the surface to produce a harder sheen. In the past, we have used a 2% solution of polyvinyl acetal resin (Alvar®) for this purpose; recently we have been relying on the waxes, where necessary. Many objects retain sufficient sheen to make a surface coating unnecessary.

All of the materials applied in these treatments are removable later, if necessary. In fact, in one case, we removed a wax coating that we had applied two years before so that an object could be polished by Japanese traditional methods. Wax fills of cracks have the additional advantage over lacquer fills that the wax can extrude from the crack if climatic change causes the crack to close up, thus preventing further spreading of the crack. All of these materials are also easily distinguishable from the original lacquer, a fact that may serve later scholars well.

Case histories: wax fills and reattachment of cleavage

1. A shop signboard (FGA #V31.77)

In 1977 we were requested to treat a large Japanese Meiji period lacquered signboard, or storefront sign panel, for a private owner. Since this was a rush job, no photographs were taken. The signboard was carved with three Japanese characters and an imitation of a seal impression. It was quite large (106 cm x 33 cm), and lacquered all over with red and black lacquer. The lacquer had extensive cupped cleavage, with flakes about 1 cm x 4 cm curling up, with both ends standing above the surface; in Western painting conservation we would call this "extensive tented cleavage." Many losses were present. The signboard had been treated before, with a thermoplastic adhesive, possibly polyvinyl acetate emulsion. Extensive remains of the old adhesive could be seen on the surface, and were quite distracting. Some of the lacquer flakes had been glued down in the wrong positions.

old adhesive could be seen on the surface, and were quite distracting. Some of the lacquer flakes had been glued down in the wrong positions.

First the old adhesive was removed by soaking with cotton swabs and a 1:1 mixture of acetone and toluene. The adhesive turned gelatinous and could be scraped off. Flakes which had been wrongly restored were removed and cleaned.

The cleavage was set down and loose fragments were reattached by the use of a wax-resin mixture. The wax-resin mixture was the same as that used for oil painting consolidation and relining, i.e., 2000 g unbleached beeswax, 1000 g dammar resin (Singapore gum dammar in lumps preferred), 1500 g paraffin, m.p. 56–58°C, and 500 g gum elemi (Bradley 1950).

To make the mixture, an aluminum pan is placed on an electric hotplate and the dammar resin is put in and melted. The melted dammar (and the whole mixture until cooled) is quite flammable; care should be taken not to overheat it, and avoid open flames in its proximity. After the dammar is melted, the beeswax is placed in the pan and stirred as it melts. When the mixture is uniform, the paraffin and the gum elemi are added, with stirring. Then the pan is removed from the hotplate and the mixture strained through cheesecloth into a shallow tray made of aluminum foil. Just after the mixture has set, it is scored with a sharp knife into blocks of a convenient size. It is then allowed to cool completely to room temperature; the blocks are removed from the aluminum foil, snapped apart and stored for use.

Incidentally, the author first encountered the use of this material for laying down cleavage in lacquer in 1965 when he was a student in the Objects Conservation Department of the Metropolitan Museum of Art, New York, where a large Coromandel screen was being repaired using exactly this method.

The wax-resin mixture was applied to the lacquer with a small, electrically heated spatula kept warm enough to thoroughly melt the mixture (60–70°C). The lacquer and substrate were thoroughly infused with the wax-resin mixture, and the tacking iron was held in the wax pool, not touching the lacquer, until the lacquer softened from the heat (usually about a minute). The lacquer was tested for softness by touching the raised portion lightly with the tacking iron or a wooden stick held in the other hand. After softening, the lacquer could be pressed back into place. A piece of glassine paper was then placed over the area, and the tacking iron used to iron the area flat, extruding any extra wax around the edges. The tacking iron was then set aside, and continuing pressure exerted on the area, either with the fingers or a cool, smooth piece of metal, until the wax had set. The glassine paper could then be peeled back and excess wax removed with ligroin or petroleum benzine. Usually only a preliminary removal of the wax was made at this stage, and work proceeded to the next area of cleavage.

After all of the loose cleavage had been tacked down, areas of loss were filled with a harder wax mixture: 38 g carnauba wax, 38 g paraffin wax, 10 g microcrystalline wax (Bareco 145®).

This was pigmented to suit with dry pigments; in this case, Venetian red iron oxide, raw umber and Mars black (ferro-ferric oxide) were used.

Any excess wax mixture remaining on the surface was removed with petroleum benzine; V.M.&P. naphtha was used here. The entire signboard was given a very light coat of the wax filling mixture dissolved in V.M.&P. naphtha, which increased the gloss, especially on the black background, making the signboard look more even and less weathered.

Figure 1. Late Zhou dynasty Chinese eared cup, red and black lacquer over wood, Freer Gallery of Art #47.24, overall view.

Figure 2. Eared cup #47.24, bottom view after repair; the repair can be seen at the upper right corner of the foot.

2. Zhou dynasty eared cup (FGA #47.24; Figs. 1 and 2)

A fragile and somewhat deteriorated Chinese lacquer drinking cup from the late Eastern Zhou dynasty, said to be from Changsha (Lawton 1982(135):183), was brought to the laboratory for repair of several small pieces of wood and lacquer which had flaked off from the foot, probably due to handling. The cup is 5 cm high and 17.1 cm long. The pieces were reattached using wax-resin mixture (see above), colored with burnt umber and vermilion. The cup was also infused with wax-resin mixture in the break area to consolidate it; some small fills were made with the same mixture to tone the area of loss. The treated area coincided with the most heavily distorted area seen at the upper right of the foot in Figure 2. An "invisible" fill was impossible, due to the heavy distortion of the surface. The lacquer of the cup was very glossy, and no general surface treatment was necessary.

This treatment was done in August 1967. When the piece was re-examined in the store of the Freer Gallery in June 1986, no change in the repaired area was visible.

3. Ming dynasty red lacquer dish (AMSG# S87.0392; Figs. 3–6)

This deeply-carved red lacquer dish, 3 cm high by 17.25 cm long, has a pattern of peonies and leaves set against a yellow background (Lee 1971:174). It was very dirty when received by the Arthur M. Sackler Gallery for inclusion in the opening exhibition in 1987. In addition to the general cracking and minor losses such as the one described below, some large areas, especially the leaves, had suffered heavy losses and had been recarved. Some loose pieces had been glued back.

The piece was cleaned with distilled water and cotton swabs. The smaller crevices were cleaned with cotton rolled on the end of sharp bamboo skewers. The cleaning evened out the color of the object and made the yellowish background evident (Figs. 3, 4). Visible remains of old adhesive were removed with acetone.

One area of loss in the background, near the right edge of the dish about one-third of the way from the top, was large enough to be quite distracting. The loss is about 0.5 cm x 0.8 cm in extent, and the black, rough ground layer is revealed. The pigments of the ground were tested by x-ray diffraction and by microscopy, and proved to be orpiment and iron oxide.

The area of loss was coated with 10% Acryloid B-72® in xylene as a separator. After this had dried, the area was filled with the wax filling mixture mentioned above (38% carnauba, 38% paraffin and 10% microcrystalline wax), colored with raw sienna, Mars yellow and indian red. As can be seen in Figures 5 and 6, the compensation of this area of loss has integrated the appearance of the object.

Figure 3. Ming dynasty Chinese deeply carved red lacquer dish with peonies and leaves on a yellow background, Arthur M. Sackler Gallery #S87.0392, condition before treatment.
Figure 4. The dish shown in Figure 3, after treatment.

Figure 5. Detail of loss in background on #S87.0392, after cleaning but before filling.
Figure 6. The same area as Figure 5, after filling.

Figure 7. Yuan dynasty Chinese multi-part inkstone box in black lacquer over wood with mother-of-pearl inlay, Arthur M. Sackler Gallery #S87.0386a-pp. This shows the inner portion of the box, assembled. Figures 8–11, opposite, are details of the small tray on the upper right, just beneath the cover. Photograph by Otto Nelson, courtesy of the Sackler Foundation, New York.

4. Yuan dynasty multiple-part inkstone box with mother-of-pearl inlay (AMSG# S87.0386a-pp; Figs. 7–11)

The object, 30 cm high by 31.6 cm long by 24 cm wide (Lee 1971:138–140; Yonemura 1987:46), is composed of an outer case with top and bottom decorated in red lacquer using the *qiangjin* technique, surrounding an inner set of boxes made from black lacquer with mother-of-pearl inlay on the outside. Figure 7 shows the inner set of inkstone boxes, assembled. The object was in precarious condition when received by the Arthur M. Sackler Gallery for inclusion in the opening exhibition in 1987.

The outer red box (not shown here) has major losses on all of its surfaces, along with extensive cupped cleavage of the remaining lacquer. Many old repairs, done in a brighter-colored red lacquer which no longer matches the original, complicated the treatment. We decided not to include the outer box in the opening exhibition but to leave its treatment for a later date.

Figure 8. One tray (#S87.0386d) showing the general condition before repair.
Figure 9. The interior of the tray shown in Figure 8. Note the general cracking, cleavage, lifting and loss of lacquer.
Figure 10. Detail of interior of the tray, before treatment, showing cupped cleavage rising 5 mm above the wood surface.
Figure 11. The interior of the tray after treatment.

The inner boxes have wooden inserts inside them to hold inkstones and water droppers, which are included with the set (Figs. 8, 9). The construction of the boxes includes a reinforcing fabric which covers the outside bottom, feet and sides of each box and wraps part-way into the interior. A sample of the fabric was removed from a large area of lacquer loss on the foot of one box and was identified as a bast fiber, probably ramie. On the inside, the black lacquer has been applied directly over the wood.

The lacquer inside the boxes was in a precarious condition, with many losses, old repairs, and much cleavage and lifting of the lacquer (Figs. 9, 10). In some cases, the lacquer has lifted 5–6 mm above the surface. Old repairs (the gray areas in Fig. 10) complicated the problem. The box was quite dirty on all surfaces, and many losses were evident both in the lacquer coat and in the mother-of-pearl on the outside. Old black lacquer repairs and reinforcements to the corners on the outside obscured some of the design; this was especially noticeable on the vertical framing pieces of the inlay towards the corners of the boxes, and can be seen clearly in Figure 7.

We made one test of complete consolidation and filling of all losses on one of the small boxes just under the top of the piece. Similar methods to those used on the street sign panel were applied, with the difference that smaller tools had to be made to work on these boxes. Pigmented wax-resin was used for the consolidation and elimination of cleavage. It was applied hot, lightly ironed, and the excess removed with petroleum benzine (ligroin). Losses on the inside were filled with the same material; removal of excess wax also removed the dirt and dust on the inside. The result is shown in Figure 11. The lacquer is now stable. Shortage of time before the exhibition forced us to set down only the worst and most precarious areas, and leave the rest of the interior untouched.

The mother-of-pearl inlay on the outside was dirty and, in some cases, loose. Cleaning and consolidation were done under a low-power stereomicroscope. An initial washing, section by section, was done with water on cotton swabs, taking care not to loosen any mother-of-pearl. When a loose piece of inlay was encountered, it was glued down with 10% Acryloid B-72® in xylene. The spots of harder dirt (which did not respond to washing) on the inlay were removed mechanically by cutting them away with a very sharp small scalpel blade, taking care not to scratch the inlay. The blade was held nearly parallel to the surface of the inlay, to lift the crusty dirt away. Similar methods were used to reveal any inlay covered by old lacquer repairs.

Work began on this object in February, 1986 and was completed in July, 1987. It took approximately six man-months full time. Difficult areas such as the back corners of the lower box (Fig. 7), where the support is no longer in plane, were treated with Acryloid B-72® for consolidation, followed by filling losses with a putty made from Acryloid B-72® and dry pigment. We considered methods for compensating the losses in the mother-of-pearl inlay, possibly cutting new pieces of inlay and setting them in with synthetic resin, or filling the areas of loss to the correct plane with wax and then inpainting the inlay with pearlescent colors. After cleaning and consolidation, however, we decided that compensation was not necessary.

Case histories: Replacement of loose pieces with synthetic resin

Synthetic resins have been used in the case histories above for setting down small areas of loose pieces, notably in the multiple-piece inkstone box just discussed. Here we show some more examples of their use.

1. Yuan dynasty deeply carved red box (AMSG# S87.0390a,b; Fig. 12)

This box is 7.9 cm high by 17.6 cm long (Lee 1971:162; Yonemura 1987:38). When received, it was quite dirty, with many small loose pieces in both the carved areas and the background. The insides of both the top and the base had been re-lacquered sometime in the past, as had the bottom of the base. The lacquer surface was cracking and lifting throughout; many of the cracked areas seemed quite stable and would probably stay in place during normal handling, but others were less secure. Some consolidation of the object was necessary to prevent further losses. Interestingly, on this object, the yellow color in the background seemed to have been applied after the red layers had been carved, as it frequently overlapped onto the vertical walls of the red carved areas. The yellow layer was quite thin and broken up. Small pieces of the yellow easily fell out: it required consolidation. The yellow was identified by x-ray diffraction as orpiment.

Figure 12. Yuan dynasty Chinese red lacquer box, deeply carved, Arthur M. Sackler Gallery #S87.0390a,b. Photograph by Otto Nelson, courtesy of the Sackler Foundation, New York.

Treatment began with cleaning with cotton-tipped bamboo skewers, moistened with distilled water. This revealed many more loose pieces, especially in the yellow background. The treatment continued, alternately cleaning and readhering loose pieces with 10% Acryloid B-72® in acetone. For the consolidation of larger areas, a hypodermic needle was used to inject Acryloid B-72® underneath a section of carved lacquer. The excess was removed with acetone. Areas of loss were left unrestored on this piece. The black areas inside the piece and on the bottom of the base were gently polished with a soft cloth to remove fingerprints. The larger cracks in the carved lacquer on the underside and at the rim were left untreated; the underlying reinforcing fabric can be seen clearly through the cracks.

Figure 13. Qing dynasty Chinese octagonal black lacquer box with mother-of-pearl inlay, Arthur M. Sackler Gallery #S87.0407a,b. Photograph by Otto Nelson, courtesy of the Sackler Foundation, New York.

Figure 14. One piece of inlay being fitted back into the surface.
Figure 15. The surface after the loose inlay has been glued into place. The color of the replaced piece is slightly darker than the rest in this photograph, possibly due to the continuous layer of adhesive behind it and to the direction of the lighting for the photograph.

2. Qing dynasty octagonal black box with mother-of-pearl inlay (AMSG# S87.0407a,b; Figs. 13–15)

This octagonal box, 24.9 cm high by 25 cm in diameter (Lee 1971:209; Yonemura 1987:47), had lost four pieces of inlay which were found in the packing surrounding the object after transport.

Some of the remaining inlay did not have a continuous adhesive behind it, as could be seen when a small area of the box was wetted with petroleum benzine (ligroin). The solvent could be seen to penetrate behind the mother-of-pearl and darken it. However, it was decided at this stage only to readhere the loose pieces and leave the rest of the inlay alone; a close watch will be kept on the security of the rest of the inlay.

The four pieces were readhered with 10% Acryloid B-72® in acetone. The most difficult part of the job was finding out where the inlay pieces went. After gluing them back, their appearance was slightly darker than the rest, due to the adhesive which holds them securely in place (Figs. 14 and 15).

3. Edo or Meiji period Japanese circular covered box (FGA #04.348; Figs. 16 and 17)

The box, or tea caddy, is 5.5 cm high and 4 cm wide (Fig. 16). It probably dates from the nineteenth century, and may have been made in imitation of designs by Korin (Yonemura 1979:87). Of the three thick pieces of mother-of-pearl inlay on the top, one had fallen off entirely and another was loose. The loose piece was removed so that a good adhesive join could be made. Under the mother-of-pearl the original adhesive could be seen; it was dark reddish-brown, and looked as if it were unpigmented lacquer (Fig. 17). A small sample was removed for solvent testing and proved to be insoluble in alcohol, acetone, 1:7 nitric acid, and 1:4 methylene chloride (heated).

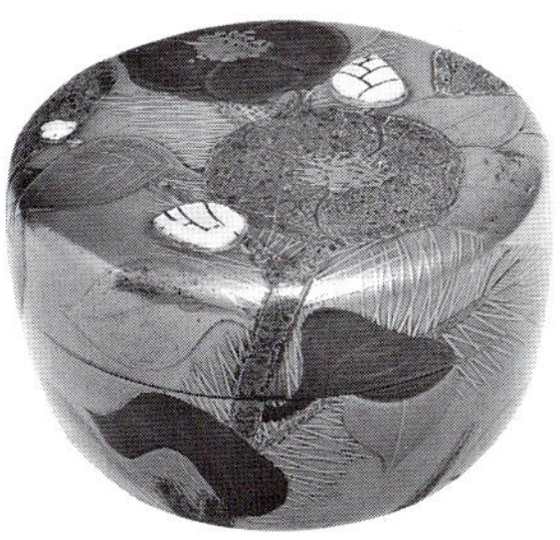

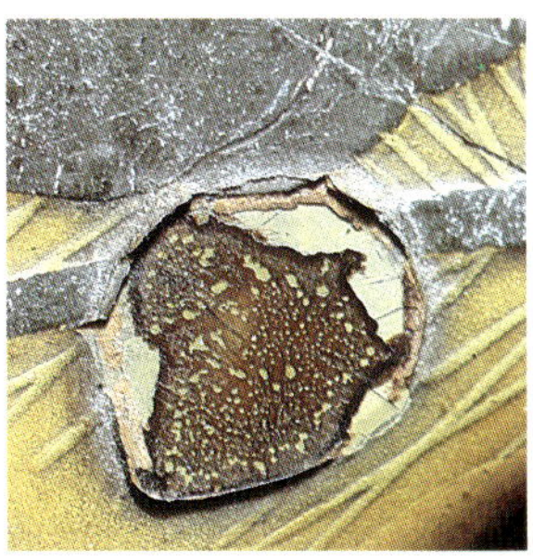

Figure 16. Edo or Meiji period Japanese circular covered box in gold lacquer over wood with inlays, Freer Gallery of Art #04.348, after treatment.
Figure 17. Detail of #04.348 with missing inlay, showing lacquer adhesive which held the inlay in place.

The mother-of-pearl inlay pieces did not fit snugly back, because the adhesive on the back had blistered. It was removed mechanically and the leaves were adhered with polyvinyl acetate emulsion (Borden's Polyco 199®, now out of production). The same adhesive, mixed with powdered gold, was used to compensate the very small losses in the lacquer around the inlay.

The lead inlay, which shows gray in Figure 16, has small, white patches of corrosion on it. The white material was insoluble in cold and hot water, toluene, and ethanol. The white corrosion product was identified by x-ray diffraction as hydrocerussite. It has been checked repeatedly since, and no spreading of the corrosion or new spots have been seen; each time it has been decided to leave it alone. Lead corrosion products on lacquer will be dealt with in more detail below.

Case histories: Structural repair with synthetic resin

1. Here the statue, shown before repair, is attributed to the Song dynasty but the date has since been revised to Tang in the Freer Gallery of Art accession records.

Table 1

	Parts by weight
Alvar 770® polyvinyl acetal (Shawnigan Chemical)	100
Acetone	130
Methanol	50
Amyl acetate	40
Xylene	100
Water	60

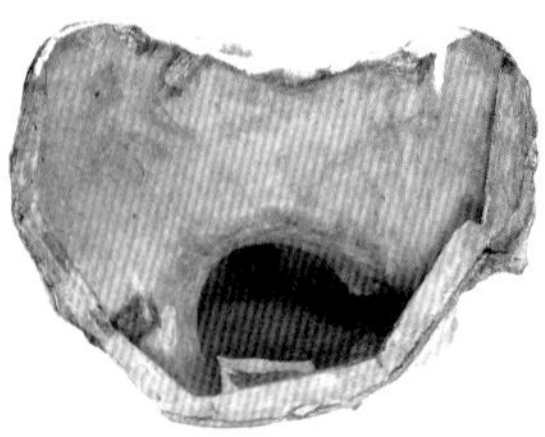

Figure 18. Tang dynasty Chinese dry lacquer statue of the Buddha Sakyamuni, Freer Gallery of Art #44.46, before repair. Note breaks and unevenness along bottom edge.
Figure 19. The interior of the base of the dry lacquer statue shown in Figure 18, after repair with AJK dough. The fabric support for the dry lacquer can be clearly seen.

1. Tang dynasty, seventh-eighth century A.D., Chinese dry lacquer statue, seated Buddha (FGA #44.46; Figs. 18–20)

This statue is nearly lifesize (height 99.5 cm, width 72.5 cm). It depicts the Buddha Sakyamuni, but the individual detail of the face suggests that it was actually a portrait of a Buddhist monk; dry lacquer portraits of similar quality were often produced during the Tang, Liao, and Song dynasties (Freer 1972:173).[1]

The statue is made in the classical dry lacquer technique, with layers of cloth saturated with lacquer built up over a clay core which is removed after the lacquer is dry; this can be clearly seen in Figure 19, a photograph of the interior of the base of the statue after treatment. Wooden reinforcements were used in the arms. The hands and earlobes are missing.

In 1959 a piece of the lower proper left front edge was pulled off the statue while it was on exhibition; an area about 5 cm x 5 cm was lost, and some of the remaining underlayers were loosened. Polyvinyl acetate emulsion was used to impregnate the area and to attach a piece of Japanese mulberry paper to hold the loose fragment in place. Impregnation with PVA emulsion was repeated in 1962, but problems of visitors removing and disturbing the broken lower edge of the statue continued.

In November 1969 we decided to fill in the skinned portion of the drapery along the lower front of the statue and even out its contour so that it would look better and not present such a tempting target for visitors. Figure 18 shows its condition at this time. First, the loose portions of the bottom and the drapery areas which were skinned were consolidated with Vinac B-15® (polyvinyl acetate) 10% in methanol, applied by repeated brushing. This rendered the areas quite strong. Next, any spots which were to be filled were brushed with an emulsion made from Alvar®, the adhesive component of the AJK (Alvar-jute-kaolin) dough which was to be used as a filler. The same emulsion was used to make the AJK dough (see Table 1).

The emulsion was thinned with xylene and water for the brush application. Then the AJK dough was made by mixing jute flock (which we had cut from commercial jute mailing twine) and kaolin (C.P.) into the emulsion to the right consistency, a little stiffer than library paste. We used about two dry measures of jute flock to one of kaolin. A little commercial whiting (chalk) was added to the later batches, to improve the surface finish in the solidified dough. Some of the AJK dough was kept sealed up wet in containers, for use as a putty, but most of it was formed into slabs about 0.5 cm in thickness and allowed to set.

After drying, the slabs were cut to fit the areas being repaired and set in place with wet AJK dough. Heat was used to bend the slabs to the correct contour before attaching them. Small spines were cut from the set dough and adhered to the surface to form the backbones of the centers of the drapery folds. The repair was allowed to dry, and more AJK dough added as putty where necessary to compensate for shrinkage. A putty was made from the Alvar® emulsion and kaolin alone to fill small holes and even out the surface. The whole area was finish-sanded to smoothness with 600-grit carbide paper, being careful not to touch the original.

The filled areas were then inpainted with Magna® polymer paints, thinned with toluene and applied with an airbrush. The fills were inpainted a uniform brownish gray so as to remain easily visible. Since these polymer paints dry quite mat, the fills were varnished with a light coat of MS2B® varnish, 25% in petroleum benzine, to bring up the gloss.

The entire statue was then lightly cleaned and polished by rubbing with British Museum wax (available as Renaissance Wax® from Picreator Enterprises Ltd, London). The wax restored some sheen and life to the lower portion of the statue, especially the drapery which had been repainted a uniform gray at some time in the past. After completion of the restoration (Fig. 20), the statue was returned to exhibition in 1969, on a new, fitted base, and remained there in a stable condition until fall, 1987, when it was moved for gallery renovation.

The restoration of this object was another case where the curator (in this instance, the late John A. Pope) and the conservator worked very closely together, and decisions were made jointly. We had thought of attempting to complete the earlobes and possibly the hands; I made a trip to the Metropolitan Museum, New York, in 1968 to measure and draw the extant earlobes on a similar sculpture. We decided that the repair to the bottom edge should be done first, and then we would decide whether any more compensation of missing areas would be done. We all felt that the statue looked much improved after the treatment described above and no more compensation should be attempted.

Figure 20. The dry lacquer statue shown in Figure 18, after repair.

Along with the conservation treatment of the piece, some of the dull blue paint from the chest area was identified by R.J. Gettens as being azurite, and possibly original. This paint was left intact in our treatment. Also, a fragment of wood was removed from the inner end of the wooden core of the proper left arm for carbon-14 dating to determine the age of the sculpture. The wood appeared to Gettens to be *Cryptomeria*, a tree that grows to some age in the Orient. The carbon-14 date was reported as A.D. 510, somewhat earlier than the seventh eighth century date now given (Long 1965). This could be either because the wood came from the inner rings of a large tree or because of re-use of an old piece of wood. At least the possible Tang date has not been disproved.

2. Japanese Heian or Fujiwara period dry lacquer statuette of a Bodhisattva (FGA #16.418; Figs. 21 and 22)

This statuette had been in the collection since 1916. It was extremely dirty and in somewhat bad repair when it was treated in 1961. It is 53.2 cm high and 40 cm wide, and was made in a dry lacquer technique similar to that of Tang statue discussed above. Figure 21 shows the statue before treatment. It was cleaned with detergent and water on cotton swabs, which removed most of the dirt. Since the black lacquer with gold surface under the dirt was very hard and compact, some of the more resistant dirt was abraded away using fine carborundum powder. Old repairs are evident in a number of places on the statue; the most visible is that on the proper left arm.

The large hole on the top of the proper right knee was filled with Japanese mulberry paper and polyvinyl acetate emulsion, probably Polyco 199®. A layer of paper was adhered from the inside first; it was then coated from the outside with the emulsion and allowed to dry. Inpainting was done with the same emulsion mixed with earth colors and whiting. Numerous other loose pieces of the lacquer (earlobes, one finger, etc.) were glued in place with the same emulsion. Figure 22 shows the statue after cleaning and filling the loss in the knee.

Figure 21. Fujiwara period Japanese dry lacquer statue of a Bodhisattva, Freer Gallery of Art #16.418, before cleaning and repair. Note hole in proper right knee. Figure 22. The statue shown in Figure 21, after repair and cleaning.

Figure 23. Late Zhou Chinese eared cup on stand, red and black lacquer, Freer Gallery of Art #49.1.
Figure 24. The interior of the eared cup shown in Figure 23, showing damage due to wood shrinkage at the center of the bowl.

3. Chinese Zhou dynasty stem cup (FGA #49.1; Figs. 23 and 24)

This cup, from the fifth–fourth century B.C., is also reported to have been found at Changsha (Freer 1972(112):177; Lawton 1982(134):181). It is 23.5 cm high and 17.9 cm wide. While the cup was in remarkably good condition, the shrinkage of the wood support had caused two problems.

First, the stem had become somewhat smaller than the base. Some unlacquered wood showed on the upper surface of the base where the cone-shaped base meets the stem, and the joint was loose and wobbly. The joint was secured by the injection of polyvinyl acetate emulsion (Jade 403®), thinned to a thick, creamy consistency with water. The object was allowed to set under its own weight; after a few minutes the excess emulsion which had extruded from the joint was removed with swabs moistened with distilled water. The cup was then allowed to set, undisturbed, for two hours. The areas of exposed wood and a few small losses around the outer edges of the base and the top were then inpainted with Magna® polymer colors thinned with toluene; a little Acryloid B-72® was added to the paint to increase its gloss so that it matched the original lacquer.

The second problem caused by wood shrinkage could be seen inside the bowl of the cup, where the horizontally-oriented wood of the bowl had shrunk with age, while the vertically-oriented wood of the stem had not, leaving an area at the center of the dragon raised above the rest of the surface. The lacquer in this area was stable; no loose cleavage was detected. A repair to make the lacquer at the center go back into the same plane as the lacquer in the rest of the bowl would have necessitated removing the lacquer and cutting down the wood of the stem, which would have been too drastic a treatment. We decided to leave this area alone for the present.

4. Chinese lacquered ewer, late Warring States-early Han periods, third century B.C. (FGA #49.22; Figs. 25–27).

This lacquer ewer is also supposed to have come from Changsha, according to a letter from J.H. Cox, dated 16 November 1949:

> The Ch'ang-sha lacquer ewer...according to one of the eye-witnesses of the excavation, came from one of the deep-hafted water-filled tombs characteristic of the late Warring States period and Early Han Dynasty. In fact there still remain traces of the white preservative clay...Such clay occurred only in these deep-shafted tombs and never in the case of the brick tombs common in the Han Dynasty.

Figure 25. Late Zhou Chinese ewer, lacquer over wood, Freer Gallery of Art #49.22, before repair.

The ewer is 15.6 cm high, 30.7 cm long and 15.3 cm wide (Lawton 1982(138):189). The black lacquer lies directly over the carved wood, with a broken layer of brown lacquer over it. The handle was detached when we received the object.

A large area of repair could be seen at the bottom, where the inset for carving out the interior of the body probably existed. The repair in the bottom was camouflaged by applications of a medium brown lacquer, which was peeling and blistering in some places, as was the brown lacquer on the rest of the object.

Peeling lacquer (especially the brown) was reattached by using a small brush to run polyvinyl acetate emulsion (Polyco 199®) around the lifting edges. The handle was reattached with new walnut pins, using polyvinyl acetate emulsion as the adhesive. The old pins (one is shown in Figure 26, which also demonstrates the thinness of the original lacquer) had shrunk badly and seemed not to have much strength. One additional pin was added at the front for extra strength where the handle had shrunk and no longer fitted the contour of the body. The repaired ewer is shown in Figure 27.

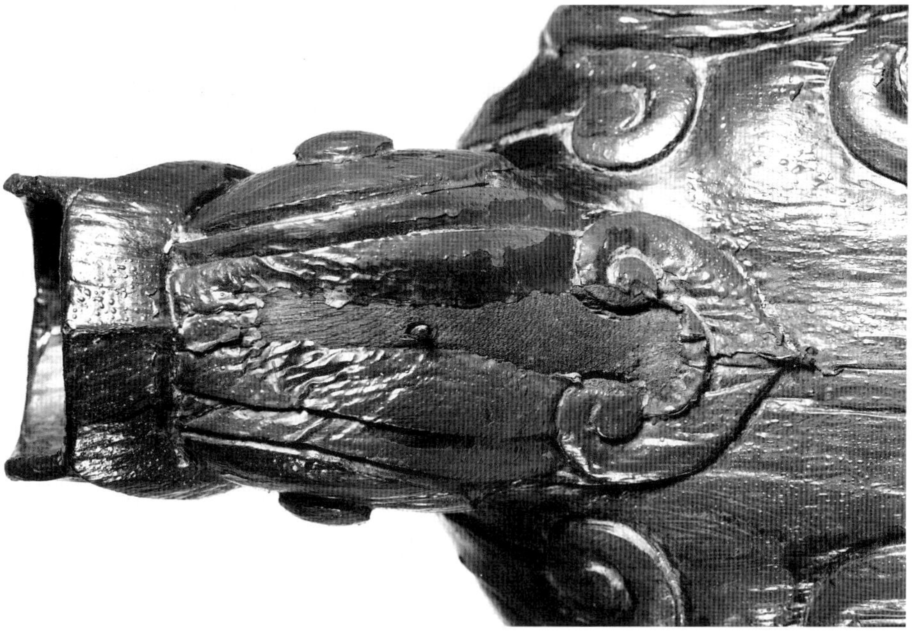

Figure 26. Detail of the handle attachment in the top spout area, from the ewer shown in Figure 25.

Figure 27. The ewer shown in Figure 25, after repair.

Lead inlay problem

Lead inlays in Japanese lacquers have posed a continuing problem in our museum, and it is a problem that we should like to investigate more deeply. In a number of cases, white, powdery corrosion has formed on the surface of inlays while the objects were on display; one object had been put into a newly-constructed exhibition case, the subfloor of which was constructed of plywood. Since the predominant corrosion product on the surface was lead formate, we assume that the urea-formaldehyde glue in the plywood had emitted formaldehyde which transformed to formic acid and attacked the lead. Lead formate is soluble in hot water, so the formate could be removed from the surface of the lead. Underneath the spots of lead formate, the lead had pitted deeply; the pitting was compensated with wax (the normal filling mixture) darkened with graphite and thinned with petroleum benzine (ligroin). Within the pits, and on one other box, hydrocerussite was also detected. This may be due to transformation of the lead formate to lead carbonate by atmospheric carbon dioxide, and the exact mechanisms of corrosion of these lead inlays would be a good subject for future study. It is clear, however, that lacquer boxes with lead inlay should be kept in a suitable environment free of formates.

Case histories: Cleaning and compensation

1. Japanese Tokugawa lacquer chest, Namban style, Momoyama-Edo period, early seventeenth century (FGA #79.50; Figs. 28–29)

This chest, of imposing size (53.3 cm high, 106.3 cm long and 45.3 cm wide), is reputed to have been sent as a gift from the second Tokugawa Shogun to Urban VIII, the Barberini Pope (Christie, Manson, and Woods 1976). Its treatment was very long and involved, and only the cleaning and compensation will be mentioned here. The surface of the outside of the chest when received looked quite yellow (Fig. 28). Longwave ultraviolet fluorescence showed a light yellow fluorescence on all the outer surfaces. This led us to suspect that the chest had been varnished, and a sample of the surface coating which was submitted to the Smithsonian's Conservation-Analytical Laboratory showed that the outer coating was mainly shellac.

The shellac was removed by exposing it to alcohol compresses and then swabbing off the soft, gelled material with cotton swabs. Some spots of repaint were removed with acetone. The cleaning considerably brightened the chest, especially the mother-of-pearl (Fig. 29).

After cleaning, losses were filled with the usual wax filler mixture and pigments. Losses in the mother-of-pearl were filled with unpigmented wax, which matches the mother-of-pearl remarkably well. As a final treatment, the chest was waxed with Renaissance Wax® (British Museum wax mixture) to restore its gloss. After many months of work, the overall appearance of the chest was much improved.

Figure 28. Tokugawa period Japanese black lacquer chest with gold paint and mother-of-pearl inlay, Freer Gallery of Art #79.50, showing the condition upon receipt, with two baroque trestle supports.

Figure 29. One end of the chest shown in Figure 28, partially cleaned. The semicircular top and the upper left corner of the body have been cleaned.

2. Chinese Song dynasty thirteenth century basketwork sacrificial box (AMSG# S87.0373a,b; Figs. 30–33)

An oblong lidded box with sides made of finely-woven basketry, this object stands 18.7 cm high. Other dimensions are: length 44.9 cm, width 19.7 cm (Lee 1971:101; Yonemura 1987:44–45). Losses in the lacquered surface had been filled with pigmented urushi during previous treatments; the old repairs had changed color and were quite visible (Fig. 32). The box was cleaned by our usual methods (Figs. 30 and 31). After cleaning, the old repairs were inpainted with acrylic polymer paints (Liquitex®), with gloss medium added to enhance the sheen. Inpainting made an especially dramatic difference at the corners, as can be seen in Figures 32 and 33. This inpainting can be easily removed with solvents, should the need arise.

Figure 30. Song dynasty Chinese red lacquer and basketwork sacrificial box with painted designs, Arthur M. Sackler Gallery #S87.0373a,b. Condition as received.

Figure 31. The box shown in Figure 30, after treatment.

Figure 32. One corner of the box shown in Figure 30, after cleaning but before inpainting.

Figure 33. The area shown in Figure 30, after inpainting.

Acknowledgments

Thanks are due to past and present members of the staff of the Freer Gallery of Art, the Arthur M. Sackler Gallery, and the Sackler Foundation. The past Directors, John A. Pope, Harold P. Stern, and Thomas Lawton were always encouraging in finding new and improved methods of treatment. Milo Beach, Acting Director of the Freer and Sackler Galleries, agreed to the publication of treatments of the objects under his care. Ann Yonemura, Assistant Curator of Japanese Art, assisted with helpful suggestions, and Takashi Sugiura, former Head of our Oriental Paintings Conservation Studio, helped in demonstrating Oriental lacquer restoration techniques.

Particular credit goes to the staff members of our Technical Laboratory. The late R.J. Gettens helped and encouraged me with the treatment of many of the pieces, notably 44–46; some of the treatment of this object (and a number of the other objects represented here) was done earlier by Elisabeth FitzHugh. Ilona V. Bene treated many of the pieces described here, including V37.72, 04.348, 16.418, and 79.50, the Namban chest, with which John Winter and Lynda Zycherman were also very involved. Jane Norman treated Sackler Gallery objects S87.0392, S87.0373a and b, and S87.0390a and b; Jane, Paul Jett, Stephen Koob, and I all worked on the inkstone box S87.0386a–pp. The skilled work, helpful advice, and excellent laboratory notes of all of them have made this paper possible.

Bibliography

Bradley, Morton C.

1950 *The Treatment of Pictures.* Cambridge, Massachusetts: Art Technology Press. Appendix C, no. 120.10, Wax-resin adhesive.

Christie, Manson, and Woods

1976 *Japanese Ceramics, Lacquers, Bronzes and Other Metalwork.* Sale catalogue, sale of Monday 22 November 1976:24, color illustration, and plate 23.

Franke, Anita

1978 "The use of beeswax and resin in the restoration of East Asian lacquer work." In *Conservation of Wood in Painting and the Decorative Arts.* N.S. Brommelle, Anne Moncrieff, and Perry Smith (eds). London: IIC, pp. 45–47

1982 "Restaurierungsbericht." In Beatrix von Ragué, *Ein chinesischer Kaiserthron: Die Pfirsche der Unsterblichkeit,* Bilderheft der Staatlichen Museen Preussischer Kulturbesitz, Berlin, Heft 40/41: 81-87.

Freer Gallery of Art

1972 *The Freer Gallery of Art, I: China.* Kodansha Ltd, Tokyo, Japan.

Lawton, Thomas

1982 *Chinese Art of the Warring States Period, Change and Continuity 480–222 B.C.* Freer Gallery of Art and Smithsonian Institution Press, Washington, D.C.

Lee, Yu-kuan

1971 *Oriental Lacquer Art.* John Weatherhill, New York.

Long, Austin

1965 "Smithsonian Institution radiocarbon measurements II." *Radiocarbon* 7: 253 (date SI–113).

Nakasato, Toshikatsu

1986 "Yamaguchiken Ryuzo-ji zo Chinkin Shitteno-zo zu Tobira no Hozon Shufuku" (Conservation Treatment of *chinkin shitteno-zo* doors, property of Ryuzo-ji, Yamaguchi Prefecture." *Hozon Kagaku* (Science for Conservation) 25: 63-82.

Umney, Nicholas

1987a "Oriental Lacquer." *Conservation News* 32: 23–25.

1987b "Oriental Lacquers." *Conservation News* 33: 13–15.

Yonemura, Ann

1979 *Japanese Lacquer.* Freer Gallery of Art, Smithsonian Institution, Washington, D.C.

1987 "The Art of Chinese Lacquer." *Asian Art* I,1: 30–49.

Some Thoughts about Conserving Urushi Art Objects in Japan, and an Example of Conservation Work

Kenichi Kitamura
Lacquer Artist and Restorer

The conservation of urushi art objects in Japan has two distinctive aspects. The first is that most urushi treasures have been restored, not by those who are specialists in conservation, but by those who are actually engaged in producing lacquerware using traditional techniques inherited from their masters. The second aspect is that experience acquired during the conservation of urushi art objects has revealed a great deal about both the techniques and the modes of artistic expression used in old urushi art objects.

In Japan today, outstanding urushi objects are designated cultural properties under the Cultural Properties Preservation Law. Conservation work is conducted under the guidance of the Agency for Cultural Affairs as part of a national program. As in the past, those who create urushi art objects are engaged in conservation. We have inherited traditions based on the foundations laid by our ancestors, but we must also try to develop new techniques for conservation by applying our knowledge of contemporary science.

This paper discusses problems that I have encountered and their solutions, together with a specific example from my experience.

Approach to conservation

It is our responsibility to recognize outstanding works by our predecessors and to hand down these cultural assets to future generations. We demonstrate this sense of responsibility in our methods of conservation and our approach to the management of cultural property during conservation work. We should attempt to learn from and to share both the spirit and the expertise of our predecessors, who created the urushi objects that we conserve.

We should try to preserve the object in its present condition and to determine what needs to be restored immediately and what should be left for future conservation work. First, we try to preserve the object in its present condition. Second, we try to determine what needs to be restored immediately; and, finally, we try to determine what should be left for future conservation work.

It is important to conduct extensive research prior to any conservation and to keep a complete record of conservation work. When deciding whether the present condition should be changed or not, we should never rely only on our personal judgment: Specialists should always be consulted.

Problems and solutions

Urushi objects are often so aged that they are on the verge of collapse. Therefore, a conservator is like a medical doctor who diagnoses his patient: He must judge which parts should be treated and which should be left alone as well as select a method of treatment. Suppose an urushi object has a few small dents on its surface. A conservator must decide whether these dents were caused by contact with other objects, or whether they are traces of *raden* (shell inlay) or gold or silver *heidatsu* (applied metal foil) patterns that have peeled off. If the latter is the case, the dents should not be treated as scars and filled in, but should be left as they are.

In order to make proper judgments under such circumstances, a conservator needs to be completely knowledgeable about changes in modes of expression and techniques. Also, he should have a mastery not only of the techniques of conservation, but also of producing replicas based on academic study.

Prior to any conservation treatment, the condition of the object should be meticulously recorded in order to prevent possible misjudgments in future generations. During conservation, the construction of the substrate and the condition of the *urushi-shitaji* (ground) may be revealed in the damaged areas. Since these serve as important evidence for studying the techniques and materials of the period in which the object was made, the conservator should try to record the findings accurately and preserve undamaged areas whenever possible.

Optical analyses using x-rays, ultraviolet, etc., are essential when there is a need to investigate internal damage and materials that cannot be detected by the eye alone.

Figure 1. Makie koto *(zither), Heian period, eleventh century.*

Occasionally, the object is restored to its original state, for example, when dirt on the surface seems likely to accelerate deterioration, or when the original urushi is obscured by past conservation work, provided that such measures are in the best interest of the object. Even if some measures seem harmless, it is necessary to ask for approval because the work may be interpreted as changing the current status of the object.

One of the most impressive lacquerware objects I have conserved is a *makie koto* (zither), a National Treasure kept at Kasuga Taisha (Fig. 1).

First, I moistened the surface using a soft brush dipped in lukewarm water. Then, I used soft Japanese tissue to remove the moisture by pressing gently. By cleaning in this way, designs that were dim became clearer. As a result, it became easier to study the *makie* technique and to determine the types of *makie-fun* (metal powders) that had been used. Unusually for this period, in addition to gold, blue-gold, and silver powders, copper powder had been used.

When objects are structurally unstable, lost parts should be replaced only after consulting specialists. New replacement parts must be recorded or have some engraved mark to identify them. To make a *shippi* (hide) box in the Shosoin (Fig. 2), raw leather was spread over the wooden form, then hardened by applying urushi, and removed from the form to make the substrate. As Figure 3 shows, nearly half the lid was lost before conservation. The ideal way to replace the lost parts would be to use the same material as in the original. But raw leather easily loses its shape and is highly unstable, so I decided to apply the *kanshitsu* technique, that is, linen cloth layered with *nori-urushi* (rice-paste and urushi), to repair this *shippi* (Fig. 4).

Figure 2. Shippi *(hide) box from the Shosoin.*
Figure 3. The shippi *box in Figure 2, before conservation. Nearly half the lid has been lost.*
Figure 4. The shippi *box in Figure 2 after restoration using the* kanshitsu *(dry lacquer) technique.*

Urushi is not the only material used to make lacquer art objects. A variety of materials, such as dyed cloth and metal clasps, and many techniques are combined to make the object. This is why it requires careful attention to avoid changing the condition of an object by applying urushi to parts that need not be touched. Take, for example, a *makie tebako* (cosmetic box) equipped with *uchibari* (decorative cloth applied inside the box) and *himokanagu* (metal fittings for cords). The *uchibari* must be removed before conservation, in cooperation with specialists, in order to prevent it being stained by excess urushi. When the conservation work is over, the *uchibari* can be replaced in its original position with *urauchi* (*washi* paper applied to the back of the cloth for reinforcement). *Himokanagu* should also be removed when possible and put back after conservation. If they are not removable, they should be covered with paper to keep them free from excess urushi. Conservators should also be extremely careful when dealing with *mitsudae* (a kind of oil painting) or *kingin deie* (painting with fine silver and gold powder mixed with animal glue), which tend to darken if they come into contact with urushi.

It is most desirable to use high quality Japanese urushi for the restoration of lacquer art objects. But if the *shitaji* (ground) is *doroji*—an urushi substitute made of *nikawa* (animal glue) mixed with *gofun* (powdered shells) and *tonoko* (clay)—it is often difficult to restore using only urushi. It is also difficult to restore solely with urushi if the entire surface of the object is colored with a mixture of *nikawa* and pigment. In these cases, use of other approved materials should be considered with the guidance of conservation scientists.

With the basic principle of maintaining the present condition, it is customary not to do any repair even if there is damage to the decoration. If a case arises, however, in which repair is needed to balance the overall structure, the first step is to seek the approval of specialists. Then a study of the original technique and material can be started to determine which techniques and materials will be most similar to the original.

According to this principle, no "antiquing" or artificial *craquelure* should be added to urushi objects. If the restored parts do not match the original and need "antiquing" for purposes of continuity, this should be limited to those areas where it is absolutely necessary. In addition, only material that can be removed later should be used.

Some urushi objects repaired in the past have had "antiquing" or *craquelure* added artificially. Those with urushi *koshoku* ("old color") and *danmon* (cracking) over the entire surface have lost their original appearance, because the painted urushi cannot be removed. Professional conservators should refrain from such repairs. Black urushi may turn brown as time elapses, so attention must be paid to the color of the repaired area. Once-matched color may become unmatched as the original color ages. A similar problem exists with *shu-urushi* (urushi colored with cinnabar) and other colored urushi. Since in-depth scientific studies of color changes have not yet been made, future research and study are awaited.

These thoughts are based on my personal experience and that of my seniors. Urushi art objects which require conservation show various kinds of damage and there is no single pattern of conservation. A conservator is always faced with contradictory problems. He must try to find the most suitable solutions and make decisions. He should never let dogmatism based only on experience dictate conservation. He must face the object with a modest attitude of exploration, and try to improve his own technique.

An example of conservation

Since 1982, I have conserved nineteen urushi art objects, made between the Sengoku period (Japan) and the Tang dynasty (China), which are kept at the Linden Museum in Stuttgart, West Germany. I went to the museum to meet Dr. Klaus J.

Brandt, head of the Oriental Department and in charge of conservation. We took pictures of the objects to be conserved, recorded their size and condition, and discussed conservation policy. After returning to Japan, I submitted a plan for the method, duration, place, and cost of conservation. I signed an agreement and started work on the objects. The conservation of one object, the lid of a *heidatsu gosu* (round box) is described here in detail (Fig. 5).

Figure 5. The lid of the heidatsu gosu *(round box).*

The conservation policy agreed upon was as follows:

- The present condition should be maintained by using urushi. Acrylic resin (Acryloid B-72®/Paraloid B-72®) can be considered for use only on the vermilion back of the lid.
- The exposed substrate can be left as it is as long as it does not cause any problem from the point of view of conservation.
- Fallen flakes should be returned to their original positions as long as these can be identified. Otherwise they should be put aside and kept as they are.
- Many points of similarity were found with four pieces of silver *heidatsu gosu* owned by the Shosoin (Fig. 6). The *gosu* owned by the Linden Museum can be restored to its original state using the Shosoin *gosu* as reference.
- The conservation process will be recorded.
- Conserved objects will be kept in double boxes made of paulownia wood.

The condition before conservation was recorded. The box is 10 cm in diameter and 1.6 cm high. The substrate is a straight-grained round plate made of maple wood, about 2 mm thick, around which 1.5 mm strips of the same material are wound approximately seven times (Figs. 8,9). It is not clear whether this ring is made of seven concentric circles or one long strip wound in a coil. It has not been determined, either, if a *rokuro* (wheel) was used for winding. The height of the lid can be measured, although only a small part of the side remains. Both the substrate and the urushi layers are so deteriorated that it is almost impossible to handle them. The structure of the substrate can be observed because the urushi layers have peeled off from the back of the lid.

The *kyushitsu* (layer sequence) consists of fine linen cloth (fifteen threads per cm^2) applied over the whole surface along the wood grain (Fig. 10). Over the thin *shitaji* (ground), black urushi is applied; the *uwanuri* (top coating) is *suki-urushi* (transparent urushi). The back of the lid consists of linen cloth, priming, black urushi, and vermilion *nikawa* (animal glue). The upper face of the lid is silver *heidatsu* (0.25 mm thick silver plate) with a design of small birds flying among trees and flowers (Fig. 11). At the center, three baby birds stretch their heads out of a nest, waiting for their parents to bring food. The fine lines of the picture suggest the skillful use of tools. The design is not made from a single sheet: silver sheets were joined together for the flower branches. Around the circumference of the lid a design called *renjumon* (a chain of round beads) is found. The sides of the lid have six-petaled flowers placed at regular intervals. In both cases, a semicircular chisel is believed to have been used to cut the designs in the silver plates. The worked silver sheets are a little higher than the surface of the surrounding urushi and there is no trace of the urushi surface having been polished with charcoal. These two facts suggest that the urushi over the silver sheet was carefully peeled off after the final coating. Parts of the silver sheets used for the *heidatsu* have become displaced because of the deformation of the substrate. Other parts have been lost together with the ground. Even the remaining parts are in danger of peeling off in the near future.

The structure of the substrate is the same as that of the silver *heidatsu gosu* in the Shosoin, which was x-rayed by Mr. Hoko Kimura (see Fig. 7). The design is also similar; the only difference is that this box is slightly smaller. The body of this

Figure 6. *One of four silver* heidatsu gosu *in the Shosoin.*
Figure 7. *X-radiograph of the* heidatsu gosu *in the Shosoin.*

heidatsu gosu has been lost, but assuming that it was similar to the Shosoin *gosu*, it would have had the structure of an *inro* lid. The sides of the lost body would probably have had six-petaled flowers of silver *heidatsu* like those on the side of the lid.

Prior to conservation work, photographs were taken of the entire lid and of individual details, in order to record the state of damage. X-rays were also taken for comparison with the Shosoin treasures (Fig. 13).

Conservation work sometimes includes fumigation with methyl bromide gas. In this case this process was omitted since there was no trace of insect damage.

Consolidation by injecting urushi from outside was not enough because of the deterioration of both the substrate and the urushi layer. In any case, use of urushi proved to be impossible because the entire back of the lid was painted in vermilion animal glue. After much thought, I attempted to remove the urushi layer from the substrate. This was done successfully by carefully using a thin, stainless steel, dental spatula (Fig. 12). The pieces of the urushi layer were placed temporarily on a distribution diagram made previously, in order to prevent any mistake in repositioning them at a later stage of conservation.

The substrate of the lid and the back of the urushi layer that had been removed were then impregnated with urushi. Impregnation was done in several steps, starting with a 30% solution and gradually increasing the concentration. In order to

Figure 8. *Diagram of the* heidatsu gosu *showing the construction.*

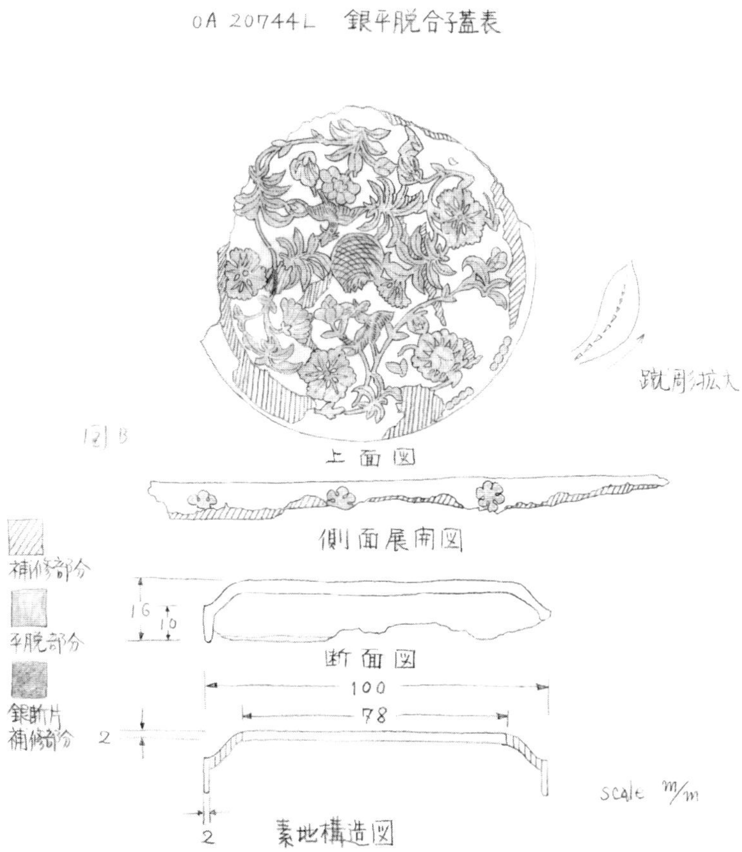

Figure 9. The substrate of the replica of the heidatsu gosu *showing the method of construction.*
Figure 10. A fine linen cloth is applied along the grain of the wood.

Figure 11. The upper face of the lid showing the heidatsu *design of birds, trees, and flowers.*

Figure 12. The urushi layer was removed from the substrate using a dental spatula.

Figure 13. X-radiograph of the lid of the heidatsu gosu *from the Linden Museum.*

Figure 14. Photograph of the completely restored heidatsu gosu *taken for record purposes.*

Figure 15. The replica of the heidatsu gosu *made for the Linden Museum.*

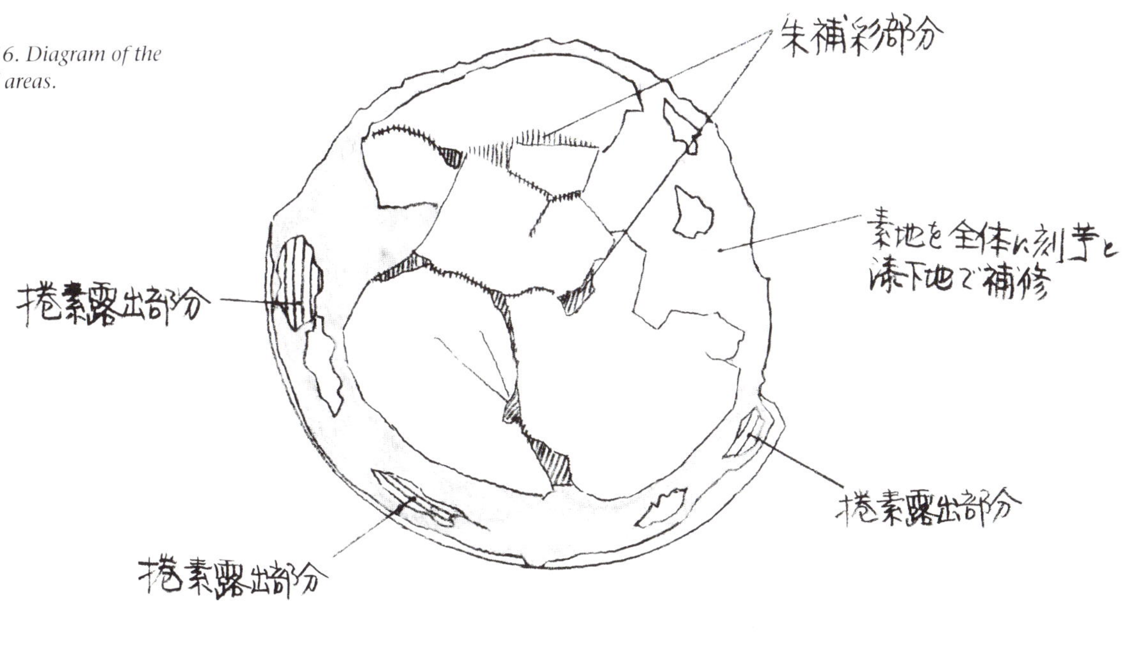

Figure 16. Diagram of the restored areas.

prevent warping of the urushi layers as they hardened, they were kept under pressure. Raw urushi was injected into the cracks on the face of the lid.

Mugi-urushi (urushi with adhesive properties made of wheat, rice paste, water, and raw urushi) was used to adhere the urushi layer in its original position. *Kokuso* (made of *mugi-urushi* mixed with burnt linen fiber and wood powder) was used to fill holes in the substrate and large cracks, using a bamboo or metal spatula; the *kokuso* surface was smoothed with a graver.

Kiriko, made of *jinoko, tonoko,* water, and raw urushi, was applied over the *kokuso* with a spatula. Then the surface was polished, varying the size of stone according to the areas of *jitsuke,* and the *kiriko* was "set" by impregnation with raw urushi.

Sabitsuke (*sabi* coating) was done by applying a mixture of *tonoko* and water to the *kiriko* base with a spatula. *Sabitogi* (polishing of the surface) was carried out using a stone which has a slightly coarser grain than that used for *kiriko. Sabidome* (fixing) was done by impregnation with raw urushi.

Depending on the size and position of the area to be painted, either a *makie-fude* or a *jinuri* brush was used to apply black urushi to form a base coat. The *shitanuri* (base coat) was polished using *togizumi* (charcoal) and water. Intermediate coating (*nakanuri*) was applied like the base coat, and polished in the same way. *Suki-urushi* (transparent urushi) was applied with an *uwanuri* brush as a top coat. The vermilion areas on the back of the lid were consolidated by applying a solution of acrylic resin (Acryloid B-72®/Paraloid B-72®).

The entire tone was adjusted. Because urushi newly applied on restored areas often stands out, it needs to be toned down for balance. A vermilion color similar to the original was applied over the cracks on the back of the lid.

Photographs were taken and records were made of the completely conserved object (Fig. 14). Diagrams were drawn to indicate restored areas (Fig. 16).

In order to prevent sudden changes in the environment, I made special paulownia wood boxes for storage. It is important to avoid low relative humidity.

I also made a complete replica of the original box, based on the information gained about the substrate and techniques of manufacture, which was sent to the museum with the restored object (Fig. 15).

Conclusion

This was the first time that I had been involved in an international conservation project. Sometimes I was troubled by the unfamiliar paperwork and the need to balance different opinions which were rooted in different cultures and traditions. Still, the conservation work was completed successfully thanks to Dr. Klaus J. Brandt, Professor Motoo Yoshimura of Kansei Gakuin University who introduced this project to me, Ms. Toshie Nakajima who did the conservation work on objects after the Meiji period, and many other supporters.

Apprenticeship and Conservation

Susanne Barchalia
The Royal Danish National Museum

Apprenticeship is a fairly common form of education in Denmark. It is still quite normal for a young person to make a contract with an older master in order to learn the particulars of a specific trade. I myself served an apprenticeship in a silversmith's workshop, so the idea of submission to a craft training was not entirely foreign to me. After four years of training as a silversmith, I was accepted by the Conservation School of The Royal Danish Academy of Fine Arts, where I trained as a conservator.

I wanted to become an apprentice again to specialize in the conservation of lacquer. The National Museum of Denmark's Department of Ethnography has a large collection of east Asian lacquer. Some of these objects were in a very poor state of preservation, so the Museum wanted a conservator to have an advanced education in Japan and then return to take care of the treatment of these objects.

As a result of the introductions of various people, a Japanese master lacquerer invited me to stay and work for a year at his workshop in the northern part of Kyoto (Fig. 1). To be taught the traditional skills of a Japanese lacquer workshop, it is very important to have recommendations from people whom the master trusts.

Figure 1. A traditional lacquer workshop in Japan.

In Japan it takes eight to ten years of apprenticeship to attain the level of a traditional master artisan. A Japanese master willing to pass on his knowledge will not accept a student whose mind is prejudiced. The openmindedness shown by the master demands a similar openness from the student, who must be willing to receive the teachings of the master without reservation.

It is usually very difficult for a Westerner to understand this total submission because we are used to learning from a critical dialogue between equals, whereas in Japan one looks, listens, and learns. So, obviously, one has to change one's behavior totally and try to achieve the reticence expected if one wishes to acquire any knowledge from a Japanese master.

My purpose was not to learn lacquer technique from the beginning, but to be taught the basic principles of lacquering in order to apply them in my daily work as a conservator. At the workshop I was introduced to techniques that had been developed by the masters over the past seven generations.

Almost all craft traditions in the world are based on knowledge that is crucial in making the works of a particular workshop a specialty. Likewise, all crafts have their own technical language, which is almost impossible to learn except through practical work. Getting to know this language is an important part of the education; it works as a kind of code or guarantee between people of the same trade.

Figure 2. Sample boards showing steps in the process of lacquering, from the basic shape (top left) to the finished product (bottom right).

The process of lacquering

The lacquer master had already planned a course of practical work for me before I arrived. His wood-turner had prepared a number of basic objects in Japanese cypress (*hinoki*) on which I was able to practice, in a simplified manner, the traditional lacquer process.

The process of lacquering can be divided into two phases: the treatment of the basic shape and the coating.

Before the grounding begins, one has to produce wood spatulas and polishing stones suited to the basic shape. The spatulas are cut with a knife from pieces of *hinoki*. Pumice stone is used for polishing.

An illustration of the lacquering process I practiced on rectangular pieces of *hinoki* is shown in Figure 2. The same process is used for objects of very different appearance and size, such as boxes, bowls, trays, and furniture.

The grounding of the basic shape

If the basic shape is fairly thick, as in this case (0.5 cm), the object will shrink and expand too much with changes in climate. To avoid this, the wood-turner makes a jointed construction designed to reduce this effect. Then the ground layer is applied with a spatula. It consists of different sorts of pulverized clay together with lacquer and water. When this layer is dry, after approximately twenty-four hours, the shape is polished with water and pumice stone. The second coating of the ground layer is the same as the first and is also polished.

When the process of grounding is finished, the coating of the lacquer can begin. The lacquer is hardened for eight to ten hours in a humidity chamber with a relative humidity above 90%.

As the grounding material consists of lacquer mixed with clay and water, it does not require high humidity to harden, but hardens under ordinary conditions.

Before the final layer of lacquer is applied, the workshop is given a thorough cleaning to remove dirt and dust. The lacquer, which is of very high quality, is filtered through fine layers of rice paper until no impurity remains.

Decoration

A Japanese lacquer workshop is usually divided into two parts: one for the process of lacquering and one in which the work of decoration is done. Here the lacquer master's wife instructed me in the three most common methods of decoration: flat decoration (*hiramakie*), relief decoration (*takamakie*), and decoration below the surface (*togidashimakie*). *Makie* means "sprinkled picture." Metal powder is sprinkled through a bamboo straw onto a wet lacquer base. The metal powders used for *makie* are usually gold and silver, which are sold in many grades and shades of color (blue, yellow, red, black, etc.).

I finished my stay at the workshop by practicing my newly learned skills on twenty-five serving trays.

The purpose of the lacquer workshop

Part of the collection of Japanese lacquerware at the National Museum of Denmark can be traced back to the establishment of the Royal Cabinet of Curiosities (Det Kongelige Kunstkammer), which was organized in 1650 by King Frederick III. These approximately one hundred objects are now the nucleus of a fairly large ethnographical collection. Today the Japanese collection contains approximately 3500 objects, most of which can be traced back to the nineteenth and first part of the twentieth centuries. The objects are mostly works made for export (Fig. 3).

Figure 3. A picnic set from the Danish National Museum (no. EAC 133), late seventeenth or early eighteenth century.

The state of preservation of the collection varies considerably, from intact objects to very deteriorated and porous materials. Very few objects in the collection had been treated before. The main causes of deterioration were:

1. Changes in climate, which meant that objects with thick wooden substrates, such as export lacquers, had changed in volume. This resulted in cracks in the wooden base and in the flaking off of the lacquer layer.
2. Ultraviolet from sunlight or from lighting in the exhibition cases.
3. Fingerprints, dirt, dust, wrong polishing materials, etc., which had left the surfaces of many objects open to deterioration.

Establishing the lacquer workshop

It has always been a basic principle at the National Museum of Denmark to treat and preserve objects in accordance with their original materials and their nature. The museum made premises available to me in Brede, just outside Copenhagen, where I established a workshop according to European traditions, but using materials imported from Japan.

Building up a lacquer workshop does not require vast financial resources. The difficulties were centered on creating a workshop that would be able to cope with any task, at any time, of whatever character. In Japan, the changing climate of the seasons is used as an important part of the lacquering process. In the dry winter season the process of grounding is done, while the last layers of lacquer are applied during the humid periods. Naturally, such considerations are not possible in a museum context, where exhibitions and loans to other museums are often arranged at very short notice.

The workshop is equipped with a humidifier so that the climate can always be adjusted according to the steps in the process of treating the objects. In order to dry, the lacquer has to go through a chemical process: hardening. Urushiol, which is the main constituent of lacquer, absorbs oxygen when changing from a fluid to a solid state. This process must be even and slow and take place in a humid atmosphere.

For this purpose a humidity chamber is used, which is a wooden box with a pair of detachable sliding doors. The best type of wood for such a humidity chamber is *hinoki*, Japanese cypress, because this type of wood both absorbs and emits moisture, without changing its characteristics. This is necessary for the hardening process. *Hinoki* is very expensive and can be difficult to find in Europe; a conifer, pine, or spruce, may be used as a substitute. The humidity chamber has to be wetted regularly with water in order to maintain the relative humidity permanently above 90%. The climate outside determines how many times the cabinet has to be moistened. The humidifier is therefore a very useful piece of equipment during dry periods.

The ideal surrounding is a south-facing room with good light and a wood floor, which can help to maintain the correct humidity. The environment should be free from dirt and dust.

The materials and tools used in the lacquering process are not available in Europe. As these things cannot be ordered through any stores, contacts with Japanese suppliers are very important. Contacts with suppliers of top-grade products are usually made through the master.

Some of the tools I use I make myself, for example, the wood spatulas for the grounding processes. The brushes for laying down paint flakes, lacquering, and decoration should be very robust and made of hair, either rat, rabbit, or human. In Japan there are workshops that specialize in making brushes for various crafts.

Concurrent with the establishment of the lacquer workshop, the exhibition and storage conditions for the Museum's collection were improved. Each exhibition case is equipped with a humidifier and a thermohygrograph; the relative humidity is controlled at 55–60% and the fluorescent lighting does not exceed 150 lux. Objects hitherto stored in antiquated conditions are now in a new, air-conditioned, dark storeroom.

A treatment process in the Japanese tradition

In order to gain some knowledge of conservation and restoration methods relating to Japanese lacquer, I visited several museum workshops in Japan. I spent the most time at the National Museum in Tokyo where the head conservator and lacquer master explained the basic principles for treating Japanese lacquers, such as fixing loose layers, mixing filling material and grounding, surface cleaning methods, and treatment of damaged surfaces. I have chosen to describe a standard treatment process for a Japanese lacquer object that is representative of the type of problems I encounter.

1. The object is documented by means of photographs, x-radiography, and infrared photography.

2. If possible, samples are taken for cross sections to permit examination of the ground and lacquer layers.

3. Flaking in the color layer is fixed with a mixture of *kiurushi* (unrefined lacquer) and ligroin (petroleum benzine, BP 80–110°C). The process of readhering flakes begins with a weak solution of *kiurushi* in ligroin and finishes with 100% *kiurushi*. After each operation the object is put into a humidity chamber for between one and three days.

4. Small cracks, insect attack and other typical damage to the wood base are improved with *kokuso*, which is a mixture of urushi, rice paste, cotton fibers, flour, and sawdust. A larger crack can, depending on the stability of the object, be filled with new wood, strengthened with paper or cloth, and then built up with the same number of ground and lacquer layers as the original.

5. The ground layers, which normally consist of urushi, clay, and water, can be porous and powdery. This may be due to insufficient urushi or to the use of animal (pig's) blood as a substitute for urushi. This condition can be stabilized with *kiurushi* and ligroin.

6. When flakes in the ground and color layers have been readhered, missing areas can be built up with the same layers of grounding as in the original, rubbing down between layers. The ground material, *sabi*, consists of clay (for example, ceramic clay fired at 800°C), urushi, and water. Each ground layer must dry for about twenty-four hours at room temperature, and then be polished with pumice and water.

7. To seal the surface, urushi is rubbed in.

8. After the grounding process is completed the lacquer used can have many different qualities and colors. The black lacquer (*roiro*) that I import is colored with iron oxide powder (Fe_3O_4). Red lacquer can be bought in many shades. I color raw lacquer myself with cinnabar because the lacquer often changes color during hardening, becoming darker. This can be corrected if one mixes the materials oneself.

9. When the lacquer layer has hardened in the humidity chamber (after eight to ten hours), it is polished with charcoal and water. I use charcoal from *hinoki*. The annual rings in the wood should be as close together as possible; if the gap between the rings is too large, the result is a porous material that can cause "stripes" in the polishing.

10. Through the years, many lacquer surfaces have lost the depth and luster that are so characteristic of Japanese lacquerwork. In the past shellac was used to revive this effect but, as Japanese urushi generally does not harden if European materials are present, it is necessary to remove them before conservation with solvents that do not damage the original lacquer. Dirt and dust can be removed with ligroin. Degraded surfaces can, after cleaning, be revived with *kiurushi*, which is rubbed into the surface and then hardened in the humidity chamber. The treatment is continued until the desired surface effect is achieved. Old fingerprints can be removed in the same manner.

11. Decorations, either *makie* or inlays of mother-of-pearl, bone, metal, etc., are rarely dealt with in my workshop.

Dry Lacquer Statues of Japan

Kyotaro Nishikawa
Tokyo National Museum

Figure 1. Gobujo (one of the eight supernatural guardians of Buddha), Kofukuji temple, Nara. Because the lower part has been lost, the technique can be observed in detail. The hemp cloth layers, the thickness of kokuso-urushi *and many other points can be studied.*

Today, the word urushi brings to mind delicate, beautiful, lustrous, compact works of art. In the past, however, urushi techniques have also produced huge and dramatically impressive works of art: dry lacquer statues.

In the middle of the sixth century, Buddhism was introduced to Japan from the Korean peninsula. Over several decades, the religion took root and many temples were erected to house Buddhist images. Together with Shintoism, the traditional religion of Japan, Buddhism began to play an important role in Japanese history.

At first, Buddhist images were made of gilt bronze or wood. But in the seventh century, two new techniques were introduced from China: *kanshitsu* (a dry lacquer technique) and *sozo* (a clay statue technique). Many dry lacquer statues were made between the eighth century and the beginning of the ninth century. Today, many works of that period can be seen in the Nara area. Because then, as now, urushi was an extremely expensive material, dry lacquer statues were made mostly for large temples in and near the capital.

According to an eighth century record, the statues of the four Deva kings erected in Daianji temple in Nara in the middle of the seventh century were created in dry lacquer. This is the earliest record of complete dry lacquer statues. However, extant wooden statues, such as the *Kudara Kannon* in Horyuji temple, which were made in an earlier period, used dry lacquer for partial surface modeling. This indicates that the technique was in use before the seventh century.

Among extant dry lacquer statues made by the beginning of the ninth century, seventeen are designated as National Treasures and thirty-one are designated as Important Cultural Properties. These are all highly important works in the history of Japanese statuary and are under the protection of the Japanese government.

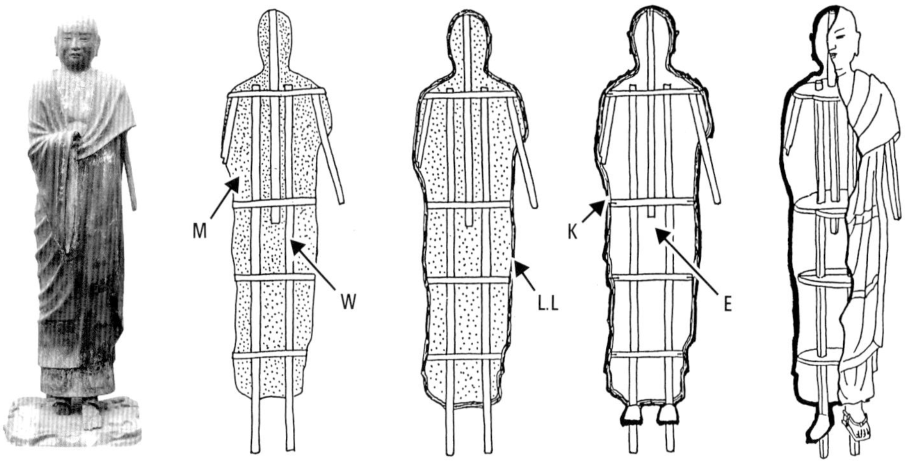

Figure 2a. Sharihotsu (one of the ten great disciples of Buddha), Kofukuji temple, Nara, height 1.55 m, hollow dry lacquer, polychromed, A.D. *734.*
Stages in the construction of the statue of Sharihotsu. Figure 2b. The basic framework (W) is made and clay (M) is put on as modeling. Figure 2c. Hemp cloth layers (LL) are applied using urushi. Figure 2d. The back of the statue is cut open to remove the clay, leaving the interior hollow (E), and kokuso-urushi *(K) is applied as surface modeling. Figure 2e. Cross section of the finished statue.*

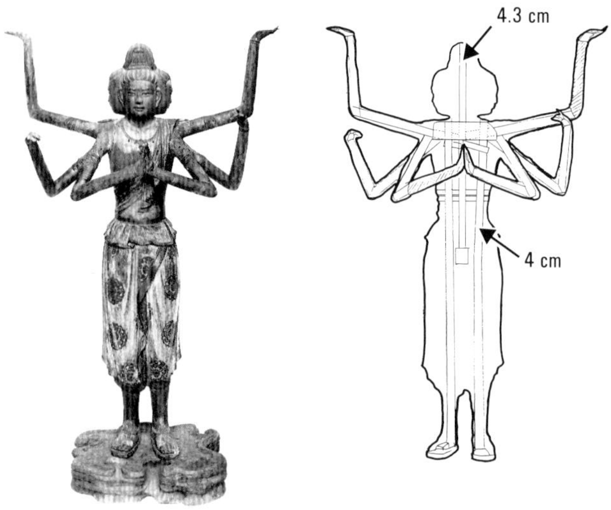

Figure 3a. Ashura (one of the eight supernatural guardians of Buddha), Kofukuji temple, Nara, height 1.6 m, hollow dry lacquer, polychromed, A.D. *734.*
Figure 3b. Diagram of the internal framework of the statue of Ashura.

Figure 4. Amoghapasa, Hokkedo hall of Todaiji temple, Nara, height 6.33 m (the largest standing dry lacquer statue), hollow dry lacquer, gold leaf pasted on, eighth century.

Figure 5a. Zochoten (one of the four Deva kings).
Figure 5b. The internal framework of the statue of Zochoten.

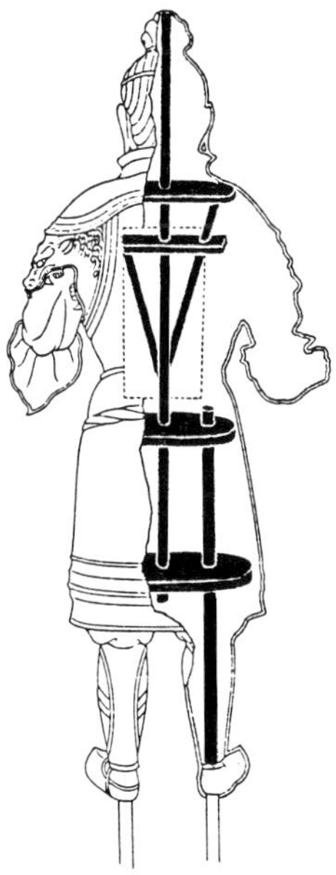

Figure 6a. Tamonten (one of the four Deva kings), Hokkedo hall of Todaiji temple, Nara, height 3.1 m, hollow dry lacquer, polychromed, eighth century.
Figure 6b. The internal framework of the statue of Tamonten.

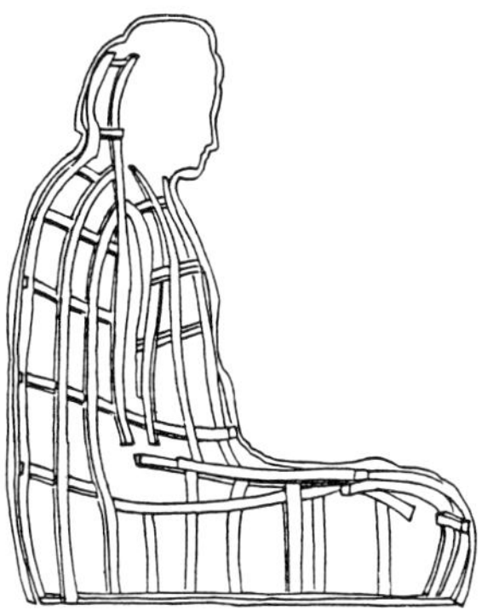

Figure 7a. Bhaisajyaguru, Saiendo hall of Horyuji temple, Nara, height 2.5 m, hollow dry lacquer, gold leaf pasted on, second half of eighth century.
Figure 7b. The internal framework of the statue of Bhaisajyaguru.

Two dry lacquer techniques

Dakkatsu kanshitsu ***(hollow dry lacquer; Figures 1-7)***

1. A simple framework is made using wooden poles. Clay is put over the framework to make a rough form. Then the structure is dried.

2. Over the dried structure, three to ten layers of hemp cloth (similar to canvas used in Western oil painting) are pasted, using urushi. After each layer, the structure is dried.

3. A rectangular opening is cut into the back of the layered structure, through which the clay and wooden framework are removed. Then, to strengthen the assembly, a wooden frame is inserted in the hollow statue and the opening is sewn up with hemp thread. In some cases, only the clay is removed and the wooden poles are left to form the inner frame.

4. Wooden arms and feet are attached to the inner wooden frame. Iron wires are used to form the core of details, such as ears and fingers. For veins on the face of the statue, hemp string is pasted on, and thin wooden boards are used as a core for fragile parts of the garments.

5. Over the surface of this basic form, *kokuso-urushi* (a paste of kneaded urushi, incense powder, clay powder, sawdust, and sometimes hemp fiber) is applied and modeled.

6. There are two choices for the final finish: either gold leaf is pasted over a black urushi coating or various mineral pigments mixed with a thin solution of animal glue are painted over a *hakudo* coating (white clay mixed with a thin solution of animal glue).

Figure 8. Head of Buddha, Toshodaiji temple, Nara, height 81.5 cm, wood-core dry lacquer, late eighth century. The area where the kokuso-urushi *layer has come off clearly shows the wood-core surface. This example shows that the details of the eyes were not carved.*

Mokushin kanshitsu ***(wood-core dry lacquer: Figures 8-10)***

1. A rough shape is carved from wood, usually Japanese cypress. Carving is done either from a single block of wood or from several blocks put together. In many cases the inside of the block is hollowed out, to prevent cracking during drying.

2. Over this carved wooden statue, *kokuso-urushi* is thickly applied as modeling. For details, iron wire is used as in the hollow dry lacquer technique, Step 4, above.

3. The application of the final finishing follows the process of the hollow dry lacquer technique, Step 6, above.

Kokuso-urushi

Since *kokuso-urushi* is a high quality modeling material, it was used even after the ninth century when dry lacquer statues were no longer made. Even today *kokuso-urushi* is used to compensate for errors in carving or as a filling for joints or knots in wood. Urushi itself was often used as a ground coating material and also serves as an important bonding agent for timber composites in carved wood statues. Until the nineteenth century, urushi and *kokuso-urushi* were important materials in the history of Japanese sculpture.

The use of urushi and *kokuso-urushi* as modeling materials made it possible to create statues with balanced style and delicate surfaces. The style itself followed that of the Tang dynasty and thus exhibits realistic, classical, and sophisticated beauty. Although the dry lacquer technique originated in China, most of the remaining works are in Japan.

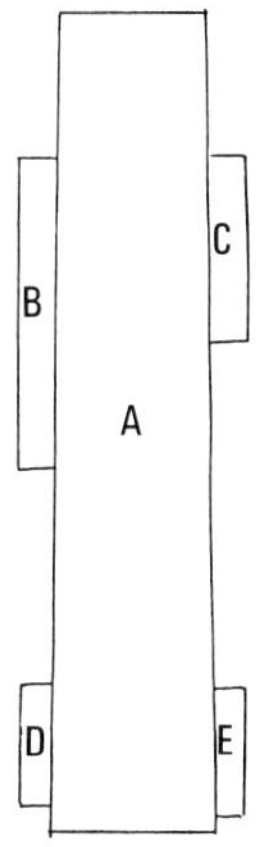

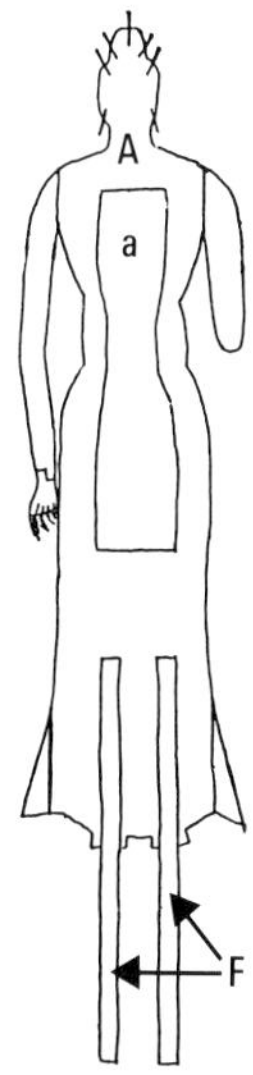

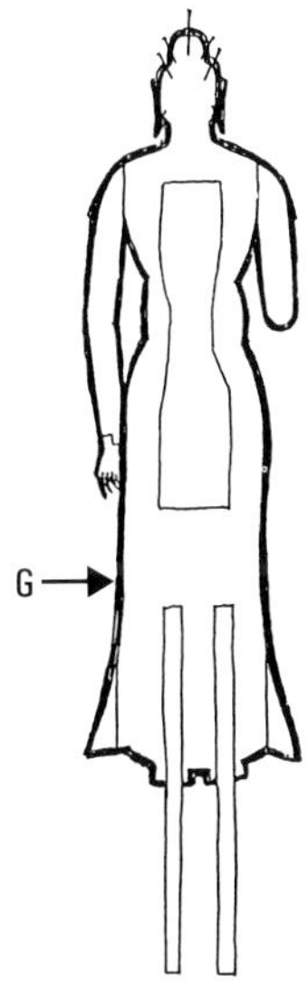

Figure 9a. Ekadasamukha (eleven-headed Kannon), Shorinji temple, Nara, height 2.1 m, wood-core dry lacquer, gold leaf pasted on, end of the eighth century. Stages in the construction of the statue of Ekadasamukha.
Figure 9b. Main wood block (A), arms (B and C), and blocks used for the hanging draperies on each side (D and E).
Figure 9c. The rough shape is carved, with a window (a) through which the interior of block (A) is hollowed out; a square wooden block (F) is joined on for the feet.
Figure 9d. Kokuso-urushi *(G) is applied to the surface.*

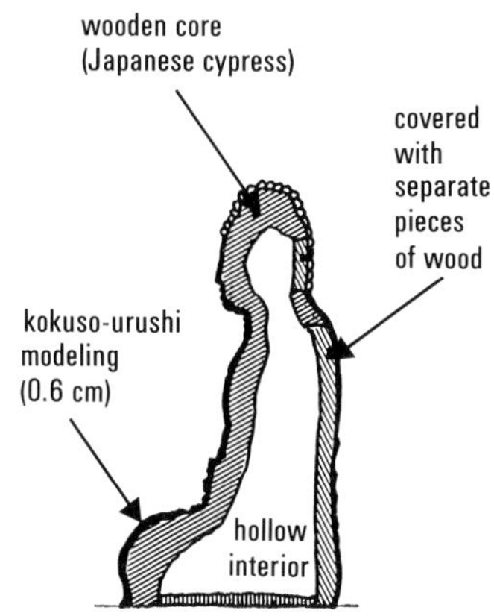

Figure 10a. Shakamuni, Saidaiji temple, Nara, height 70.8 cm, wood-core dry lacquer, gold leaf pasted on, late eighth century.
Figures 10b and 10c. The statue of Shakamuni showing the wood-core construction.

Chinese Lacquer

Alena Skálová
Private Conservator

The application of natural lacquer is one of the oldest techniques in Chinese craftsmanship. It is one of the techniques that influenced the characteristic features of fine art not only in China itself but in the Far East in general.

Lacquer is made from the sap of the *qi shu* (lacquer tree, *Rhus verniciflua*), cultivated in central and southern China. Originally, it was probably used to reinforce implements and utensils and to glue parts together. For these purposes the sap was used in its natural form, in the cold state. In the course of time, however, the properties of hot processed lacquer came to be used to decorate objects made of other materials. Finally, implements and utensils were made of lacquer. This lacquerware served ceremonial (or ritual) purposes, and it was used in everyday life by the upper classes of Chinese society. Here, the decorative possibilities of the medium were developed to their full potential.

The characteristic properties of lacquer—durability and workability—were utilized for various purposes. Its strength and durability provided an excellent protection for the materials used as a base, such as wood and metal, against the effects of weather or insects. Its workability made possible grinding; polishing; carving in deep and shallow relief; coloring first with black and red dye, and later with other colors; layering in various thicknesses; and inlaying with various materials.

Characteristic of the Neolithic period is the black earthenware found in Wujiang, Jiangsu province and in Yüyao, Zhejiang province, decorated with material similar to lacquer. Vessels made from lacquer alone have been found in excavations dating from the sixteenth to fourteenth centuries B.C. (Shang dynasty). During the first millenium B.C. the range of lacquerware objects increased as new decorative techniques were introduced, for example, inlaying with precious stones, or wood (*xiang qian*), or mother-of-pearl (*luo tian*). This last technique underwent a complicated development. While the oldest specimens (Western Zhou period, tenth to eighth centuries B.C.) consist of thick layers of lacquer inlaid with entire cubes of mother-of-pearl, products characteristic of later periods are very delicate: thin layers of lacquer are inlaid with finely cut slices of mother-of-pearl and polished to a glossy luster. Very often mother-of-pearl of different colors was used for an artistic effect. Later on, another technique was developed: applied decoration with gold and silver, both sheets and pieces of wire. After the fourteenth century A.D. (Ming and Qing dynasties) applied decoration alone was rarely used; more often it was combined with inlaid mother-of-pearl. This combined technique, which was at first used only to decorate the backs of mirrors, gradually developed into an independent form of decoration.

Applied decoration combined with inlay is very similar to the technique of incised decoration filled with gold (*qiangjin*) or silver (*qiangyin*) where the motifs used for decoration such as landscapes, figurative scenes, etc., conform to the spirit of Tang paintings.

Sometime around the fourth century A.D. a new technique called dry lacquer (*tuo dai qi*) arose. The center for dry lacquer under the Tang dynasty (A.D. 618–907) was in Sichuan. To manufacture dry lacquerware, a piece of cloth is fixed to a clay or earthenware core over which layers of lacquer are applied. Thus the completed product is very light. This method was mainly used for making bowls, cups, trays, toilet boxes, vases, etc. But during the third to fifth centuries A.D. an earlier version of this technique was used to make Buddhist statues.

Another widespread technique was carving in lacquer (*diao qi*). In this method, many layers of lacquer (from thirty-eight to over two hundred) were applied to a substrate of wood or metal. The lacquer was then carved in either deep or shallow relief. Typically in the Ming period, the whole surface was ornamented with motifs of flowers and fruits. The technique evolved into complicated pictorial compositions of landscapes, architecture, and figurative scenes.

A variation of the carved lacquer technique is through-cut lacquer (*tixi*). Layers of lacquer of different colors, at first black and red, and later also other colors, were applied successively. The decoration was then carved, uncovering the individual layers according to a deliberate pattern. The oldest known specimen of this technique is a fan handle from the Southern Song dynasty (thirteenth century A.D.). This technique probably evolved from the manufacture of long leather shields during the Tang period and the epoch of the Five Dynasties (A.D. 907–960). The cores of these shields were formed of leather (most frequently camel leather) over which seven alternating layers of black and red lacquer were applied. The decoration was created by uncovering the individual layers to form the desired pattern. This technique reached perfection in the latter half of the Song dynasty.

Besides the traditional decorative techniques, whose variety can hardly be treated in detail in this brief introduction, there is another combined technique that has no Chinese name. In the West it is known as "Coromandel lacquer," after the part of the Indian coast from which it was shipped. Coromandel pieces were made chiefly for export and the oldest known specimens date from the late seventeenth century (the Kangxi period). This technique, combining carving, lacquer, and polychromy, was used in south China in workshops producing some small toilet boxes and similar minor items, but mainly large pieces of furniture: tables, dressers, wardrobes, cabinets, and, especially, screens.

Generally, a screen has an even number of panels (four, six, eight, or twelve) 177 to 275 cm high. Both sides are usually decorated. As a rule, the front has figurative scenes while the reverse bears motifs of flowers, birds, or cultural objects. I should like to emphasize here the artistic qualities of Coromandel lacquer and its connection with painting in colored ink. The contour lines of each motif, not only the general outline, but also the details of attire, features of faces, etc., were cut into the surface layer of black lacquer. Thus, the fine black lines of the lacquer, combined with the polychromy of the exposed underlying layer, resemble a fine ink painting, in a different and more durable medium.

In China today there are three principal places where lacquerware is produced. Beijing is the center of lacquer carving. Fuzhou in Fujian is the center of lacquer painting, especially of modern techniques using multicolored lacquer, gold dust, and applied decoration with mother-of-pearl and eggshell, and this technique is slowly growing into a new medium of modern painting. Finally, Sichuan remains the traditional center of dry lacquer manufacture.

As an epilogue to this introduction, I would like to point out that the qualities of genuine Far Eastern lacquer contribute to the overall artistic quality and aesthetic effect and do not change in the course of time, unless treated unsuitably or damaged. They cannot be replaced by other materials or different technologies.

Restoration of a Chinese screen

After study at the Tokyo National Research Institute of Cultural Properties, followed by ten years of practice in Europe, I was given the job of restoring a Chinese screen.

This screen is in the collection of the Residenz Museum in Munich. It dates from the Kangxi era (1662–1722) and consists of ten panels, each 184 cm high and 47 cm wide, decorated with carvings in lacquer depicting fantastic landscapes painted in the Tang style. There are mountains, and lakes with boats floating on them. Architecture is represented by pavilions, studios, and ornamental bridges. Figurative scenes show the lives of intellectuals.

At the edge of each panel of the screen there are borders with a design of "fortunate" motifs and objects. The first border from the center contains stylized flowers: lotuses and tendrils. There follows a wide band of cult objects: vases, fumigators, vessels with offerings, symbols of immortality such as Buddha's hand and the mushrooms of immortality, musical instruments, the mythical three-legged frog, coins, fans, palm leaves, and gongs. In the third marginal band there are flying cranes in colorful clouds, also a symbol of immortality.

When I was given the job of restoration, the screen had already been repaired several times. The first repair was done in the traditional way with original materials, probably in China. This repair can be detected only by minor changes in the color of the lacquer covering the repaired area.

In the second and third repairs, European materials were substituted for the original Chinese ones. The areas where the black lacquer was damaged had been filled with chalk combined with glue. After polishing, the surface had been coated with black oil paint. These repairs respected the form of the original lacquer carvings.

For the fourth repair, red wax had been used to fill chips, large cracks, and irregularities in the lacquer caused by the warping of the wood. This red wax covered not only the damaged areas but also the surface of the lacquer around them, so that the sharp contours of the relief were obscured. Indeed, in some places the carvings themselves were entirely obscured. A coat of black oil paint over the red wax had been intended to reproduce the rich gloss of lacquer, but it had dulled in the course of time. The original chalk layer with delicate shades of polychromy was overlaid with thick, harsh-looking oil paint. Visible traces of brush strokes on the surface of the screen contributed to the general disharmony.

Figure 1. A detail of the screen (a) before restoration, (b) after removal of later repairs.

Figure 1a shows a detail of the screen before restoration. In the center panel, traces of the repair using red wax are visible as is a test area in the polychromy. After consultation with the owner (Bayerische Verwaltung der Staatlichen Schlösser, Gärten und Seen, Munich), the second, third, and fourth repairs using unsuitable European methods and materials (oil paint, wax) were removed. The surface was then consolidated using traditional materials. Only the oldest Chinese adhesives, those complying with the overall aesthetic effect (even if the color of the lacquer had changed), were left in place (Fig. 1b).

Figure 2. Areas filled with chalk (a) after removal of the oil paint and (b) after restoration.

Figure 3. After the removal of black oil paint.

Figure 4. After the removal of the red wax.

Figure 5. After repair, and removal of oil paint from the polychromy.

Figure 6. Red wax adhesive applied over the original relief.

Figure 7. Black oil paint that had penetrated under the surface (a) was removed, revealing the earliest repair (b) made with traditional materials.

Figure 8. A test area in the polychromy reveals the original coloring.

After the oil paint was removed, the areas filled with chalk could be seen. Figure 2a shows that the form of the original carvings was respected. The chalk combined with glue used in earlier repairs was unsuitable due to its high moisture content. Because of this it was necessary to replace the chalk with material similar to that used originally, while respecting the form of the carvings (Fig. 2b).

The condition after the black oil paint had been removed can be seen in Figures 3a–c. The red wax spoiled the aesthetic effect and covered a large undamaged area around the repair. Figures 4a–c show the condition after the red wax was removed from the surface of the black lacquer as well as from the cracks and fissures, and the areas filled with glue. The real extent of the damage to the surfaces of both lacquer and wood could now be seen.

Figures 5a–c show the condition after repair, including removal of the oil paint from polychromed areas. Test areas showed the insensitive application of red wax adhesive over the original relief (Fig. 6). The mechanically damaged black lacquer surface had been covered with black oil paint that at first concealed the defects; later, however, it penetrated under the surface and dulled the lacquer (Fig. 7a). This heterogeneous layer had to be removed entirely, thus uncovering the earliest repair, which had been done in China in the traditional way (Fig. 7b). After consultation with the owner, this earliest repair was left as it was. A test area in the polychromy after the oil paint was removed showed the original coloring (Fig. 8).

This ten-panel screen was restored from 1984 to 1986. The problems we encountered with the repair indicate the need for a unified approach to the restoration of Asian artifacts. The materials commonly used in the West can never provide the qualities found in Far Eastern cultural objects: color, shine, luster, and overall aesthetic effect. Moreover these Western materials usually penetrate the surface around the damaged area, which makes it difficult (if not impossible) to redo the repair at a later date in the traditional manner with materials similar to those used to create the original.

Acknowledgments

Translated from the Czech by Dr. Zlata Cerná, Curator of the Národního Muzea—Náprstekova Muzea, Prague, Czechoslovakia, who also translated the Chinese articles cited in the bibliography.

Bibliography

Herberts, Dr. K.

1959 *Das Buch der Ostasiatischen Lackunst.* Düsseldorf.

Garner, H.

1979 *Chinese and Associated Lacquer.* London.

Wang Shi-xiang

1973 "Zhong guo gu dai qi gong za shu." *Wen-wu* 3.

Fu Zhen-lun

1982 "Xüan-li duo cai di Ming Qing dai dian qi qi." *Guogong bowuyüan yüankan* 1.

Essay on Shells That Became Cherry Blossoms

Yoshikuni Taguchi
Lacquer Artist

Shells, whose origins are in the wide oceans of the earth, have many uses. Some shells are a source of food, some shells have monetary value, and some shells are appreciated solely for their beauty. In the art of urushi, which uses shells as a material, the beautiful colors of the shells are appreciated all the more since they cannot be adequately reproduced in urushi.

The appreciation of beauty is something of which only human beings are capable. Fish, animals, and birds cannot feel the beauty of things around them. Even human beings cannot appreciate beauty unless their stomachs are full. In earliest times, shellfish were merely a source of nutrition. This is proved by the various shell mounds, consisting of many layers of discarded shells, that have been found in different places in the world. It took people a long time to notice the beauty of shells because they were too busy surviving. Ancient hunters decorated themselves with flowers, shells, and animal fangs only after their hunger was satisfied.

When prehistoric peoples started to create beautiful objects, they used shells and urushi as decoration. Red urushi combs and ornamental hairpins dating from the Jomon period (pre–200 B.C.) have been excavated at Torihama Kaizuka. Bracelets and knives made of shells were found at Kosaku Kaizuka in Funabashi, Chiba prefecture. Scarlet ornamental hairpins from the first half of the Yayoi period (200 B.C.–A.D. 250) were found in the Aman ruins. Earrings made of shell have been found in various other places.

Shells were also used as money. For example, during the Ying period in China, *takara* and *koyasu* shells were treasured as coins. It may be hard to imagine shells being used as money. However, even now shells are sold as material for urushi in shops and thus have monetary value. In European casinos, mother-of-pearl discs have been used as coins for roulette; the sound made by these shells is similar to that of silver coins.

Having been fished from the ocean to be eaten, shells are further used for decoration because of their beauty. Therefore people who handle shells should show great appreciation. Shells conceived in beautiful places in the ocean serve as food, architectural decoration, and craft materials on earth. Mother-of-pearl, *awabi*, and *hakuchogai* are used mainly for craftwork.

Shells are also used to decorate altars. They comfort worshipers with their beautiful colors, seeming to shine with the mercy of God. One may say that this is the most precious use of shells.

Figure 1. Scales on the ryuto raden ryoka-gata bon *(tray) from the Yuan dynasty in China.*

The colors of shells match the colors of the ocean as well as the colors of the fish living there. The motion of fish swimming catches the mind of an artist. To express such beauty, I use silver leaf, platinum leaf, and gold leaf on the back of a shell. There are many designs that vividly represent fish scales using shells. In the *ryuto raden ryoka-gata bon* (a tray with a design of dragon and waves inlaid in *raden*) of the Chinese Yuan dynasty (1279–1368) shells were used for the body of the dragon (Fig. 1). First, they were cut into pale red or blue scales and inlaid. Then they were engraved in *kebori* technique to express the scales dynamically. The fine ripples were expressed with *aogai*. Shells were also used for patterns of trees, flowers, and birds on other art objects. All are examples of how effectively the colors of shells were used and how exquisitely the artist's images were expressed.

There are many kinds of shells. The spiral shells that are used for *shikki* (lacquerware) include mother-of-pearl, *hakuchogai* (a bivalve), *kichogai*, and *kokuchogai*. *Awabi-gai* is a spiral shell, although it looks like a univalve, for only a small part of the shell has a spiral. Mother-of-pearl and *awabi-gai* have been used since ancient times because they are hard enough to cut patterns easily. When collecting *awabi*, the women divers take *kuro-awabi* first, for it is quick to flight and sticks hard to the rock. It is oval and the inside of the shell is green and lustrous. *Ezo-awabi* is a thin shell with uneven "pleats." The inside is mostly green. It is good for making fine patterns on *shikki*. *Madaka-awabi* and *mekai-awabi* are more red than blue; the pink or purple parts are rare and valuable. The color of *sazae* resembles mother-of-pearl but is greener. This shell is small and suitable for making sharp curves.

When *raden* is inlaid on a column in a building, the shells are used whole. First a design is made on Japanese paper and placed over the shells. Then the shells are cut with a fretsaw on a stand to fit the curvature of the column. The natural curves of the shells are retained in cutting and the shells are polished. The flower patterns on the ends of the roller, which is thinner than a column, of the *shitan raden-jiku* (hanging scroll) in the Itsukushima shrine were made in the same way. The same method was also used for the national treasure, *ho-o raden kagari tachi* (sword) of the Heian period (794–1185).

The shells used since ancient times have reflected the aesthetic preferences of each country. Although a very beautiful *awabi* existed in Japan, it was not used for a long time. Instead, mother-of-pearl was used in the Nara, Heian, Kamakura, and Muromachi periods (710–1568), under the influence of the Tang culture. Mother-of-pearl appealed to the Japanese because of its luster, which was an apt metaphor for the life of people who enjoyed sunshine though all seasons. *Awabi* began to be used in the Muromachi period (1392–1568). In the Momoyama period (1568–1600), new patterns of gorgeous maples in autumn and beautiful cherry blossoms in spring became popular. Since these patterns were made on *kin-makie, enashiji, aokin* (all techniques that make abundant use of gold) and on silver, the blue and red colors of *awabi* matched them well.

Some of the *raden* methods use *atsugai* (thick shell) and make a flat surface. In one of these methods, when applying *nakanuri* (intermediate coating), shells are inlaid so that they are slightly higher than the *nakanuri* surface. When *roiro-shiage* (polishing) is done, the entire surface becomes even. When making *ikakeji* (lacquerware thickly coated with gold dust and polished to look like gold), shells are inlaid still higher than the urushi surface. Tin foil is pasted over the entire surface and dried. Then the *raden* is ground with a whetstone until the entire surface becomes level. After that, the tin foil on the urushi is peeled off. As a result the *raden* will be slightly higher than the *nakanuri* surface. When *ikakeji* is then done on the *nakanuri*, the entire surface becomes level.

The two *atsugai* techniques mentioned above are quite often seen nowadays but were not so popular in the Heian and Kamakura periods. Outstanding examples of *atsugai* technique include the Chusonji Konjikido *raden* and octagonal *shumidan* (altar) from the Heian period and the *magaki-giku raden suzuribako* (ink-stone box with a fence and chrysanthemum design) of the Tsurugaoka Hachiman shrine from the Kamakura period (Fig. 2). In all these examples, which are designated national treasures, the impression given by the *raden* is soft, even though the material itself is hard.

In the *usugai* technique, thin shells are placed on the *nakanuri* and urushi is applied three times to raise the surrounding urushi area to equal height with the shells. Then the urushi covering the surface of the shells is scraped away. The *suhama-u raden suzuribako* (ink-stone box with cormorant design) of the Heian period is a masterpiece of *usugai* technique. When *raden* is applied over the surface of a *shikki* in such a way that the *raden* forms a pattern, then the shells are kept a little higher than the surface of the urushi. In the final coating of such a design, care should be taken with the handling of the brush so that no dust remains on the final coating and no extra urushi is deposited around the shells. In my *nuritate-shiage* technique, in which only one coating of urushi is applied over the *nakanuri*, a knife or a spatula is used to scrape urushi off the shells. Thus the shells theoretically become slightly lower than the urushi *uwanuri* (final) surface. Such a condition is also seen in *ikakeji* of the Heian period. It is this condition that makes the outline of the *raden* uniquely soft.

Figure 2. The magaki-giku raden suzuribako *(ink-stone box) from the Tsurugaoka shrine, Kamakura period.*

Figure 3. Meigetsu-wan *(Momoyama period). Also called* warigai *cherry-blossom bowl set, it was given to Meigetsu-in in Kamakura by Urakusai Oda, a younger brother of Nobunaga.*

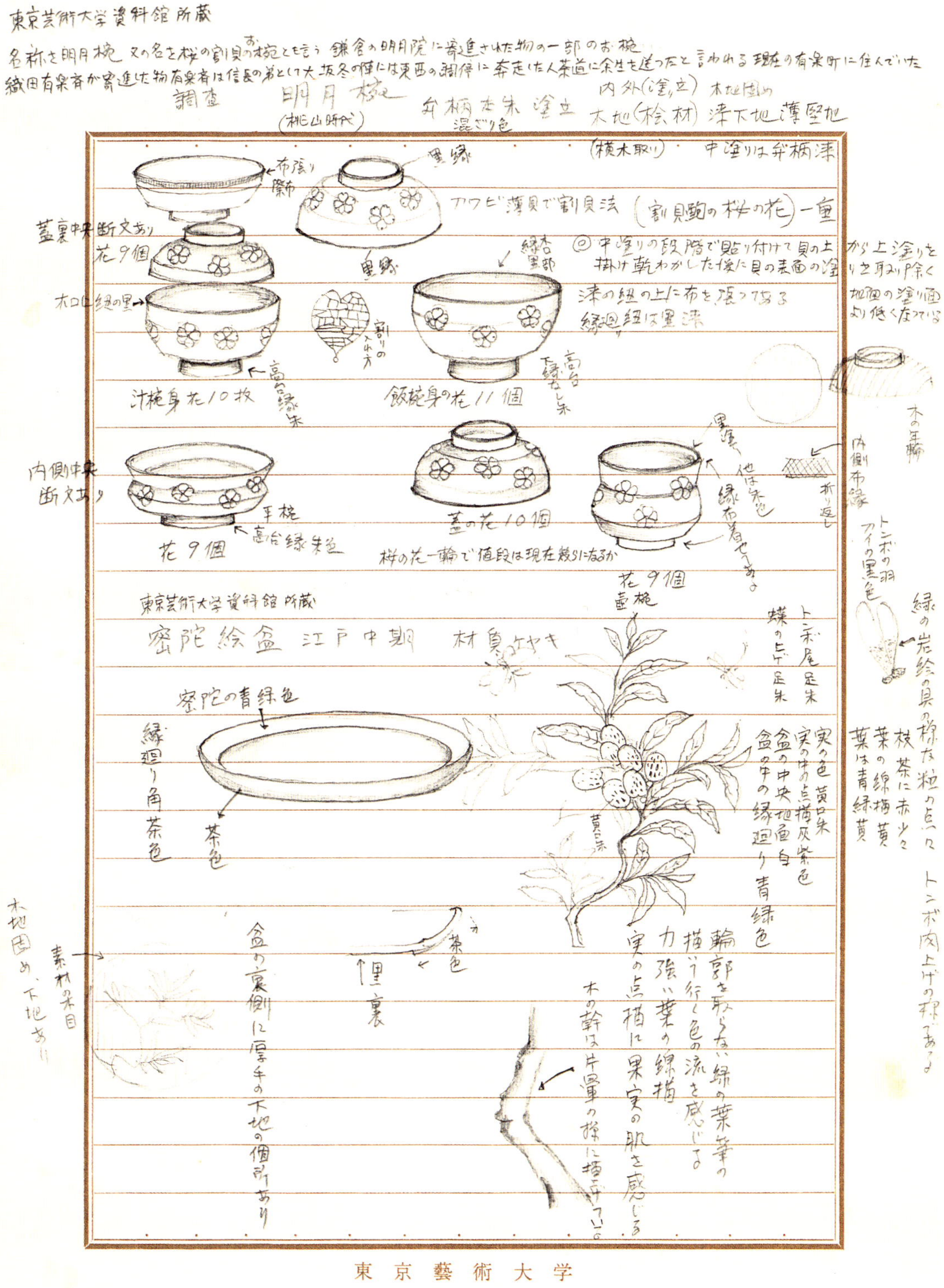

Figure 4. Meigetsu-wan *research data.* Shiruwan *(soup bowl), height 7.5 cm, diameter 13.2 cm;* meshiwan *(rice bowl), height 9.9 cm, diameter 14 cm;* hirawan *(flat bowl), height 5.4 cm, diameter 13.2 cm;* tsubowan *(vase-shaped bowl), height 7.3 cm, diameter 11.1 cm.*

Figure 5. Hira-natsume *(tea-powder container) using* warigai *technique, made by Yoshikuni Taguchi. Silver leaf and gold leaf on the reverse of the shells create a unique variation. Gold dust sprinkled on the side of the* natsume; *horizontal lines scratched;*

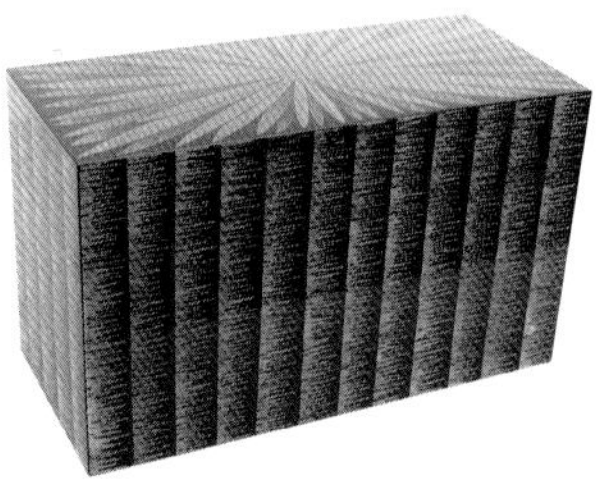

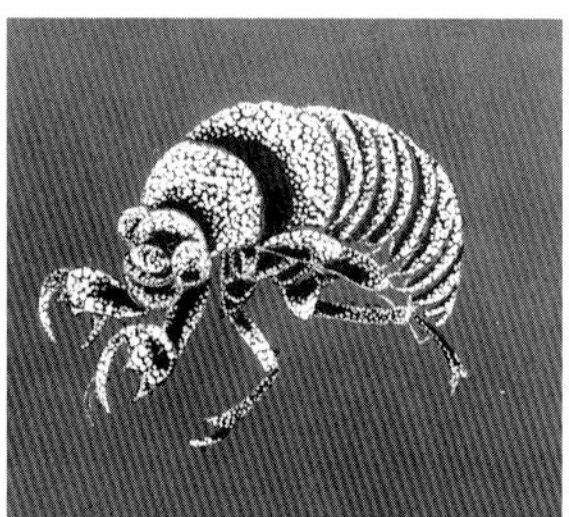

Figure 6. Okigai *design by Yoshikuni Taguchi. Shells are placed one by one on the side, thus creating a variety of patterns. Gradated* kin-hirame-fun *on the surface of the lid;* togidashi.
Figure 7. Silver tsubuoki *sample board made by Yoshikuni Taguchi. The shell particles are of the same size as* gin-hirame.

Meigetsu-wan, a set of bowls, in Meigetsu-in in Kamakura, are also known as "cherry blossom bowls" (Fig. 3). The cherry blossom designs are made from shells from the ocean. Urakusai Oda, a younger brother of Nobunaga who lived in Yuraku-cho, Tokyo, donated this bowl to Meigetsu-in in the Momoyama period. Urakusai worked as a mediator between Nobunaga and the Toyotomi family. He then devoted the remainder of his life to the tea ceremony. Some of the bowls now belong to the collection of the Tokyo National University of Fine Arts and Music.

After a long chilly winter, spring comes and new leaves sprout on the trees in Tokyo at the beginning of April. Cherry buds grow in the sunshine day by day. In the middle of April, the cherry trees come into full bloom. According to Japanese custom, people gather under the cherry trees, drink *sake*, enjoy themselves, and relax. As they go home, they walk on numerous cherry blossom petals on the road. Five of these petals placed on the hand make a cherry flower. It is from this image that the famous *Meigetsu-wan* cherry blossoms on vermilion urushi bowls were born. They look like cherry blossoms in the moonlight, suitable for Meigetsu-in (literally, the temple of the bright moon). There are eleven flowers on the *meshiwan* (rice bowl) and ten flowers on its lid; there are ten flowers on the *shiruwan* (soup bowl) and nine on its lid. The *tsubowan* (vase-shaped bowl) has nine flowers and the *hirawan* (flat bowl) has nine also. They are all single-petaled cherry blossoms. The wooden substrate is Japanese cypress. Hemp cloth was pasted on the edges and *urushi-shitaji* (priming) was applied. *Bengara-nuri* was used for the *nakanuri* (intermediate coating). Shells were then placed in position and the final urushi coating of a mixture of *bengara* (red ocher) and vermilion was applied. Lastly, urushi on the surface of the shells was scraped away with a spatula or knife.

With regard to the *raden* parts, *awabi-gai* was cut to make the cherry blossom petals. Japanese paper was pasted over the five-petaled cherry blossoms made of shells, using paste that is soluble in water. On a wooden stand with a thick cover, the shells were cracked finely, through the paper, with a knife that had round edges and would not cut, or with rectangular sticks of wood or bamboo. This is known as *wari-gai* technique. Positions for the shells were marked on the *wan* (bowls). After that, *roiro-urushi* or *kijiro-urushi* was applied with a brush to the reverse of the shells. They were then placed on the burnished *nakanuri* surface. After drying, the Japanese paper was dampened so that it could be removed from the shells with a small brush. Then urushi consisting of vermilion and *bengara* was applied and, after drying, urushi on the surface of the shells was scraped away (Fig. 4).

In general, *warigai* technique is used to paste flat pieces of shell on a curved surface (Fig. 5). The cracks that appear on the shells give a new impression to the flat surface of the shells just as wrinkled Japanese paper gives an entirely different effect compared with unwrinkled paper. The lines separating *raden* and *ikakeji* found on urushi objects of the Heian and Kamakura periods are similar to the fine lines on cloth that has been tie-dyed. The *botan karakusa raden-bako* (inlaid box) is an example of *warigai* from the Muromachi period. Since the surface of the box itself is flat, cracks in the shells create a unique contrast.

Other techniques similar to *warigai* include *okigai* in which small pieces of shell are used as in mosaic (Fig. 6) and *tsubuoki* in which still smaller pieces are used to make patterns (Fig. 7). Eggshells were also used instead of shells on sword sheaths in the Edo period (1600–1868).

Mugi-urushi is generally used to paste *atsugai*, but the amount of starch in the mixture depends on the individual urushi craftsman. Pure flour is used. First, the flour and water are kneaded until they have the consistency of an earlobe. *Kiurushi* (unrefined urushi) is added gradually so that the mixture becomes sticky. This *mugi-urushi* is used to paste not only shells but also other materials.

When *atsugai* is inlaid on urushi, it should be placed deep into the *nunokise*; if it is not embedded deeply enough, it may peel off. To inlay *atsugai*, urushi is baked onto the reverse of the shell for two hours at 120°C, so that it adheres well. After that, the baked urushi surface is given a slightly mat finish with charcoal.

A much stronger way of pasting *atsugai* is to use *nikawa* (animal glue). *Sanzen-bon nikawa* is soaked in water overnight; then it is boiled to a thick consistency over low heat. Finally, it is filtered through a cloth strainer. During use, it should be kept heated over a water bath. Adding this thick *nikawa* gradually to *kiurushi* strengthens the adhesive power of the urushi. I think now that it is better to use this for *atsugai*. In some cases, hemp fiber and powdered spruce may be mixed with urushi to make it easier to dry.

When pasting *usugai*, *roiro-urushi* or *kijiro-urushi* is used to paste small square pieces of shell. Cut patterns should be pasted when the *roiro-urushi* or *kijiro-urushi* is half dry, or *nikawa-urushi* should be used. It is also much better to have a thin coating of urushi baked on the reverse of the shell before it is used. The temperature is the same as for *atsugai*. I use a domestic iron for this purpose because the temperature can be easily adjusted.

The national treasure, *nashiji kiri-mon raden koshigatana* (short sword) in Itsukushima shrine, made in the Nambokucho period, uses the *warigai* technique. There are several horizontal rough cracks which give an impression of vigor. It is a very rare example in which *awabi* was used.

It is more difficult to make *warigai* from *Mexico-gai* (compared to *aogai* and mother-of-pearl) because *Mexico-gai* has brilliant patterns like the scales of a pine tree trunk which easily peel off when the shells are cracked (Fig. 8).

Urushi began to be used as an adhesive in Japan from the middle of the Jomon period. An earthen vessel was excavated from Juno Deitanso in Saitama prefecture. The vessel had been repaired with urushi so that it could be reused. I was struck with admiration to realize that urushi had been used for this purpose since ancient times.

People often place great value on an old tree of historical interest. Urushi is an inconspicuous tree. It cannot become any other kind of tree, no matter how much it wishes to do so. However, the urushi tree has valuable sap, which may be likened to human blood, and we should make good use of it. Significant works of art are national properties in any country. They should be well preserved and kept in good repair. Each country has its own remarkable materials. Each should make the best repair by using such materials. I myself think my life is like that of an urushi tree.

Figure 8. Hira-natsume *(tea container) using* warigai *of* Mexico-gai, *made by Yoshikuni Taguchi. The lines on the lid are* kin-hirame-fun *and* kin-hira-makie.

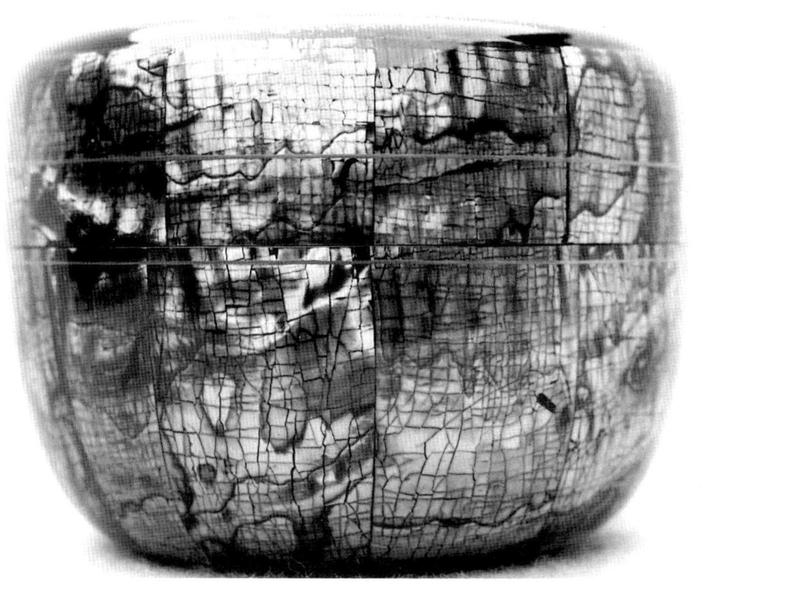

***Appendix: shells used in* raden**

Aogai	*Patelloida* (*Chiazacmea*) *striata* (Quoy et Gaimard), from the Ryukyu islands; literally "blue shell."
Awabi-gai	General name for *haliotis*; *gai* and *kai* mean shell.
Ezo-awabi	*Haliotis discus hannai* (Ino), a sub-species of *kuro-awabi* (q.v.).
Hakuchogai	*Pinctada maxima* (Jameson), also known as *shirochogai*; *haku* and *shiro* mean white.
Kichogai	The yellowed edge of an aged *hakuchogai* (*ki* means yellow).
Kokuchogai	*Pinctada margaritifera* (Linne), more correctly known as *kurochogai*; *kuro* and *koku* mean black.
Koyasu	Trivial name for *takara* (q.v.).
Kuro-awabi	*Notohaliotis discus* (Reeve); "literally black *awabi*."
Madaka-awabi	*Haliotis (Nordotis) gigantea* (Gmelin).
Mekai-awabi	*Haliotis (Nordotis) gigantea sieboldi* (Reeve).
Mexico-gai	Resembles *madaka-awabi*; the main part is deep blue-green and the rest is spotted.
Sazae	*Batillus cornutus* (Lightfoot).
Takara	The cowrie (genus *Cypraea*).

Urushi Technique in the Prehistoric and Antique Periods in Japan

Toshikatsu Nakasato
Tokyo National Research Institute of Cultural Properties

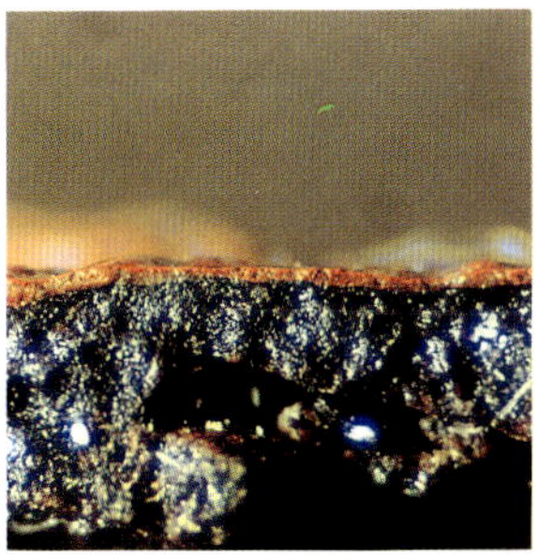

Figure 1. Cross section of rantai-shikki *(lacquerware with a woven bamboo substrate) found in the ruins of Korekawa.*

Urushi began to be used in Japan very early in the prehistoric period, about 5000 B.C. It was greatly influenced by the superb works and techniques introduced from the mainland in the sixth century A.D. (when Japan had just begun to develop as a nation) and in the seventh and eighth centuries. After the ninth century, the relationship with the mainland was severed and urushi art developed in its uniquely Japanese style until the end of the Antique period in the twelfth century.

Almost all the fundamental techniques employed in present-day urushi work are believed to have developed during this Antique period. However, over the centuries many changes in technique occurred as the art form developed. As a result, techniques used today are not exactly the same as the techniques used in the Antique period.

This paper will discuss urushi techniques from the Antique and even earlier periods with special emphasis on *urushi-shitaji* (priming), *raden* (shell inlay), and *makie-fun* (sprinkled metal powders) and will describe the differences between earlier techniques and those used today.

Urushi-shitaji *(priming)*

Urushi-shitaji is a general term for *sabi* and *jinoko* and is the ground coating. *Sabi* is *tonoko* (finely ground baked clay) mixed with water and kneaded with an equal amount of raw urushi; *urushi-jinoko* is *jinoko*, which is coarser than *tonoko*, kneaded with raw urushi.

In the early technique of urushi work in Japan, around 5000 B.C., urushi was applied directly on wood in layers, and *urushi-shitaji* was not essential. However, towards the end of the Jomon period, around 500 B.C., *rantai-urushi* wares, which employed woven bamboo as a substrate, were introduced. Thus, *shitaji* became an important technique for covering the mesh surface.

A *rantai* (woven bamboo) vase, found in the ruins of Korekawa, dating from the end of the Jomon period, has a clear, dark brown *shitaji*, and includes transparent quartz particles along with a sandlike substance (Fig. 1). Although no scientific investigation has been done, urushi is likely to have been used as a binder.

Rantai combs were also made in this period, and *urushi-jinoko* was used to give shape to the combs. Among the ten combs excavated at Terachi, some were shaped by *kokuso* (a paste made of urushi, clay powder, sawdust, etc.) and others by *urushi-shitaji*. The *urushi-shitaji* of this period are extremely solid and have scarcely any cracks.

Figure 2. Hemp and shitaji *(priming) from the Hachimanyama tumulus.*
Figure 3. Urushi-shitaji *from the Hachimanyama tumulus.*

Figure 4. Urushi-shitaji *used to bind cloth in a* shippi-bako *(leather box) from the Horyuji treasures.*
Figure 5. Black urushi-jinoko *used in a stage mask from the Horyuji treasures.*

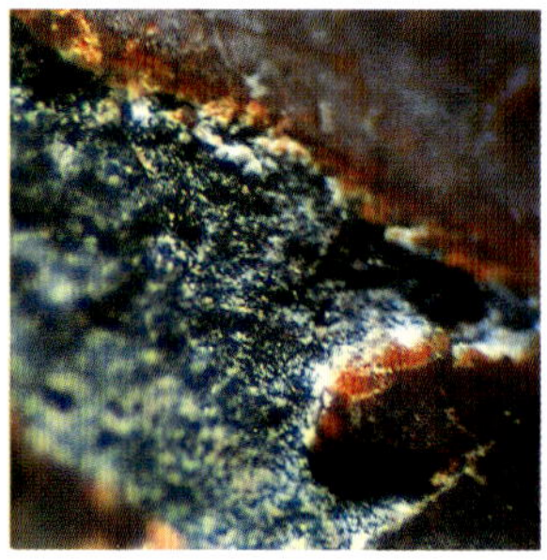

Urushi techniques around A.D. 400–500 show retrogression. The technique was greatly simplified: in most urushi articles of this period only black urushi was applied and *urushi-shitaji* was seldom used.

From the years 700 to 800, when relations with the mainland became active, the technique of *kanshitsu* (dry lacquer) was introduced, and along with it the technique of *urushi-shitaji* reappeared. Some articles of this period include *kanshitsu-zo* (statues), as well as *kanshitsu-men* (stage masks), *soku-hachi* (bowls), *soku-bako* (boxes), *shippi-bako* (leather boxes), and *shikkan* (coffins). Among techniques known for making coffins, *urushi-shitaji* is found in both *kanshitsu-kan* (dry lacquer coffins) and *toshitsu-kan* (coffins with only urushi applied). Examples of *kanshitsu-kan* were found in the Kegoshi-zuka tumulus and the Anpuku-ji collection and of *toshitsu-kan* in the Takamatsuzuka and Hachimanyama tumuli.

In some of these coffins, *urushi-jinoko* was used for its original purpose of simply coating the hemp cloth (Fig. 2), but in others, *urushi-shitaji* was used as an adhesive between layers of cloth. In the latter case, each layer of *urushi-shitaji* was made 1–1.5 mm thick because the silk cloth used for the coffin is very thin. In both cases, the *urushi-shitaji* is usually coarse and not much different from present-day *jinoko*. However, it is interesting to note that the *urushi-shitaji* of the Hachimanyama tumulus, in particular, is uniformly round and finer (Fig. 3).

As mentioned earlier, *urushi-shitaji* was originally used to coat hemp cloth. However, there are other uses of *urushi-shitaji*, for example, to adhere cloth as on a *nunokise shippi-bako* (leather box; Horyuji treasures; Fig. 4) using hemp, or to shape objects such as the urushi stage mask (Horyuji treasures; Fig. 5).

Types of *urushi-shitaji* can be distinguished from each other according to the materials used and by their particular characteristics: they may be *urushi-jinoko*, black *jinoko*, or clear brown *shitaji*. *Urushi-jinoko* is seen only in large-scale works such as coffins. Black *jinoko* was used for the shaping of boxes, for example, those excavated from the Nishinoyama tumulus, and wooden stage masks, but the material cannot be identified. Clear brown *shitaji* was used in leather boxes and stage masks, but these materials also cannot be identified. In Chinese sculpture of the same period, we find *nerimono* (a molding substance). Again, the materials used are not specified, but the *urushi-shitaji* is similar in appearance to those previously described.

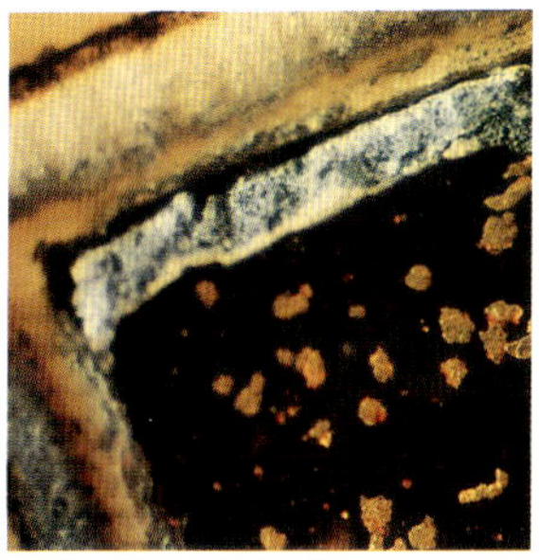

Figure 6. Katawaguruma makie-tebako *(cosmetic box with waterwheel design; twelfth century) with black* urushi-shitaji.

The *urushi-shitaji* of the twelfth century is represented by the so-called *Chusonji-jinoko* of Chusonji Konjikido. This *urushi-shitaji* was used for filling spaces between *raden* (mother-of-pearl inlay) and is characterized by its ability to make a 2 mm layer in a single application. *Urushi-shitaji* diluted with water seems to have been used when narrow spaces between intricate *raden* designs were filled. This *urushi-shitaji* is similar to present-day *urushi-jinoko*, being light brown in color with coarse particles. There is no evidence of the use of *tonoko* in the twelfth century.

It should be noted that in the interior decoration of the Konjikido, only the left and right *shumidan* (altars), which were worked on at a later date, have black *urushi-shitaji* between the *raden* designs. This special *urushi-shitaji* is found in almost all the *raden* objects of this period (Fig. 6) and may have been an essential technique for decorative works. However, what the material was and what sort of effect it achieved have not yet been discovered. This technique was handed down and used until the early thirteenth century, when it became extinct.

Raden

In the Antique period, only *yakogai* (*Lunatica marmorata* Linne) was used for *raden*. The process is still a mystery. Considering other techniques of the Antique period it is easy to figure out how the shells were made into sheets; however, the tools used to cut out the designs are not familiar. Since these tools have not been passed down, and since there have been no related archaeological discoveries, the only clue is the carved surface of the *raden* itself.

Investigations of the marks on the surface of *raden* (Figs. 7,8) belonging to the twelfth and thirteenth centuries reveal an interesting fact. Presumably something like a wire with a round, narrow, cross section was used to make the cuts. In order to employ a wire as a saw, one only needs to make incisions. The result will not be much different from today's wire-saw blade. The cross section of the tool is thought to have been round because the *irizumi* (cut edge) of the *raden* fits the tool, as is shown in Figure 9.

How this blade was fixed to its wooden frame is unknown, but saws similar to present-day steel saws have been found in excavated ruins of the eighth century. Furthermore, it is known that a more typical bow-shaped saw was used in China. Hence, we can imagine the shape of the *raden* tool then in use. The instrument used for cutting agate in the Izumo district proves that by sprinkling emery, a hard rock can be cut with simple twine.

Even in the *raden* of the Tang period (A.D. 618–907), traces of a wire blade with a round cross section are found. This technique developed steadily and was used in the twelfth century during the elaborate project at the Konjikido. The *raden* of the Konjikido amounted to a total of several tens of thousands of pieces, so these works could not have been completed by only a few skilled artists in a short period of time. Therefore, the work of producing *raden* must have been done by a simple method requiring little skill and carried out by a large number of people. In the thirteenth century this method was still in use and reached its peak.

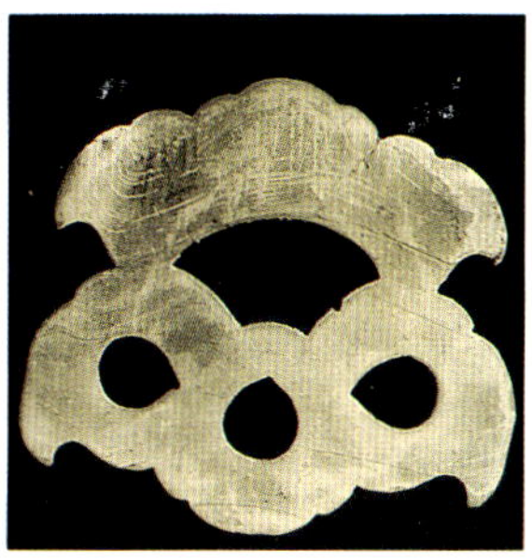

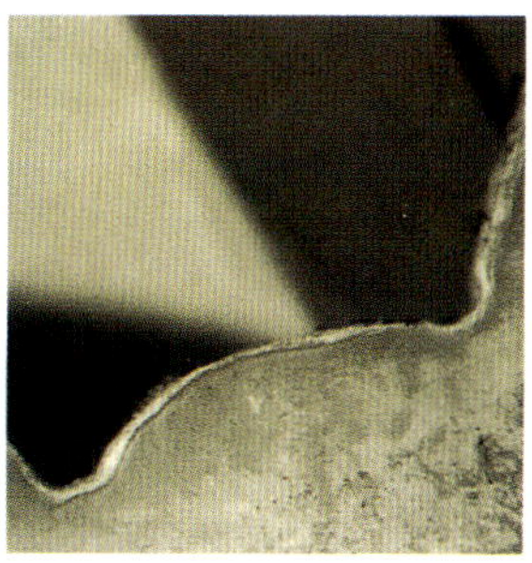

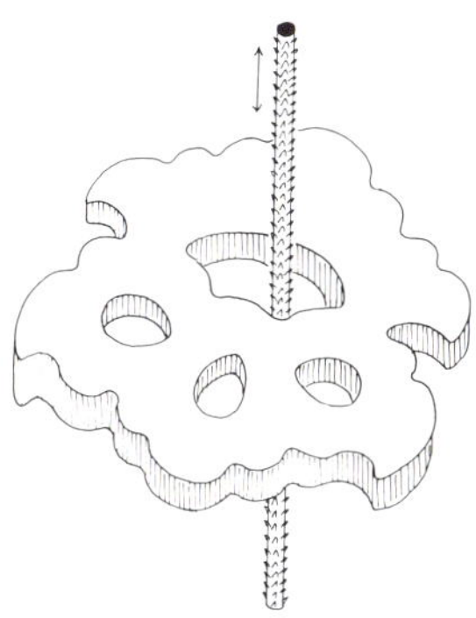

Figure 7. Raden *(shell inlay) from Chusonji.*
Figure 8. Detail of the raden *in Figure 7.*
Figure 9. Method of cutting raden *with wire.*

Figure 10. Raden *from the Shigure* kura *(saddle; thirteenth century) showing the marks of a blade with a round cross section.*

Among the principal remains from this period, the Fusenryo *makie-raden tebako* (cosmetic box) in the Suntory Museum, the Shigure *raden kura* (saddle) from the Eisei Bunko, and the *sakura-raden kura* in the Tokyo National Museum are well known, and there are other excellent works, especially *raden* saddles. The *raden* of the Shigure saddle shows traces of a blade with a round cross section and such exquisite technique that it cannot be imitated today even with a wire-saw (Fig. 10).

The general standard of *raden* technique in this period was high, due not so much to the skill of the artists as to the introduction and improvement of tools. In fact, one *raden* product demonstrates the process of improvement in the blade: the Matsukawabishi *raden kura* (Konda Hachiman). Interestingly, this saddle shows traces of square blade marks similar to the wire-saw blade of today. However, since this kind of blade is found only in this one example among the remains from the thirteenth century, it may be an exceptional work.

Traces on the cut surface of *raden* should also be considered. For example, one *raden* in the Konjikido shows a U-shaped cut surface that gives a smooth outline and creates its peculiar atmosphere. The *raden* of the *makie koto* (zither; Kasuga Taisha), also dating from the twelfth century, possesses sublime expression from having the sides of the *raden* sliced almost vertically.

There are several possible reasons for such great differences in the process of cutting *raden*. First, the *raden* of the Konjikido was mass-produced whereas that of the *koto* was carefully created as an offering for a god. Secondly, there was difference in the precision of the tools. In mass-production the tools are inevitably less accurate. Trying to cut by force with a blunt blade will naturally make a U-shape. Thirdly, the skill of the artists may have been different. For the creation of the Konjikido many artists were gathered from Kyoto, but not all these *raden* artists had superior skills. The *makie koto* and the Konjikido display contrasting characteristics in the *raden* works of the twelfth century. The *raden* of the *makie koto* is far superior to the *raden* technique of the Shosoin and it is considered to be one of the best *raden* relics in Japan.

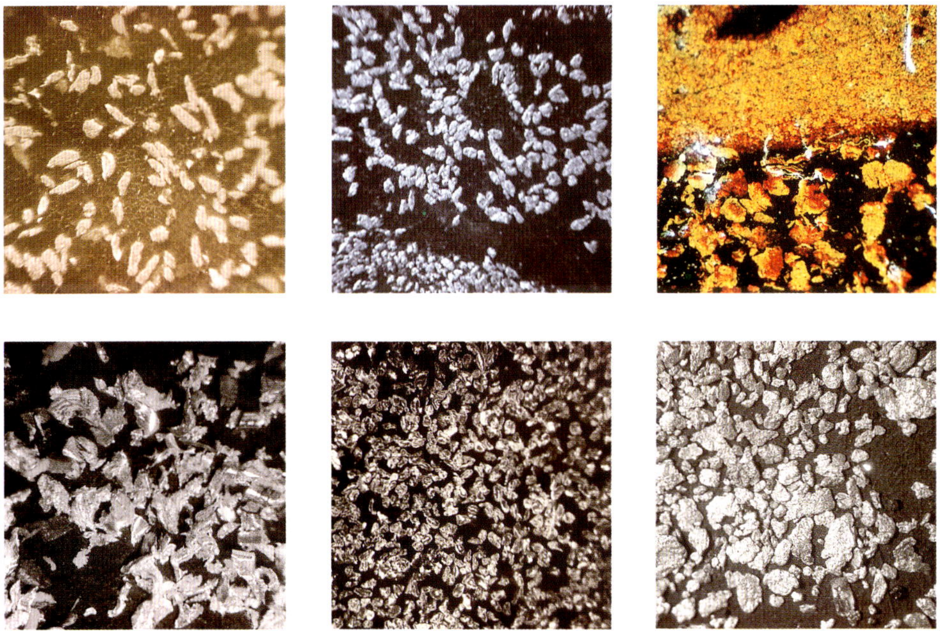

Figure 11. Makie-fun *(metal filings): (a) rice-grain shape (twelfth century), (b) circular shape (twelfth century), (c)* hirame-fun *(thirteenth century), (d) Filing shapes, (e) present-day* maru-fun *(round shape), (f) gold dust.*

Makie-fun

Japanese *makie* dates back to the eighth century and early *makie-fun* may well have used more natural materials such as sand gold. The shape of *makie-fun* changed from a rice-grain shape, to an oval shape until the thirteenth century when it developed into a round shape.

Makie-fun which were generally very coarse were called *yasuri-fun*. This name comes from the Edo period (1600–1868) and has been brought into common use. However, the ancient *makie-fun*, which were finer than *yasuri-fun*, are better called "ancient filings."

Ancient filings are those *makie-fun* used from the eighth to the twelfth centuries. Recent investigations show that these ancient filings have different characteristics from the *makie-fun* used from the thirteenth century onward. The most significant characteristic of the ancient filings is the long, narrow, rice-grain shape. From the thirteenth century the shape gradually becomes more circular.

Besides the rice-grain shape, the filings before the twelfth century can be broadly grouped as follows: wedge-shaped, triangular, and whisker-shaped. Different metals yield different shapes of filings: for gold, rice-grain shape and for silver, wedge-shape are the most common; for greenish gold, triangles and wedge-shapes are found. Such categorization merely notes the tendency and is not absolute. There are many *makie* remains dating from the eighth to the thirteenth centuries, yet each example of *makie-fun* possesses different types of unique filing shapes; hence it is difficult to find similar shapes of filings even among these remains. The significance of this is still unknown but perhaps the making of filings was originally done by individuals. *Shokunin tsukushi-e*, which pictures many artisans at work, depicts a man making *makie-fun* next to a man engaged in *makie*.

Hirame-fun (metal flakes, literally "flat-eye dust"), found in remains from the latter part of the twelfth century, became popular in the thirteenth century. Apparently some kind of significant change took place in the *fun* manufacturing method. The method described in the contemporary *Wakan Sansai zu-e* shows the process of first making *hirame-fun* and then *sai-fun* (a very fine powder). This suggests that the process goes back to the thirteenth century when *hirame-fun* was popularly used.

Photomicrograph (x16) of thin section of Chinese lacquerware object, fifteenth century.

Scientific Approach to Traditional Lacquer Art

Toshiko Kenjo
Tokyo National Research Institute of Cultural Properties

Traditionally, lacquering techniques have been handed down from fathers to their sons and from the sons to their own sons, and so on—knowledge gained from accumulated experience. However, the scientific basis for these techniques has never been clarified. After World War II, the old apprentice system was abolished and the duration of training was considerably shortened. This interrupted the traditional passage of lacquer art from father to son. As a result, young lacquer workers today are likely to seek new materials and, particularly, there seems to be some confusion in the handling and evaluation of "good" vs. "bad" urushi. Consequently, to enable us to preserve the traditional art of lacquering for the future, I attempted a scientific study.

At the beginning of the study, I visited some skillful *urushikaki* (sap collectors) and lacquer artists to hear about ways of collecting the sap of the lacquer tree and applying urushi. Then I tried some fundamental experiments to explore the scientific significance of their techniques.

Time of collecting and properties of the sap

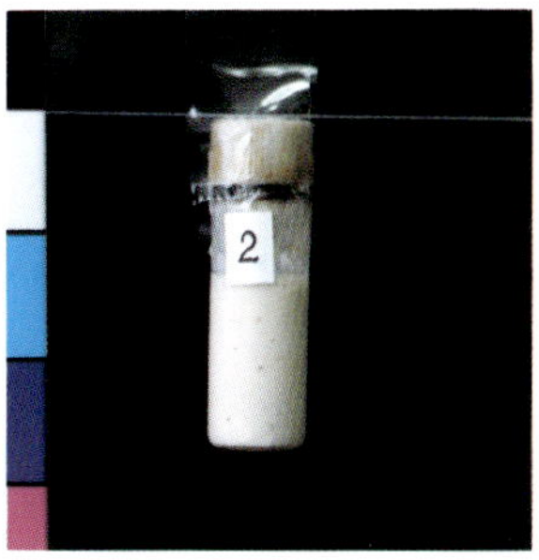

Figure 1. Urushi sap immediately after collection (hatsu urushi).

The best time for the first incision to collect the sap is said to be about fifty-five days after the cherry trees have blossomed, when the lacquer trees are in full bloom. The sap collected from the first six incisions is called *hatsu urushi*; during July and August it is called *sakari urushi*, and in September it is called *ura urushi*. These terms mean the first-, peak-, and last-collected sap respectively.

Since the sap produces bubbles on its surface, it is covered with a sheet of *shibugami* (Japanese paper) and stored in a cool, dark place; the foam collected on the paper is removed every day. As time passes, the foam gradually reduces in volume and the sap changes from milky-white to a lighter color. The top layer is the most transparent (like seed oil) and brown sediments settle on the bottom of the container. This condition is called *shusse-shita* (matured) by the sap collectors. The sap requires about one month to reach this stage.

The specimen of *hatsu urushi* that we received for study in our laboratory was whiter than we had ever seen before, of very low viscosity, and had a small amount of transparent liquid at the bottom of the test tube (Fig. 1). As soon as the cap was loosened, it popped off and the liquid foamed out. We assume that such copious foam was caused by the high activity of the enzyme in the sap; this was subsequently proved by measurement using the triethanolamine method. As expected, the enzyme activity proved to be much higher than in ordinary raw urushi, i.e., matured urushi sap.

Table 1. Analysis of urushi

	Date	Water content	Urushiol	Gummy matter	Nitrogenous compounds	
1984 immediately after collection	July 31	40.98	47.56	9.20	2.26	*Sakari urushi*
	Aug. 31	26.92	63.65	7.29	2.14	*Sakari urushi*
	Sept. 24	34.29	56.32	7.61	1.78	*Ura urushi*
	Oct. 4	34.20	54.25	8.99	2.60	*Ura urushi*
	Oct. 28	49.35	39.31	9.88	1.64	*Ura urushi*
1985 approximately one year after collection	May 20	23.80	67.15	6.60	2.45	*Hatsu urushi*
	May 11	20.11	70.74	5.70	2.45	*Sakari urushi*

Immediately after collection the sap has a 27–50% water content (Table 1), indicating a different balance of water, nitrogenous compounds, and gummy matter as compared with ordinary raw urushi; about a year after collection it shows a similar composition to ordinary raw urushi. This aging is one of the most important things to be considered in traditional lacquering techniques. Both *hatsu urushi* and *sakari urushi* are low in viscosity at first; as time elapses, white precipitates and a transparent aqueous phase are deposited on the bottom of the test tube. The milky white fraction left in the upper part of the tube increases in viscosity. At about the same time the enzyme becomes less active: no foam is produced but tiny bubbles are seen moving about in the milky white sap.

High performance liquid chromatography (HPLC) of the sap immediately after it was received, when it was producing copious foam, showed only one peak (attributed to urushiol), but as time passed several additional peaks appeared. These peaks are believed to be attributable to urushiol dimers or trimers formed as a result of the oxidative polymerization of urushiol.

The pH of the sap immediately after collection was about 6.0 for *hatsu urushi* and about 5.5 for *sakari urushi.* Both figures have a tendency to decrease with time. Ordinary raw urushi has a pH of about 4.5. It is said that the optimal pH value for enzyme activity is between 5.6 and 6.8. High pH values of the sap in its early stages can be considered important for the maturing of the sap rather than for the hardening of urushi.

From the fact that the sap contained a large amount of water immediately after collection, it was supposed that the aqueous phase might constitute the external phase. To test this, a mixture of water-soluble yellow dye and oil-soluble blue dye was added to the sap. As expected, the external phase was tinted yellow and the internal phase blue, indicating the external aqueous and internal oily phases as shown in Figure 2.

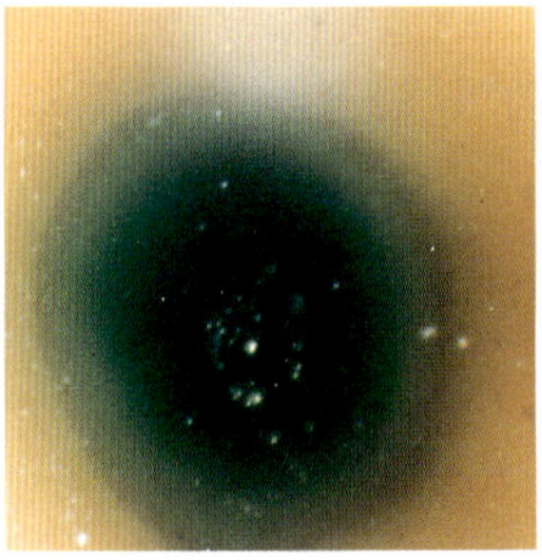

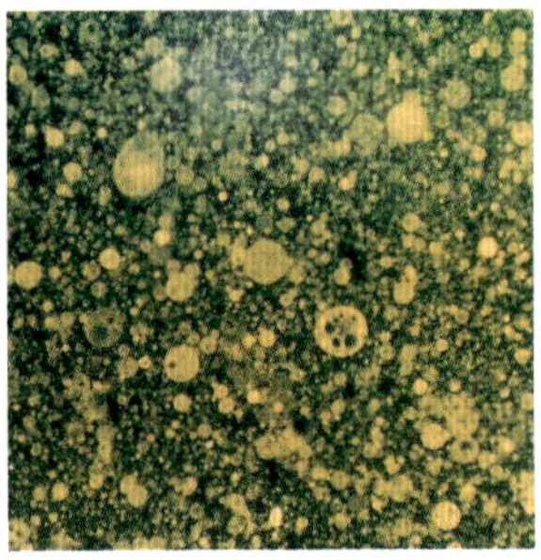

Figure 2. Photomicrograph of urushi sap (40x).
Figure 3. Photomicrograph at higher magnification of the internal phase of Figure 2.

With higher magnification, the internal blue region was found to contain a dispersion of small yellow particles (Fig. 3). Thus the sap of the lacquer tree immediately after collection is a double emulsion of water/urushiol/water.

In actual practice, the sap collected is put into a wooden bucket with a lid; the foam that collects on the *shibugami* (Japanese paper) cover is removed daily (a process called *awa keshi*). Although *awa keshi* seems trivial, it is very important because during the short time of disengaging the cover and removing the foam, the sap is exposed to air and gradually matures into raw urushi.

Figure 4. Changes in weight of lacquer film with time.

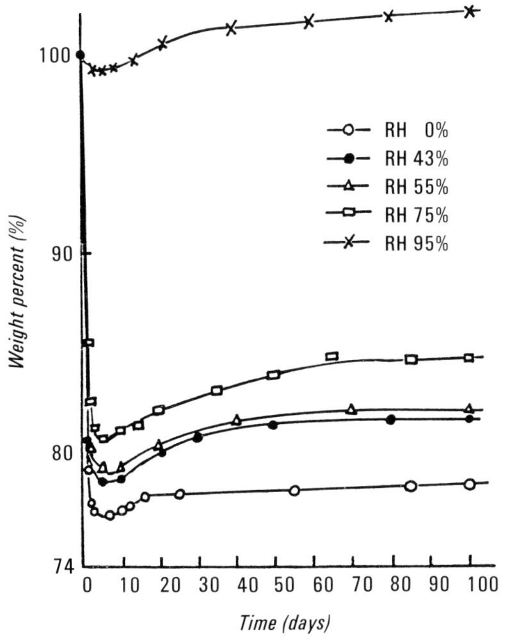

Hardening of raw urushi and its use as an adhesive

Water content and the ambient relative humidity have a considerable effect on the hardening of raw urushi. The polymerization and hardening of raw urushi at an early stage proceeds with the aid of the enzyme it contains, in direct proportion to the amount of water.

Generally, raw urushi is used as an adhesive for *shitajigatame* (ground consolidation). Raw urushi can form a hardened film varying in appearance and properties according to the ambient relative humidity. The higher the RH, the more opaque and darker the film. Below 55% RH the film is more transparent and lighter brown and has a smooth surface. Figure 4 shows the change in the weight of raw urushi films with time, exposed to several different relative humidities. The weight decreases first at all levels of RH, reaches a minimum, and then increases. The minimum value is reached generally between the fourth and the seventh day. In our experiment, one hundred days after the application of the urushi, when the films were considered to have absorbed almost the maximum amount of oxygen and when the water content had reached equilibrium with the surrounding atmosphere, the water content of the films was determined by drying them at 105°C to constant weight. Water content thus obtained is shown in Table 2, which also shows the hardening time and the appearance of the films. Films having water content below 1.5%, corresponding to those hardened at relative humidities below 55%, are transparent and light brown in appearance. However, at relative humidities below 33%, the films cannot harden completely. Depending on the water content, it can reasonably be assumed that a relative humidity of about 55% will allow a film to harden to "tack free" in a period of one day.

Table 2. "Tack free" time, appearance and water content of urushi films

Relative humidity	*"Tack free" time*	*Appearance of film*		*Water content of dried film*
		Color	*Gloss*	
%	*hr*			*g(H_2O)/100g*
33	72	Brown	+	1.10
55	24	Brown	+	1.49
75	17	Dark brown	–	4.07
95	15	Dark brown	–	22.60

Figure 5. Changes in oxygen and water contents of thin and thick urushi films at three different relative humidities.

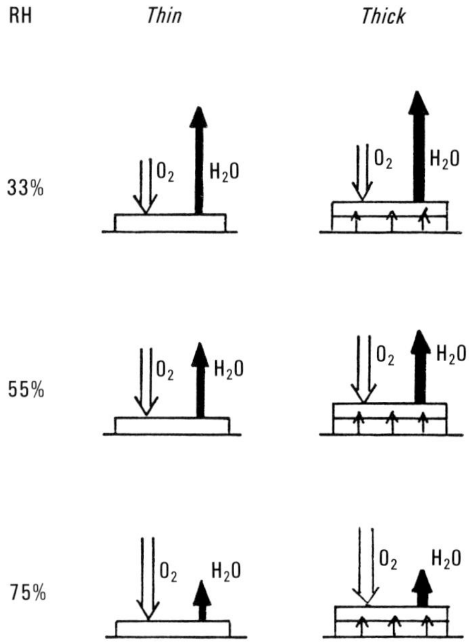

As shown in Figure 5, water evaporates more rapidly from the surface of an urushi film at a low RH than at a high one. With a thick film, at the lower relative humidity of 33%, there is a greater gradient of water concentration leading to faster diffusion of water towards the surface, so the water concentration in the film becomes uniform more quickly. However, because the water content of the film is lower, the film absorbs oxygen more slowly, thus causing slower hardening. On the other hand, at a higher relative humidity (75%), water evaporates so slowly from the surface of the film that the gradient of water content is much less, making water diffuse more slowly in the film. As a result, because of its high water content, the film absorbs oxygen faster, resulting in an opaque, nonuniformly hardened film, with a significant amount of water remaining. At an intermediate relative humidity of 55%, the rate of water evaporation from the surface harmonizes well with the hardening rate (i.e., the oxygen-absorbing rate) to form a uniform, transparent, light-brown, hardened film in a reasonable period of time, about twenty-four hours.

When the film is thin enough to disregard the question of water diffusion within it, the hardening of urushi depends only on the amount of moisture in the environment: the higher the relative humidity, the quicker the rate of hardening. However, since water remains in the film, the latter will become opaque. To keep the film as thin as possible while preventing it from becoming opaque, it is necessary to place it in an environment with higher relative humidity than for a thick film, but not too high. For example, if a thick urushi film hardens in a given length of time at 55% RH, it will be necessary to keep a thin film at 75% RH for it to harden in the same time. It is important to realize that a thin film will harden more slowly than a thick one and therefore require a higher relative humidity if it is to harden in a reasonable period of time. To make a film very hard and smooth like stone, old lacquer artists used to apply many layers of thin urushi at fairly high relative humidity, as in *urushiburo*, a technique for hardening urushi which uses a wooden box (*urushiburo*) kept permanently moist by spraying it with water. This technique can be considered very reasonable from a scientific point of view.

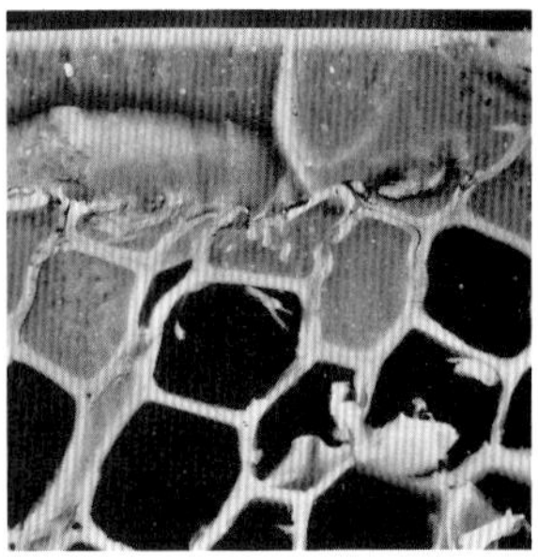

Figure 6. SEM photomicrograph of a hardened urushi film on a wooden substrate (1000x).

The photomicrograph (Fig. 6) of an urushi film on a wooden substrate hardened at 55% RH shows that the urushi has penetrated into the micella of the wood. However, the urushi film hardened at 75% RH is not uniform and the film hardens so fast that there is no penetration of urushi into the micella.

Figure 7. Change with time in adhesive strength of urushi films kept at different relative humidities.

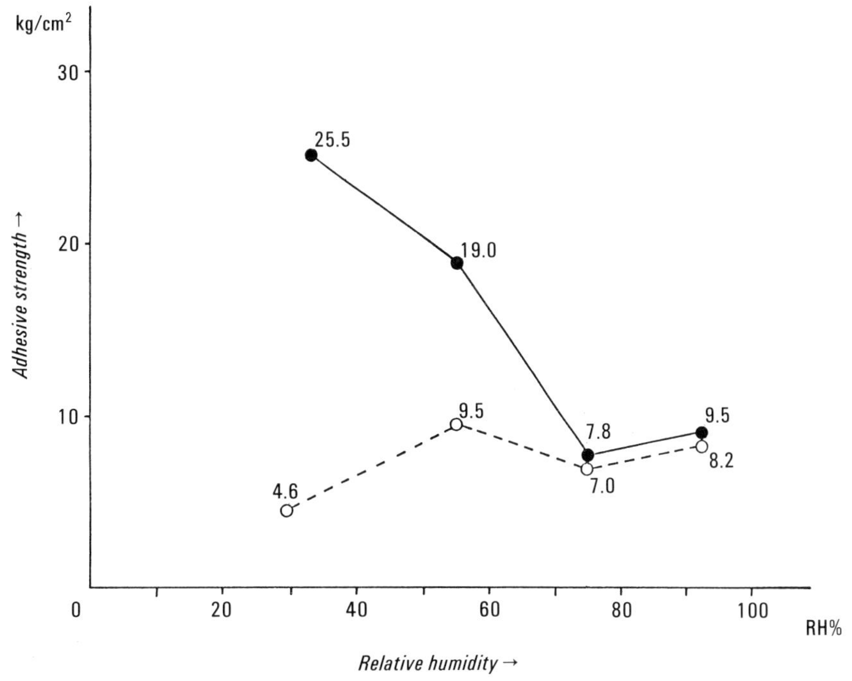

The results of measuring the adhesive strength of urushi films hardened at various relative humidities are shown in Figure 7. A comparison of the adhesive strength of the films one month and six months after application shows that it does not change significantly with the lapse of time at higher relative humidities (75% and 92%); however, it increases over time at lower relative humidities (33% and 55%). In particular, at 33% RH, the strength six months after application is five times greater than that one month after application. These results prove the old lacquer artists' saying that the longer the period of time required for hardening, the greater is the adhesive strength obtained.

In order to discover the effect of the water content of the substrate on the hardening of raw urushi, urushi was applied to boards made of Japanese cedar (*Cryptomeria japonica*) and Japanese cypress that had been equilibrated to atmosphere at different relative humidities (92% and 35%, respectively) and kept at various relative humidities (33%, 55%, and 75%). The boards measured 65 x 75 x 5 mm and the thickness of the urushi film was approximately 0.01 mm. On the boards equilibrated at 92% (average moisture content 12.2%), only opaque, dark-brown, hardened urushi films could be obtained, even at the lowest RH (33%), while on the boards equilibrated at 33% RH (average moisture content 7.2%), transparent, light-brown films were obtained at lower relative humidities (33% and 55%). These results show again that moisture content should be kept as low as possible in order to obtain well-hardened urushi films. This generally refers to thick films but, if the time required for hardening is disregarded, well-hardened urushi can always be obtained at the lowest possible RH.

Shitajigatame *(ground consolidation)*

Japanese traditional lacquering techniques include the *shitajigatame* process, which consists of two subprocesses, *nunokise* and *jigatame*.

Nunokise is a process for protecting the substrate and serves as a *komai* (base) for the subsequent process. Cloth is attached to the substrate with *nori-urushi*, a mixture of rice paste and raw urushi. Skillful lacquer artists have told us that rice paste is the best for making *nori-urushi*. The rice paste and raw urushi should be blended until the mixture can be pulled out to make a string about 50 cm long.

Figure 8. Photomicrograph of (a) hemp (200x), (b) flax (200x), (c) ramie (200x).

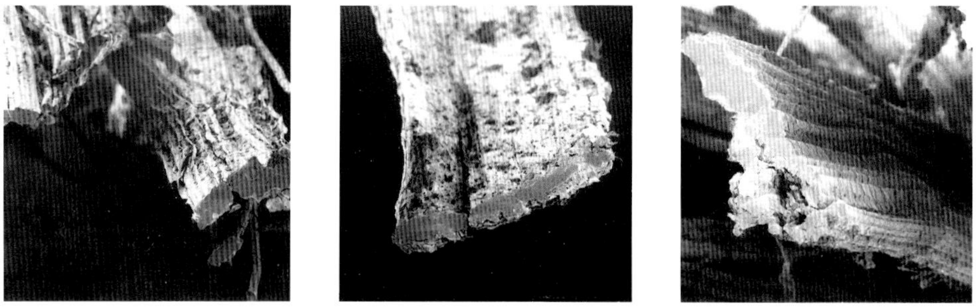

Figure 9. Hemp from makie-bashira *in Chusonji temple* (300x).

Table 3. Weight gain in saturated water vapor of several fabrics

Sample	*Weight of dry sample (g)*	*Weight gain in saturated water vapor (g)*			
		4 hr	*8 hr*	*12 hr*	*15 hr*
Hemp	100	118.3	130.1	142.3	148.7
Flax	100	118.6	129.2	140.8	147.5
Cotton	100	120.3	126.8	136.7	144.5

Table 4. Water and oil absorption of typical clay samples for use in *shitajigatame*

Sample	*Water absorption* (g)*	*Oil absorption* (g)*
Wajima-jinoko	8.6	9.8
Yamashina-jinoko	4.5	3.2
Hirasawa-tonoko	5.4	5.1

*The amount of the vehicle required for the sample to be pasty.

Table 5. Water content at each stage of urushi priming (one week after application)

Relative humidity (%)	*Substrate (%)*	*Solidification (%)*	*Cloth application (%)*	*Grounding (%)*
55	8.3	8.3	8.5	8.3
75	8.5	10.8	11.9	12.5

Various samples of *nori-urushi*, with different proportions of rice paste and raw urushi, were made and examined in a microscope to assess the blending. The results showed that only the blend with a ratio of six parts rice paste to four parts urushi gave a uniform dispersion. The blend with this critical ratio was also found to have the stringy consistency mentioned above. Due to its water content, the paste accelerates the hardening of the urushi.

Today, hemp, silk, and cotton cloths are used for *nunokise* but it is said that in the old days hemp cloth was mainly used for this purpose. A scanning electron microscope (Figs. 8 a–c and 9) showed the *nunokise* of the *maki-bashira* (a pillar with *makie* decoration) in Chusonji temple to be hemp.

Which is more suitable for *nunokise*, hemp or cotton? Data of absorption and evaporation of water for hemp published by the Sen-i Association (Table 3) show that both rates are greater than for cotton. Therefore hemp sticks to the substrate and also assists the hardening of *nori-urushi* by quickly absorbing excess water from the *nori-urushi* and evaporating it.

In the case of dry lacquer (*kanshitsu*), *nori-urushi* is made by mixing flour into raw urushi, producing what is called *mugi-urushi*. Unlike ordinary *nori-urushi*, in *mugi-urushi* the flour affects the hardening rate of the raw urushi by absorbing water, giving conditions similar to those for hardening at lower relative humidities. Therefore the film hardens more slowly but the hardened film has greater adhesive strength and stiffness than when *nori-urushi* is used.

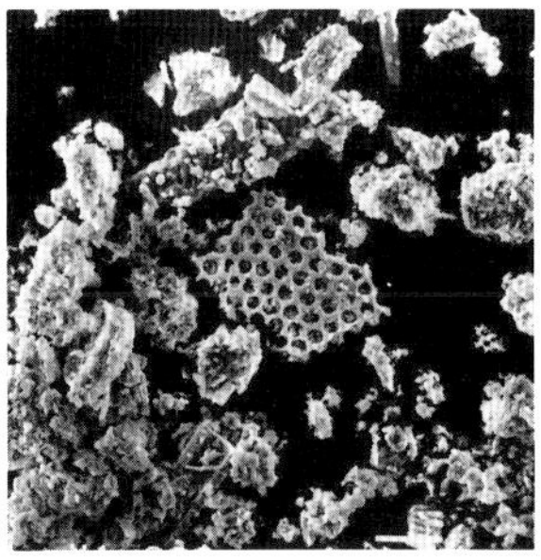

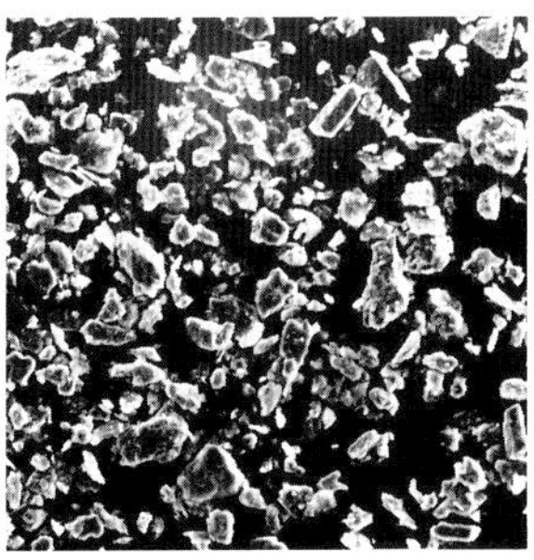

Figure 10. SEM photomicrograph of Wajima-jinoko *(1000x).*
Figure 11. SEM photomicrograph of Yamashina-jinoko *(1000x).*

Jigatame serves to protect and thermally insulate wooden substrates. Typical *jinoko* (fired, ground, and screened clays) used for *jigatame* include *Wajima-jinoko* and *Yamashina-jinoko*, named for the places of origin of the clay. Data for water absorption—the amount of water required for making a given *jinoko* paste—are shown in Table 4. This table shows that *Wajima-jinoko* is capable of absorbing a lot of water, so that it requires much more water to make a paste, whereas *Yamashina-jinoko* needs much less water. From photomicrographs of these *jinoko* (Figs. 10, 11) it can be seen that *Wajima-jinoko* contains diatomaceous earth having a porous structure like honeycomb and that *Yamashina-jinoko* is a conventional kind of clay.

Traditional lacquering techniques have skillfully utilized these two kinds of *jinoko*. For example, *Wajima-jinoko* is used by mixing it with rice paste. In this case, the paste serves to accelerate the hardening of the urushi, while the *jinoko* absorbs excess water from the paste to contribute sufficient stiffness and adhesion for the purpose of *jigatame*.

In order to discover the effect of ambient relative humidity on the water content of the material in each *shitajigatame*, wooden boards were subjected to consecutive treatments of *kijigatame* (solidification), *nunokise*, and *jigatame* after having been equilibrated to 55% RH and their water content measured. After each treatment they were kept for one week at 75% and 55% RH and then their water content was measured. The results (Table 5) show that the water content of the substrate kept at 55% RH did not change whereas that kept at 75% RH steadily increased in water content with successive treatments.

From these results it can be seen that the traditional lacquering technique of *shitajigatame* places great emphasis on removing the water content of the urushi film. The technique has been refined in practice, based on the accumulated experience of many artists.

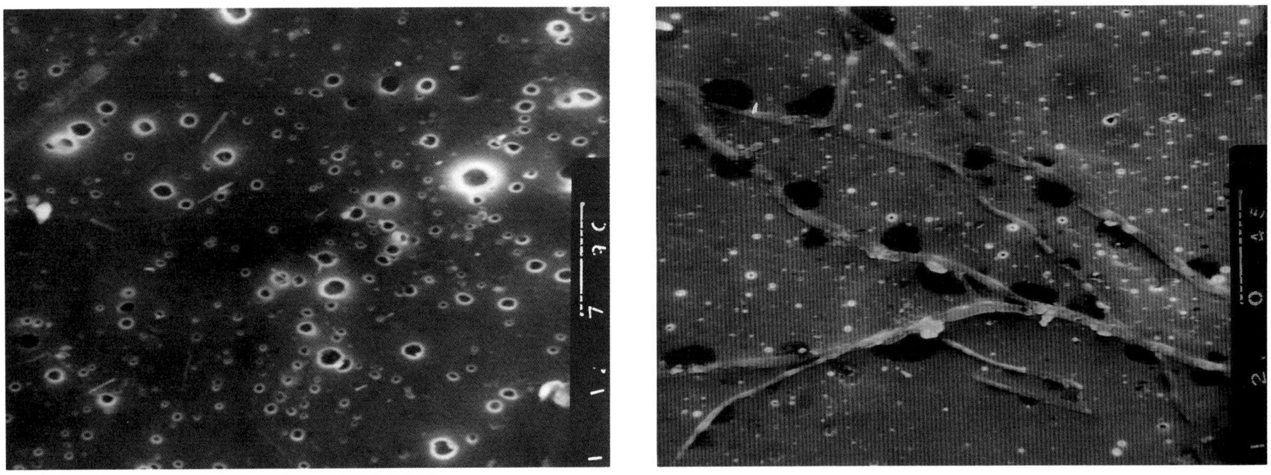

Figure 12. SEM photomicrographs of urushi film after exposure to UV radiation of different wavelengths: (a) 1000x, (b) 3000x.

Effect of ultraviolet radiation on hardened urushi films

Although hardened urushi films are excellent coating materials because they are lustrous and very stiff, their one drawback is that they are sensitive to light, especially ultraviolet radiation. When the film is exposed to UV radiation, part of the polymerized urushiol is decomposed and will volatilize, leaving the film in a heterogeneous condition. A photomicrograph (Fig. 12) of a film exposed to UV shows many black spots surrounded by lighter regions.

Since ultraviolet radiation is very harmful to other works of art besides lacquerware, non-UV fluorescent lamps are now widely used for exhibition purposes.

As a fundamental experiment to discover the effect of non-UV fluorescent lamps, hardened urushi films were exposed to a non-UV fluorescent lamp at several different light levels: 100, 200, 350, and 500 lux for two weeks. Films exposed at 350 and 500 lux showed deterioration similar to that in Figure 12, in which many black spots were seen, whereas those exposed at 100 and 200 lux exhibited no such deterioration. These results show that hardened urushi films are sensitive not only to ultraviolet but also to visible radiation. Therefore, even if non-UV fluorescent lamps are used, works of art, especially lacquerware, should be exhibited below 200 lux and ideally for short periods only.

Conclusion

Japanese traditional lacquering techniques were studied from a scientific point of view. It was found that urushi sap differs in composition depending on the time when it is collected. It is a double emulsion of water in urushiol in water immediately after collection but, as time passes, the water-in-urushiol phase, which is used as ordinary raw urushi, separates from the external aqueous phase. It was also found that oxidation intermediates of urushiol, including urushiol quinone and dimers, which are essential for the hardening of urushi, are formed in raw urushi through a kind of maturing process called *awa keshi*, done by sap collectors during the period of time between collecting the sap and handing over the raw urushi for distribution. This maturing process is important to obtain good urushi films.

In the process of *shitajigatame*, it was found that *nori-urushi* accelerates the hardening of urushi, *mugi-urushi* enhances the adhesion of urushi to the substrate, hemp is used for the quick absorption and desorption of water in *nunokise,* and *jinoko* is utilized to regulate water content. Therefore *shitajigatame*, one of the traditional lacquering techniques, is quite reasonable from a scientific point of view in that it tries to remove as much water as possible from urushi films for their optimal hardening and adhesion.

Technical Studies of Chinese Lacquer

A. Burmester
Doerner Institut

The title for this article has been chosen both to recall and to extend the work of Sir Harry Garner which was published almost twenty-five years ago. Garner, a collector of Far Eastern art, was the author of several pioneering publications, particularly on Oriental lacquer. He was the first person in the West to employ scientific techniques to try to explain the technical and technological peculiarities of Oriental lacquer. He combined a scientific approach, resulting from his education as a mathematician and astronomer, with a knowledge of Far Eastern art, evidenced both in his publications and by the works of art he donated to the British Museum and to the Victoria and Albert Museum in London. His interdisciplinary interests can be seen in his "Technical studies of Oriental lacquer," published in 1963. He pursued these interests until his death in 1977 and they are found again in the second chapter of his book *Chinese Lacquer* that appeared posthumously in 1979 and that also deals with the technological aspects of urushi.

The beginnings of the present article go back to 1976 and my first thoughts were based on Garner's preliminary work (1963). The aim of these investigations was to find out to what extent scientific examination could help to answer questions of dating, provenance, or identification of Chinese lacquerware. At the same time I intended to extend Garner's work, which concentrates on two pieces from the Ming dynasty (1368–1644), by introducing a series of additional objects to cover a greater timespan, i.e. from Zhou (1027–256 B.C.) to Qing (1644–1912). These investigations and their results are the subject of this article.

Garner's literary estate

It was a happy coincidence that, while a member of the Urushi Study Group in Japan in 1985, I learned that Garner's literary estate was in the possession of the Victoria and Albert Museum in London. Thanks to the generous cooperation of that museum, I was able to study the papers while preparing this article. Garner's papers confirmed that he had tried repeatedly with the help of the natural sciences to solve the technical questions which had occupied him greatly, in and beyond his publications of 1963 and 1979.

At the start of his work there are written notes of his visual appraisal of lacquer objects mainly carved from the sixteenth to eighteenth centuries.[1] The first concrete analytical results soon arrived[2] and are included in Table I of his publication of 1963. They cover not only qualitative but also quantitative emission spectral analyses. Unfortunately, it was difficult to isolate a layer for emission spectroanalysis and so it was necessary to repeat some of the measurements.

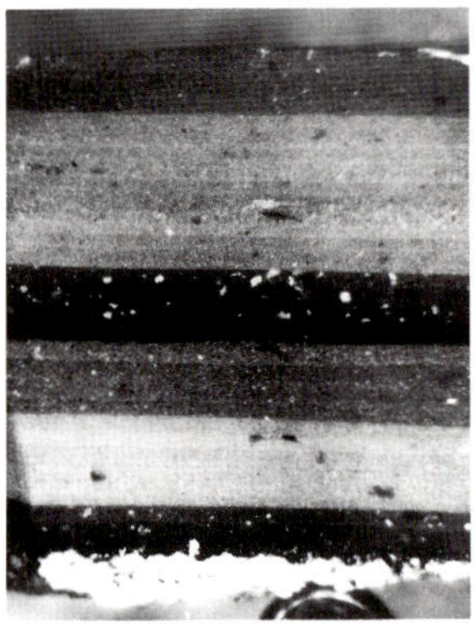

Figure 1. Photomicrograph of a cross section from a Chinese lacquerware object, late sixteenth century, from the estate of Sir Harry Garner (Victoria and Albert Museum, London), taken by A.R. Sollars, Cranfield (magnification x50), supplied with the following commentary on the layer construction: "band 1: 2 layers brownish-red, moderately fine with large inclusions; band 2: 5 layers light red, fine; band 3: 4 layers bright red, coarse; band 4: 1 interface layer, discontinuous, fine layer, grayish; band 5: 1 layer dark gray or black, structureless compared with normal pigmented layers; band 6: 3 brownish-red, similar to band 1; band 7: light red, fine, similar to band 2; and band 8: bright red, coarse as band 3."

At the end of 1962 Garner received five fabric analyses[3] that, together with the cross section and the analyses already mentioned, provided the data for his publication of 1963. This work was received with interest and goodwill.[4] In the same year R.J. Gettens gave Garner additional experimental advice on a piece of carved lacquerware from the late sixteenth century. He recommended the use of a "microscope with vertical illuminator and dark field stop"[5] and magnification of 100–200x. It is interesting that one of the dark layers in this investigation was described as "transparent and non-pigmented." This observation will be discussed later. In reference to the same cross section, A.R. Sollars[6] informed Garner that he had followed the usual procedures for preparing cross sections for microscopy and photomicrographs and that he agreed in general with Gettens' findings. One of the photomicrographs he sent is reproduced here (Fig. 1). This demonstrates the desirability of color photographs for the complex layer construction; Garner's publication (1963) suffers from the absence of color. Apart from these practical issues, Garner was looking for further samples[7] or other analytical procedures. The first attempts to subject urushi to pyrolysis gas chromatography appear to have been successful.[8] Furthermore, scanning electron micrographs increased the knowledge of the pigments used (Hornblower 1962) and Garner received radiographs of earlier lacquerware samples from Gettens.[9] However, Garner also had unanalyzed samples returned to him[10] and these were not the only failures. A manuscript prepared at his suggestion on the chemistry of urushi (Moss 1967) remained unpublished at the request of the author, and many scientists who had hitherto been of great use to him in his research denied him assistance because too great a commitment seemed to be required.

From his correspondence, Garner was also interested in questions of the pigmentation of the black layers,[11] the composition of the black ground used in carved lacquerware,[12] the choice of a suitable method of analysis,[13] correct terminology for the Chinese sources of urushi,[14] the lacquer tree that supplies urushi,[15] and its spread over Asia,[16] and the identification of the fabric support in the lacquerware of the Han era.[17] At the same time he attempted to persuade various colleagues to study a translation he had prepared of the travel diaries of d'Incarville of 1760, which in fact appear to have been one of the most reliable contemporary reports.

Among Garner's papers, evidence can be found of two more experimental initiatives on additional lacquerware. Figure 2 shows the cross section of a carved lacquerware piece that has a construction typical of the Ming dynasty and will be discussed later.[18] A further thin section shows for the first time the construction of a piece from the Han dynasty (206 B.C.–A.D. 220; Fig. 3). It has a thick layer of ground and a thin covering layer over four layers of fabric support.[19]

Figure 2. Fifteenth century, heavy black layer. A new photograph of a thin-section from Garner's estate in transmitted light (16x). From bottom to top: wooden base; heavy black layer (mostly broken away); very thin transparent lacquer layer (looks orange); dark, yellow ground layers; black guide line (Garner 1963) looks transparent orange (see text); red ornamental zone.
Figure 3. Thin section of a Han dynasty lacquerware object from Garner's estate.

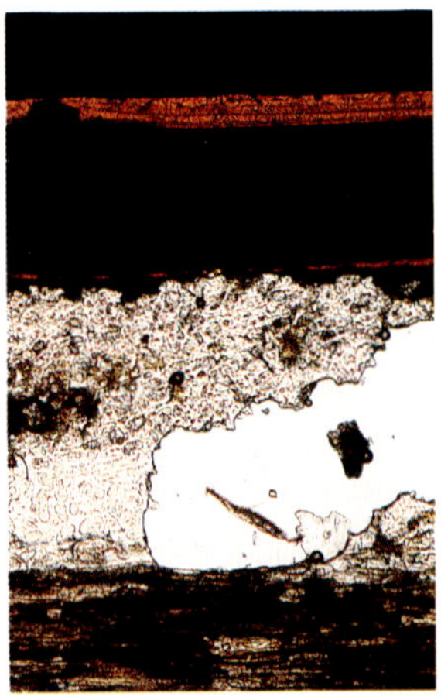

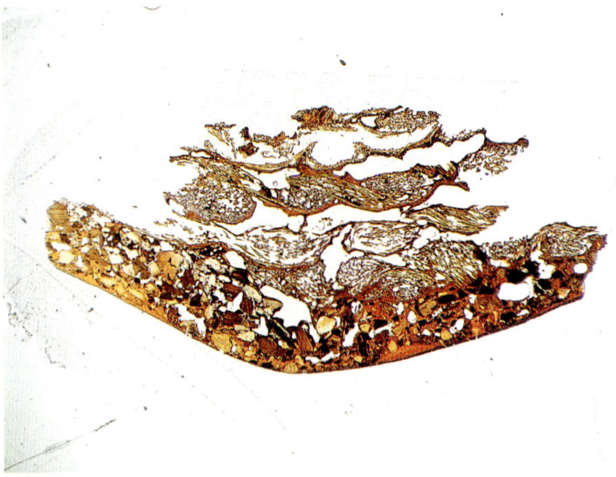

The urushi debate

In order to clear up some misunderstandings and misinformation, we shall start with a discussion of some of the main aspects of debate.

Terminology

The first of these misunderstandings is the question of terminology, to which Garner applied himself in an attempt to discover a common approach in the natural sciences and the humanities to the origins of urushi (Burmester 1985). One glance at this theme in books and catalogs clearly demonstrates the need to establish an agreed terminology.

Identification of the lacquer tree as *Rhus verniciflua* or *Toxicodendron verniciflua* is in keeping with current botanical usage (see note 15; Zander 1972; Hora 1981; Schulze 1966). Hopefully, in the interests of simplicity, *Rhus verniciflua* will be adopted as the exclusive term.

It should be stressed—and this is also clear from Garner's correspondence—that there is no information concerning the spread of *Rhus verniciflua* in the last three or four millenia. Analytical examination of the organic constituents of urushi itself may provide the answer to the question of whether *Rhus verniciflua* has always been the source of raw lacquer. The sap of other trees of the same genus, for example, *Rhus succedanea*, may also have been used.

The extent of the uncertainty about the chemical composition of lacquer can be seen in the range of ambiguous and even false designations that have been used. This is because modern scientific knowledge has hardly ever been applied to the relevant literature. An example of this is the use of the term "urushic acid"(see note 14), meaning the main ingredient of lacquer, which is also called "urushiol." After its first description as "urushic acid" by Yoshida in 1883, it was described as "urushin" and "oxyurushin" in the paper of Tschirch and Stevens in 1905; these authors clearly wanted to demonstrate that the compound in question had nothing to do with acids in the chemical sense. In 1906 Miyama was the first to describe it as a phenol; then finally, in 1908, he used the designation "urushiol." Since then it has been correctly and exclusively referred to in chemical literature as urushiol (for example, Majima 1909). Urushiol is in fact a mixture of phenols and can be taken as a trivial name for an entire group of chemically related compounds.

History and technology

It would appear reasonable to place Oriental lacquer within the range of known binding agents. Urushi can be considered as a colloidal system of a water-in-oil emulsion (Gettens and Stout 1966). Drops of water, containing polysaccharides and an enzyme, are dispersed within an oily phenol phase in which glycoproteins act as emulsifiers (Kumanotani 1979, 1983a, 1983b).

Chemical and physical analysis

As Garner and other scientists have already demonstrated, the natural sciences offer a starting point for the solution of technological problems associated with Oriental lacquerware, although Garner's attempts to analyze the binding agent urushi were not successful. A study of the relevant literature shows that urushi does not form part of the classic analysis of binding agents which is concerned with oils, resins, waxes, gums, proteins, etc.[20] The analysis of urushi—and this is a complicating factor—must take account of its high degree of polymerization and cross-linking, and the addition of inorganic fillers and pigments and organic substances such as drying oils or glues. Although, thanks to the findings of Kumanotani (1979, 1983a, 1983b), our knowledge of freshly polymerized systems is relatively precise, there is no information on the effects of aging. Ideally, urushi should be analyzed after separation into its components; this is always the first step in the analysis of binding agents and in the case of unhardened urushi. Unfortunately this is prevented by the insolubility of the polymeric system. The limited information supplied by infrared spectroscopy (Kumanotani 1979, 1983a, 1983b; Kenjo 1978) from systems which have not been separated shows that this step cannot be completely ignored. These and other analytical attempts show that our objectives must be modified to pursue analysis "in a quite empirical way, not even identifying components but trying to get distinguishable patterns."[21] Here the possibility of thermal decomposition of the matrix comes to mind. This is achieved by breaking down the polymeric framework and permits analysis of the resulting fragments. But any conclusions about the matrix prior to decomposition still remain problematical.

There is a series of analytical possibilities associated with thermal decomposition of the polymeric framework, which promise to provide more information. The first experiments with pyrolysis gas chromatography carried out for Garner demonstrate this (see note 8). More extensive investigations have shown that pyrolysis mass spectrometry of Oriental lacquer (Burmester 1983a, 1983b), as well as of other natural polymers (Meuzelaar 1982), promises to produce interesting results. However, an interpretation is difficult beyond the level of accepting the results of analysis as a pattern. The introduction of multi-variable statistical evaluation has made progress in pattern recognition (Burmester 1983a, 1983b) but, without extensive preparatory investigations, comparative studies and relevant experience, the results must be treated with caution.

Along with the results already published, limited conclusions may be drawn from the mechanism and regularities of decay of the lacquer sample from its pyrolitic fingerprint in low resolution mass spectrometry (Burmester 1982, 1983c). In relation to questions posed by Garner, some interesting conclusions can be drawn from this experimental starting point. For example, the decomposition of urushi in the case of post-Tang (after A.D. 906) specimens seems to be very well defined. The expected decomposition pattern (Occolowitz 1964) is found in each of the pyrolysis mass spectra, but only in samples of similar age. This is in contrast to earlier lacquerware samples which obviously decompose in a much less defined manner.[22] This may be due to alteration resulting from aging of the lacquer matrix or because the matrix was never structured in the same way as that of more recent carved lacquerwork. This immediately raises the question of whether a different method for preparation of urushi from *Rhus verniciflua* (*kurome* or *nayashi*) was used in earlier times or, indeed, whether another latex was used.

A more precise evaluation of the results can be achieved by improving the system of analysis (a switch from low- to high-resolution mass spectrometry). Computer-supported interpretation of these experimental results permits the identification of individual pyrolysis or ionization products which, in turn, in individual cases, give a clue as to the matrix before thermal decomposition. The experimental conditions for high-resolution mass spectrometry are similar to those already published for low resolution (Burmester 1983b). In order to establish results from this series of measurements on the same basis as those of previous investigations, measurements of samples already published were repeated (Burmester 1983b). The dates of the twenty-six samples, all from lacquer objects of the Linden-Museum (Stuttgart, German Federal Republic), ranged from Zhou to Qing and included a few modern ones. The information obtained so far is summarized.

If the high intensity peaks are filtered out from the mass spectra and related to the most probable combination of elements according to their mass (in this case only C, H, O) it is apparent that certain fragments[23] can be observed in each of the samples investigated. These appear to be quasi "index fossils" that should permit recognition of urushi as a binding agent. As a limiting factor, I must admit that I had no opportunity to examine other saps of the genus *Rhus*.

It is striking that pre-Tang (before A.D. 618) specimens have a substantially higher proportion of carbohydrates (fragments of type C_xH_y), while later samples comprise primarily oxygen-containing fragments (of the type $C_xH_yO_z$). This supports the differences already mentioned in lacquerware according to its age.[24]

Technological examination of Chinese lacquerware

In this section the construction of lacquerware is examined with the help of cross sections, and an attempt is made to determine the pigments used. Garner's publication of 1963 was a pioneering work on both of these points. However, the scope has been greatly increased in the present work.

I am indebted to the Linden-Museum, for generously allowing me to examine an extensive group of Chinese lacquerware as well as many pieces of carved lacquer. All the pieces investigated belong to this museum; they were purchased from the Fritz Löw-Beer Collection in 1978 (Burmester and Brandt 1982, Brandt 1982, 1986). The extensive scientific research project examined over forty objects, twenty-eight of which are presented here (Burmester and Brandt 1982).

Methodology

Small samples were taken from all the objects for the preparation of cross sections. By using laser microanalysis (Moenke, 1968; Roy 1979; Schrön 1983) it was possible to obtain an overall view of the pattern of elements detectable in the individual layers.[25] In this way it was possible to avoid the difficulty, to which Garner referred, of preparing single layers for analysis. In association with microscopic evaluation of the cross sections[26] it was possible to obtain an indication of the pigments used. Parts of the lacquer samples were pulverized for further investigation using mass spectrometry and x-ray diffraction analysis.[27] Here, both Debye-Scherrer and vertical goniometric diffractograms were taken.[28] Under suitable conditions, it was possible to identify the pigments used.

The Appendix contains the results obtained: the macroscopic construction of the pieces, the function of the individual layers, and their pigmentation. In keeping with Garner's method (1963), also reproduced along with the description of the object is the sequence of bands and layers,[29] the color of the bands and their probable function, their approximate thickness (measured from the cross sections) and the pattern of elements obtained by laser microanalysis.

Early Chinese lacquerware between Zhou and Song

The group of early painted lacquerware pieces under investigation includes objects from the Zhou dynasty (nos. 1–5, see Appendix), the Han dynasty (nos. 6-10), the Tang dynasty (A.D. 618–906; nos. 11, 12), and the Song dynasty (no. 13; see also Figs. 4–18). These objects are discussed later in relation to their construction and function and also, in association with carved lacquerware, their pigmentation.

Most of the early painted lacquerware objects have a simple construction. In the case of object No. 1 from the Zhou era (Figs. 4,5) there is only one layer, which is considerably darkened on the surface, over a very thin ground layer. It is very difficult to determine if this was applied on purpose or whether, as has been observed in other cases, it results from residues of abrasives used.

The construction of the objects from the late Zhou and Han eras onwards can be schematized as follows: a ground mixed with coarse minerals (mostly with black additives), followed by one or more covering layers, and on the surface an ornamental, colored layer, usually red or black (Figs. 7,9,11,12,13,15). It is striking that none of the pieces under discussion display any evidence on the surface of the layers of having been worked upon. This indicates that after the drying process the films of lacquer were not rubbed down with abrasives. Plant fibers are often found in the individual layers (Fig. 5, yellow layer; Fig. 7, layer over the black ground). This suggests that the urushi was not filtered.

All the black grounds contain traces of minerals. Laser microanalysis has detected high intensities of silicon, calcium, iron, magnesium, titanium, manganese, and aluminum, while x-ray diffraction has identified the presence of quartz (SiO_2), gypsum ($CaSO_4 2H_2O$), calcium-magnesium or calcium-aluminum silicates, and calcium phosphates ($Ca_5(PO_4)_3OH$).

Further study provides the answer to the question of whether or not the typical brown coloring of the covering layers (visible in the cross sections of no. 2 in Fig. 7, no. 4 in Fig. 9 and no. 10 in Fig. 15) was pigmented.[30] Laser microanalysis demonstrates that practically no elements other than calcium and magnesium are to be found in the brown covering layers of Han era objects. For the moment the question must remain open as to whether the top layer was altered due to aging of the binding agent or was intentionally modified.

Finally, we come to the lacquer painting of a golden bird on a black ground from the Tang dynasty (no.11; Figs. 16,17,18; Mänchen-Helfen 1937a). This painting demonstrates several peculiarities in its pattern of elements and its cross section. It was not possible to explain the presence of gold in the dark layers of the black ground that appeared unpigmented under the microscope. The actual ornamental layer (Fig. 18) contains orpiment. The particles of metal visible in Figure 18 are tin, whereas silver foil could be detected in box no. 12 (Tang dynasty).

Chinese carved lacquerware from Yuan to Qing

Although taking samples from carved lacquerware is substantially more difficult, it was possible to examine a number of pieces, one from the Yuan dynasty (1279–1368; no. 14; Figs. 19–21), ten from the Ming dynasty (nos. 15–25; Figs. 22–39) and three from the Qing era (nos. 26–28; Figs. 40–42). In the case of nos. 16 and 18 sampling was restricted to the stand ring. Here, because of the ever-present possibility of overpainting, information on pigmentation could only be assessed with reservation.

While I am aware of the dangers of schematization, it appears feasible in the case of the objects examined here (mainly from the Ming dynasty), which had a number of features in common. The construction of these pieces can be described by a simple system of three functional zones: the base, a two-part ground, and an ornamental zone above that. On top of the base there is usually a thick layer of coarse, black ground. On top of that, there are three or four layers of a much harder and finer ground which is often yellow, or in any event a color different from that of the ornamental zone. Over this lie numerous lacquer bands, often with different pigmentations, which form a comparatively inelastic ornamental zone (Figs. 20,23, 26,29,31,32,34,39).

One might expect conservation and restoration problems similar to those found in panel painting and polychrome sculpture due to the conflicting properties of the mobile base and the immobile lacquer layers above. However, carved lacquerware is far less sensitive to variations in temperature and humidity than one would think (see section 5).

Figure 4. No. 1, wooden pedestal, China, end of Zhou dynasty (fifth–third century B.C.), Linden-Museum, Stuttgart (OA 20.739LA).
Figure 5. Cross section of no. 1 (Fig. 4; x160).

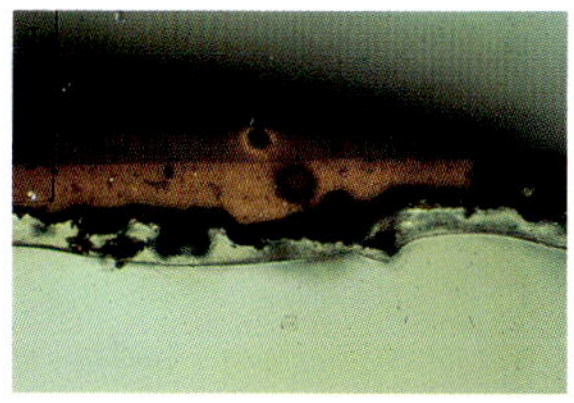

Figure 6. No. 2, two-eared cup, China, Zhan Guo era (fourth–third century B.C.; LMS OA 20.731L).
Figure 7. Cross section of no. 2 (Fig. 6; x32). The two laser craters are clearly visible here, as in the following photographs. The position of the craters permits the pattern of elements to be matched to its layer.

Figure 8. No. 4, round lidded box, China, late Zhou, beginning of Western Han dynasty, (c. second half of third century B.C.; LMS 20.720.L).
Figure 9. Cross section of No. 4 (Fig. 8; x32).

Figure 10. No. 8, two-eared cup, China, Eastern Han dynasty (100 B.C.–A.D. 100; LMS OA 20.736L).
Figure 11. Cross section of no. 8 (Fig. 10; x160).

Figure 12. Cross section of No. 9 (not illustrated), two-eared cup, China, Han dynasty (100 B.C.– A.D. 100; LMS OA 20.735L; x80).
Figure 13. As for Figure 12, but in polarized light. The sequence of layers is clearly visible (x80).

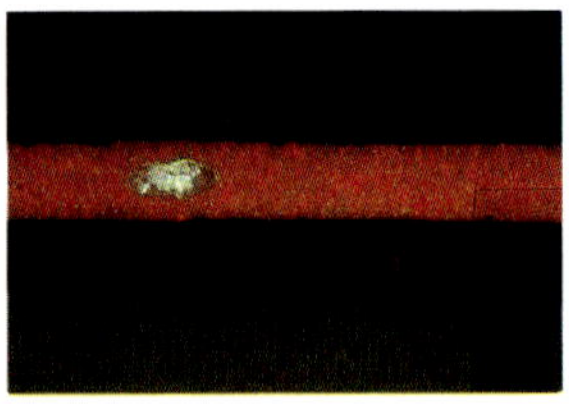

Figure 14. No. 10, bowl, China, first half of the Eastern Han dynasty (first century; LMS OA 20.730L).
Figure 15. Cross section of No. 10 (Fig. 14; x63).

Figure 16. No. 11, lacquer painting, China, Tang dynasty (possibly later; LMS OA 20.848 a–gL).
Figure 17. Cross section of No. 11 (Fig. 16) in partly polarized light (x16).
Figure 18. Part of the cross section of no. 11 (Fig. 17), showing the uppermost layers (x200).

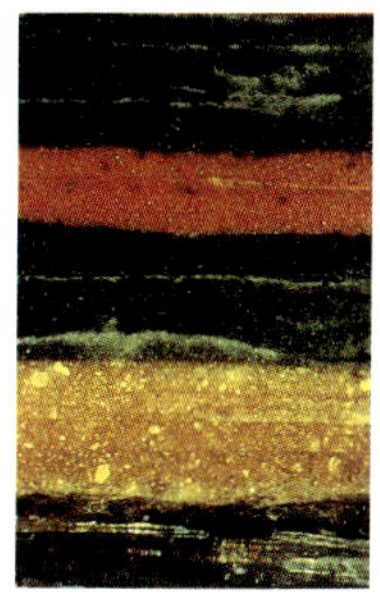

Figure 19. No. 14, bowl on stand, China, Yuan dynasty, probably fourteenth century (LMS OA 20.825a–bL).
Figure 20. Cross section of No. 14 (Fig. 19; x16).
Figure 21. Part of the cross section of no. 14 (Fig. 20) showing the lowest layers (x32).

Figure 22. No. 15, plate, China, Ming dynasty (late fourteenth–early fifteenth century; LMS OA 20.824L).
Figure 23. Cross section of No. 15 (Fig. 22; x32).
Figure 24. Part of the cross section of No. 15 (Fig. 23), showing the yellow ground layers (x63).

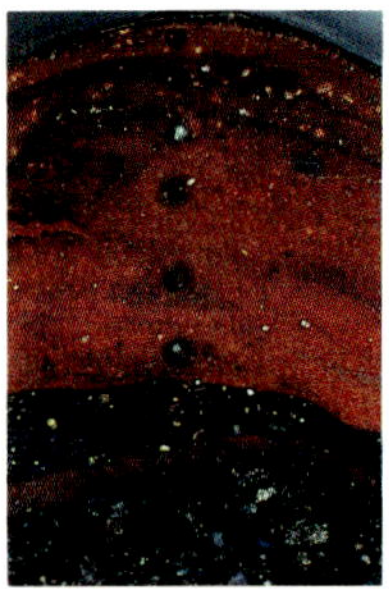

Figure 25. No. 19, large chest, China, Ming dynasty, Jiajing era (1522–1566) (LMS OA 20.761L).
Figure 26. Cross section of no. 19 (Fig. 25), upper layers (x16).
Figure 27. Cross section of no. 19 (Fig. 25), lower layers (x40).

Figure 28. No. 21, plate, China, Ming dynasty, Wanli era, dated 1592 (LMA OA 20.756L).
Figure 29. Cross section of no. 21 (Fig. 28; x32).

Figure 30. No. 22, box, China, Ming dynasty (second half of sixteenth century; LMS OA 20.763L).
Figure 31. Cross section of no. 22 (Fig. 30; x16). The position of the laser crater has been made visible with a white substance.
Figure 32. Part of the cross section of no. 22 (Fig. 31), showing the yellow ground layers and above them the black guide line (Garner 1963; x63).

Figure 33. No. 23, box, China, Ming dynasty, Wanli era (1573–1619; LMS OA 20.819L).
Figure 34. Cross section of no. 23 (Fig. 33; x32).

Figure 35. No. 24, box, China, probably seventeenth century (LMS OA 20.765L).
Figure 36. Cross section of no. 24 (Fig. 35; x16).
Figure 37. Cross section of no. 24 (Fig. 35) in polarized light. The individual layers are clearly recognizable (x16).

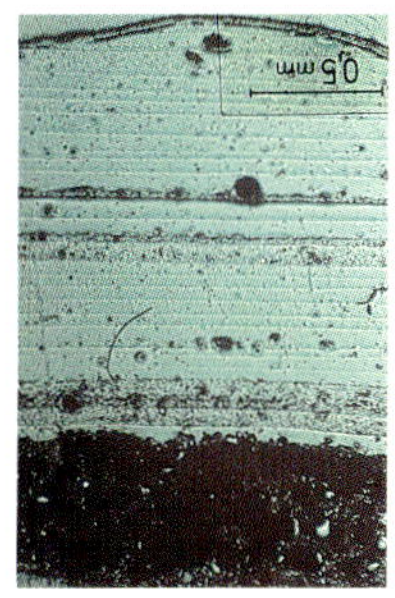

Figure 38. No. 25, box, China, probably from the end of the Ming dynasty (sixteenth–seventeenth century; LMS OA 20.817L).
Figure 39. Cross section of no. 25 (Fig. 38; x32).

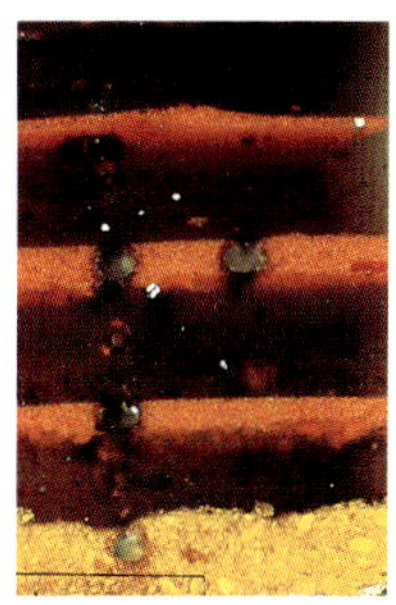

Figure 40. No. 28, box, China, Qing dynasty, dated 1775 (LMA OA 20.787L).
Figure 41. Cross section of no. 28 (Fig. 40; x16).
Figure 42. Part of cross section of no. 28 (Fig. 41); here the red layers were pigmented with vermilion (x63).

1. Bases and supports

This study can say little about the nature of the base, that is, the type of wood used, even though x-rays appear to give very interesting results (Shosoin Office 1975). This is because cinnabar (the predominant pigment in the later objects) strongly absorbs x-rays (Rees-Jones 1975) and therefore the graining of the wooden base is very difficult to identify. Nonetheless, since x-ray investigations (unlike many older methods) do not damage the objects, they should not be forgotten, and an x-ray survey of the pieces is planned. The same reservation applies to the fabric often laid on top of the base (see Fig. 21, just visible at the bottom of the picture), which on intact pieces cannot be visually examined and sampled.

2. Black ground: construction, function, and pigmentation

The lowest black ground contains mineral additives. High intensities of silicon, calcium, manganese, aluminum, titanium, iron, and magnesium indicate the use of dark earths, ashes, or clays.[32] According to experimental results, their proportion must be very high. The thickness of this layer is very striking and is often mentioned by Garner in his correspondence. This must have extended the drying time,[33] as well as prevented the formation of an urushi matrix without defects. However, it also leveled out any unevenness in the base below. Microscopic investigation does not reveal how many layers of this black ground were applied.

3. Yellow ground: construction and function

In the series of pieces investigated from the Yuan and Ming dynasties, a further zone of ground is found above the thick black one. This zone consists of several layers; it is usually yellow (nos. 14,15,21,22,23,25 and Figs. 20,21,23,24,29,32, 34,39) or occasionally another color that in any case contrasts with the ornamental bands on top (brown-black in no. 19, Figs. 26,27, and black in no. 24, Figs. 36,37).

In contrast to the zone underneath, the proportion of binding agent is much higher in comparison to that of pigment and other material discussed later. This should result in increased stability. From a visual appraisal, it seems as though the surfaces of the layers have been very carefully worked (Figs. 21,24,32). This cannot be observed to such a great extent in the black zone beneath, and anyway is difficult to see because of the greater coarseness.

The first function of the yellow ground is to provide an additional leveling out of any unevenness in the black, coarse ground. Without this zone the layers above would follow any unevenness, which would greatly disturb the aesthetic impression, especially in carved lacquerware with a different pigmentation of the ornamental zone.

Secondly, these layers absorb mechanical shearing forces between the wooden base and the ornamental zone. (The black zone below actually fulfils this function to an even greater extent.) The fact that these shearing forces often occur is mentioned in manuscripts from the fourteenth and fifteenth centuries which tell of the damage typical of carved lacquerware.[34]

The third function lies in conveying an optical impression of depth and at the same time in emphasizing the contours of the ornamentation because, in general, the yellow ground is exposed during carving in works of the Ming era, giving the impression that the ornamentation is painted on. The absence of this effect of depth on Qing dynasty piece no. 28 (Fig. 40) may account for its flatness and lifelessness, qualities also observed in pieces from the Ming era which are in need of cleaning.

A fourth and final function of the contrasting coloration of this zone was to aid the craftsman by giving him an indication of the maximum permitted cutting depth. After application and drying of the various layers, carving had to be done with extraordinary precision, as corrections were hardly possible. Therefore, it was of great importance that the depth of carving was uniform; the increased number of yellow layers permitted a certain, if limited, tolerance. The rubbing down of the surface of the black and yellow layers, already referred to, helped to maintain a uniform cutting depth.

4. Yellow ground: pigmentation

There is evidence of arsenic[35] as a color-producing pigment in the yellow layers: diffractometric results on nos. 14,15,18, and 21 all clearly prove the use of orpiment (As_2S_3). The presence of calcium, magnesium, silicon, iron, mercury, aluminum, and manganese (listed here in order of their frequency, and most conspicuous in no. 21) point to a natural origin (Wallert 1984; Schafer 1955), whereby all these elements can be correlated to associated minerals (Schröcke 1981; Doelter 1926). Therefore it seems unlikely that synthetic orpiment (Wallert 1984; Schafer 1955) was used on the objects cited above. The presence of mercury can be explained by the natural admixture of cinnabar (Wallert 1984).[36] The earliest evidence of orpiment, within the limits of this investigation, can be traced to piece no.11 from the Tang era. In the case of carved lacquerware the evidence starts with no. 14 from the Yuan era and continues through to the Qing dynasty.

The literature shows that orpiment was used in Chinese painting from the ninth–tenth centuries A.D. (Winter 1984). Before that, yellow ocher and natural organic yellow pigments in China (Winter 1984; Gettens 1938–39) and also in Japan (Goffer 1980:177-179; Yamasaki 1979; FitzHugh 1979; Gettens 1976:241–252) were dominant from the beginning of the first century A.D. In Japan, orpiment was used frequently after the middle of the seventeenth century. Yonezawa (1956) gives some references to historical sources from the Six Dynasties onwards and refers to inland deposits in Southern China;[37] in fact the same regions in which there are large urushi plantations today. These key dates may, in association with the results of the present investigations, help to date and perhaps even to identify imitations.

5. The ornamental zone: construction and function

The ornamental zone shows a huge range of styles as can be seen in Figures 20,23,26,31,36,39,41. Because of the slightly inhomogeneous distribution of pigments in the binding agent, the microscope reveals the differentiation between individual layers of lacquer, also clearly visible in the illustrations.

The condition of the surface of the layers in the cross sections sometimes suggests that very thorough treatment took place after the drying process. Exceptions are corners, edges, and heavily rounded sections, where working would be naturally more difficult and less regular. It is rare to encounter remains of abrasive materials such as in no. 14 (Fig. 21) which appear to have become embedded in layers which were not quite dry.

There is a great diversity in the literature on Chinese carved lacquerware about the number of layers involved. Estimates—and this is what is being discussed here—vary between a few and over 300 (Burmester and Brandt 1982). Garner attempts greater precision when he writes "the thickness of individual layers is in the order of 0.3 mm or less" and "between one and two hundred layers" (Garner 1979). Microscopic investigation of the pieces examined here permits a much more exact idea. If all the lacquer bands are included in the calculation, this gives an average thickness per layer of approximately 4µm.

Because of the many and various techniques which are dealt with in standard works on the subject, the question of further functions of the ornamental zone is not discussed here; it is obvious that this zone is open to visual examination (Garner 1979; Lee 1972; Kuwayama 1982).

The well thought out construction described above prevents variations in the moisture content of the base. Gain or loss of moisture is made extremely difficult due to the homogeneous, polymeric, and hydrophobic character of the lacquer layers. This explains the surprisingly good condition of some Chinese carved lacquerware. If a piece sustains mechanical damage, however, moisture can enter or leave, rapidly causing extensive damage because of the movement[38] of the base against the immobil ornamental zone.

6. The ornamental zone: pigmentation of the red layers

In the pieces examined—the early lacquerware[39] and the carved lacquerware[40]—mercury is found in the red pigmented layer almost without exception. This, together with the results of x-ray diffraction of samples from objects nos. 2,4–6,8, 9,12,13,15,16,18–26, and 28, proves that cinnabar (cinnabarite, beta-HgS) was used as a pigment.

Apart from mercury, other elements present (in order of frequency of occurrence) are calcium, magnesium, iron, silicon, manganese, arsenic, aluminum, barium, titanium, copper, chromium, and lead. These can be correlated to minerals associated with cinnabar (Schröcke 1981; Doelter 1926). The detection of this type of natural admixture using diffractometry proved to be possible on a few objects. However, an unambiguous matching was not achieved because it was not possible—for experimental reasons—to isolate only the red layers, that is, those which had been pigmented with cinnabar.

While it is possible to assume, from the impurities detected, that natural cinnabar was used in the majority of the pieces (nos. 15–23), this is not valid for the late Ming and Qing pieces (nos. 24–28). Apart from mercury, only calcium and magnesium could be detected, and these can be attributed to the binding agent itself. The absence of accompanying elements proves the application of vermilion, a synthetic cinnabar produced by repeated sublimation.[41] In relation to the dating of the pieces examined, this would indicate a change in the use of this important pigment from the seventeenth century onwards, and may be significant in providing further help in dating lacquerware. In pieces nos. 26 and 27 arsenic is also found, here certainly as orpiment. It is conceivable that in the purification of cinnabar by repeated sublimation, orpiment was carried through as an impurity. This is only possible for orpiment because it sublimes easily; it does not apply to other naturally associated minerals.

Data from the literature are very variable concerning the vermilion made from mercury and sulfur in the so-called dry process. Gettens (1972) says that the Chinese may have invented this process and that the earliest evidence for it is in the seventh century A.D. According to Needham (1976), the first written reference can be found as early as the fifth century B.C. From his comments, based on numerous sources, it is evident that cinnabar and mercury (often obtained from cinnabar) were the mainstays of Chinese alchemy and the related world of ideas. Therefore, it seems plausible that even before the seventh century, alchemists had made the observation that in the cooler areas of the experimental apparatus (ovens, furnaces, etc.) there were traces of a red substance—vermilion, resulting from sublimation. When this developed into a systematic process remains an open question.

Yamasaki (1979) reports vermilion being produced in Japan in 1609. According to Oguchi (1969), the synthetic pigment was widespread during the Edo period. These dates are compatible with the findings of the present investigation. Clear evidence of synthetic cinnabar in no. 17 supports the doubts expressed by art historians over dating the piece to the fifteenth century and identifying it as Chinese. It is more likely to be a later Japanese imitation of Chinese carved lacquerware.

In none of the cross sections was there a blackening of the red cinnabar when transformed into metacinnabar as discussed later on (Gettens 1972). Urushi appears to stabilize the pigment as well as an oily binding agent would do. The fact that this transformation can be observed is shown by the black coronae around the laser craters in Figures 23, 36, and 41. Black metacinnabar has formed here.

Compared with cinnabar, the other red pigments play a wholly subordinate role. Red iron oxide (haematite) was used as pigmentation in no. 14, in no. 19 in both upper layers, and in no. 26.[42] In nos. 14 and 26, the evidence could be secured by diffractometry. The choice of iron oxide was certainly due to its dull red tone, which differentiated it from the red of the surrounding layers derived from the presence of cinnabar. The pattern of elements in these dull red layers is especially pure. In particular, the pattern of no. 14 indicates very pure iron oxide, which was applied mixed with cinnabar in the other two cases.

The evidence of chromium in nos. 22 and 24 (Figs. 31,36, uppermost dark red layers) is puzzling; chromium pigments have only been known since the nineteenth century, so maybe this can be explained by restoration work.

7. The ornamental zone: pigmentation of the black layers

The pigmentation of the black layers of the lacquerware examined here requires more detailed consideration.

Once again, laser microanalysis offers clues as to the type of black pigments used. In the black layers there are remarkably frequent instances of iron,[43] mercury,[44] or spectra that are very low in elements and intensity.[45]

In the case of opaque pigmentation of the black layers (nos. 22 and 23), the evidence indicates that iron must be present as black iron pigments (as oxides or hydroxides), although it was not possible to support this by diffractometry.

For nos. 1, 8, 12, and 19 black metacinnabar could have been used in the opaque black layers. This is indicated by the presence of mercury, found by laser microanalysis, and proved additionally by x-ray diffractometry of nos. 1 and 19.[46] This result is very interesting because, as far as I know, metacinnabar has only been found once and confirmed by diffractometric analysis in a grave from the Han dynasty (Winter 1984).

It is obvious from the genesis of cinnabar and metacinnabar, as well as from the description of their deposits, that they are found associated as natural minerals. Furthermore, to the naked eye they are almost impossible to tell apart. If cinnabar is present as large crystals or crystal aggregates, then it is dark red, almost black. Only after grinding will it become "cinnabar red." However, this means that a proportion of the minerals found in the natural deposits, that is, metacinnabar, remains black on grinding. Metacinnabar becomes red on heating, and mercury can also be obtained this way, so metacinnabar certainly had the same importance in alchemy as cinnabar or mercury. For these reasons, metacinnabar would have been an obvious choice as a black pigment; this is confirmed by the present findings.

However, in the remaining three pieces, nos. 14, 24, and 25, the black layers are transparent and, in addition to a similar brown-black coloration, only rarely contain small aggregates of black pigment (for example, no. 14, Figs. 20,21). The piece pictured in Figure 2 from Garner's estate shows a layer of similar appearance which, from its function, is referred to as a black guideline (Garner 1963, 1979; Burmester and Brandt 1982; see also Fig. 32). Garner's cross section in transmitted light shows the transparent character particularly clearly. In all these cases, the addition of carbon black can be assumed; this would explain the weak concentration of elements detected by laser microanalysis. Garner had already guessed as much and it can also be seen in many other objects of Far Eastern art (Winter 1984). Along with the use of soot or lamp black (Gettens 1938–39), the admixture of Chinese ink should also be considered. Thus Chiang-Yen (Yonezawa 1956) refers to Chinese ink as a dye as early as the fifth century, and Winter (1974) introduces other sources. Gettens (1976) and Yamasaki (1979) both name Chinese ink as the only black pigment in Japanese paintings. It is very interesting that Chinese ink (or more appropriately the glue used in its preparation) and urushi have a very similar pH (4–5; Reinhammar 1970). This has a positive effect on the compatability of the ink with urushi. Because the hydrophobic character of soot would make it unsuitable as the only additive, it is better to mix it into hydrophilic glue which then stabilizes the colloidal system (Winter 1974). In connection with this there is an interesting reference in a source manuscript of Shen Chi-sun who reports the production of Chinese ink by combustion of urushi lacquer (Vaultier 1957).

Pigments used in Chinese lacquerware and their relationship to Chinese philosophy

A summary of all the pigments used in Chinese lacquerware investigated here gives for yellow, orpiment (As_2S_3) and for red tones, cinnabar (beta-HgS) and occasionally haematite (Fe_2O_3). For black hues there are metacinnabar (alpha-HgS), black iron pigments and carbon black as Chinese ink. Silver and tin have been detected as metal inlays. It is remarkable that none of the known blue and green pigments has been used.

In Winter's summary (1984), reference is made to the frequent application of other pigments in Chinese objects with other binding agent systems: for white, chalk ($CaCO_3$) and "lead white" ($2PbCO_3.Pb(OH)_2$ and $PbSO_4$), for blue, azurite ($2CuCO_3.Cu(OH)_2$) and for green, malachite ($CuCO_3.Cu(OH)_2$) and copper chloride ($Cu_2(OH)_3Cl$ paratacamite and atacamite), though the presence of $PbSO_4$ and $Cu_2(OH)_3Cl$ was exceptional.

Why are all these very common pigments not found in Chinese lacquerware? A few experiments, as well as reports from craftsmen, have demonstrated that chalk, lead white, azurite and malachite decompose in the binding agent urushi, that is they are unstable. The reason lies in the more acidic pH value of urushi: approximately 4.6 (Kenjo 1976) and 4.2 to 5.3 (Oshima 1985). This more acidic medium causes a chemical reaction which decomposes the pigment and gives a more basic pH value. This in turn deactivates the enzyme laccase (Reinhammar 1970), disturbs the equibibrium in the phenol-water emulsion and therefore hinders the drying of the urushi film (Kenjo 1976). The incompatability of urushi with a large range of common pigments results in an inevitable limitation of colors used in Chinese lacquerwork.

I have already mentioned that cinnabar has a function as a reagent (Needham 1976) as well as a provider of color. It occupied a central role in the alchemy of ancient China, in daily life as well as in the Chinese spiritual world. Schafer (1955) assigns a similar role to orpiment. Because both cinnabar and orpiment can be repeatedly sublimed, they are bound closely to the concept of material immortality (Needham 1976). Objects made with urushi were regarded in the same way, as shown both by the great value placed upon them by the highest levels of society—including the imperial court—and by the fact that their production was, for a time, under state control (Garner 1979). Furthermore, lacquerware objects found in graves from the Han dynasty lasted until the Ming era and longer. This must have invested urushi with the mystical qualities of material immortality. The "mystery" surrounding urushi still disrupts the debate between east and west in matters of conservation and restoration. This combination of circumstances certainly did not go unnoticed and accounts for the unchanging high regard for urushi, cinnabar, and orpiment over a very long period.

Acknowledgments

For their cooperation I wish to thank S. Baier, W.T. Chase, Dr. M. Hansen, Dr. G. Höhne, S. Holzer, K. Junghans, Dr. T. Kenjo, K. Kitamura, Prof. Dr. J. Kumanotani, Dr. C. Müller, Prof. Dr. J. Müller, Prof. Dr. H. Murawski, Dr. H. von Sonnenburg, and N. Umney. The experiments were financed by the Volkswagen Foundation in the years 1979-1983. High resolution mass spectrometry measurements were commissioned by the Getty Conservation Institute. My thanks to the Victoria and Albert Museum (J. Earle, C. Clunas) for their kindness during my research into Sir Harry Garner's papers. Also to the Linden-Museum, Stuttgart for releasing the color photographs of the objects discussed here. These photos were all taken by Frau U. Didoni. Dr K.J. Brandt (Linden-Museum, Stuttgart) granted permission to examine the museum's collection of Chinese lacquerware, and Prof. Dr E.-L. Richter conducted the laser microanalyses. Finally, thanks to my wife for her everlasting patience and support.

Appendix

In the following section the results of microscopy and laser microanalysis are combined. For details of the methods of sampling see an earlier publication (Burmester and Brandt 1982). The cross sections, most of which are shown in Figures 4–42, were systematized in the following way. Each cross section contains lacquer bands (see note 29) A,B,C ... (first column), which are on either a wooden or a fabric base (T). Each band can incorporate n lacquer layers (see note 29) and is then referred to as nx (for example, 3D in the event that band D has three layers). In the second column a crude color is mentioned, in the third the function of the band and in the fourth its approximate thickness in µm (if not measurable, marked as *). The last column gives the pattern of elements provided by laser microanalysis; the subscript numbers indicate the estimated intensity of the spectral lines from "only recognizable with difficulty" to "stronger than normal" and "auto reversal." In cases where the laser beam did not penetrate exactly into a band, comments as to the positioning are given (for example, 80%A, 20%B, if 80% of the laser crater lies in band A). For technical reasons, memory effects could not be excluded in all cases.

Chinese lacquerware

1.	Wooden pedestal, China, end of Zhou (fifth-third century B.C.; Linden-Museum, Stuttgart, West Germany, OA 20.739A)			
A	black	ground	*	$Si_5, Fe_5, Mg_4, Al_4, Ca_4, Ti_4, Hg_4, Mn_2$
B	sand color			$Ca_3, Fe_2, Cu_2, Hg_1, Mg_1, Si_1$
C	black		<12	
2.	Two-eared cup, China, Zhan Guo era (fourth-third century B.C.; Linden-Museum, Stuttgart OA 20.731L; Brandt 1982, 1986).			
A	black	ground	*	$Ca_3, Si_2, Ti_2, Mg_1, Fe_1$
B	sand	ground?	*	Ca_3, Mg_1, Si_1, Fe_1
C	brown		60	Ca_3, Mg_1
D	brown		<10	
3.	Lid of a vessel, China, end of Zhou (fourth-third century B.C.; Linden-Museum, Stuttgart OA 20.738L; Brandt 1986)			
A	black	ground	*	Ca_2, Fe_2, Cu_2, Mg_1
B	brown			$Ca_3, Fe_3, Mg_2, Si_2, Cu_2, Hg_1$
C	sand			$Ca_3, Mg_2, Fe_2, Si_1, Cu_1$
4.	Round lidded box, China, late Zhou/beginning of Western Han (*c.* second half of third century B.C.; Linden-Museum, Stuttgart OA 20.720L; Brandt 1982, 1986)			
T				
A	brown-black	ground	*	$Ca_3, Al_2, Mg_1, Fe_1, Hg_1$
B	brown		>50	90%B, 10%C: Ca_3, Hg_3, Mg_1
C	red	ornament	15	Hg_1
5.	Lid of a wine beaker of the "lian" type, China, probably Qin (227-206 B.C.; Linden-Museum, Stuttgart OA 20.719L)			
T				
A	sand	ground	75	10%A, 90%B: $Ca_4, Mg_2, Si_2, Fe_2, Al_1$
B	brown		17	$Ca_4, Mg_2, Si_2, Fe_2, Al_1$
6.	Internal construction of a box, China, Han dynasty, probably second half of second century B.C. (Linden-Museum, Stuttgart OA 20.721C)			
C	black	ornament	17	20%C, 70%B, 10%A: Ca_3, Hg_3, Si_1
B	red		25	
A	brown-black	ground	*	
T		wooden core		
A′	black	ground	*	Ca_5
B′	red	ornament	*	Ba_4, Ca_3, Cr_3, Hg_3
7.	Cylinder-shaped container, China, probably Western Han (200-100 B.C.; Linden-Museum, Stuttgart OA 20.722L)			
T				
A	brown	ground	*	
2B	brown		36	Ca_3, Fe_3, Mg_1, Hg_1
C	brown		30	
D	red		16	
8.	Two-eared cup, China, Eastern Han (100 B.C.–A.D. 100; Linden-Museum, Stuttgart OA 20.736L)			
A	black	ground	*	80%A, 20%B: $Hg_4, Si_3, Ca_3, Mn_3, Fe_3, Ba_3, Mg_2, Al_1, Ti_1, Cu_1$
B	brown		30	
9.	Two-eared cup, China, Han dynasty (100 B.C.–A.D. 100; Linden-Museum, Stuttgart OA 20.735L)			
A	black	ground	*	
B	black		>20	60%B, 40%C: $Ca_4, Hg_4, Fe_3, Mg_2, Si_1, Mn_1$
C	red		46	$Ca_4, Hg_4, Fe_3, Mg_2, Mn_2$
D	black		6	
10.	Bowl, China, first half Eastern Han era (first century; Linden-Museum, Stuttgart OA 20.730L; Brandt 1986)			
T		wooden core		
A	black	ground	*	
B	brown		50	20%C, 80%B: $Ca_4, Mg_3, Mn_3, Fe_3, Ba_3, Hg_3, Si_1$
C	red	ornament	55	$Ca_4, Ba_3, Hg_3, Fe_2, Mg_1, Si_1, Mn_1$

11.	Seven lacquer paintings with gold-colored bird on a dark background, China, Tang dynasty, possibly later (Linden-Museum, Stuttgart OA 20.848a-gL; Brandt 1986)			
	from the surface of the painting:			
A	black	ground	*	Si_4, Ca_4, Fe_4, Mg_3, Ti_3, Mn_3, Al_2
B	clear lacquer		70	Ca_4, Sn_3, Mg_1
C	clear lacquer		50	
from the painting:				
A	brown-black	ground	*	Si_4, Ca_4, Fe_4, Mg_3, Ti_3, Al_2, Mn_2
4B	brown		170	Au_2, Ca_1
C	sand	ornament	90	Sn_5, Ca_4, As_3, Mg_1, Si_1, Fe_1
		metal part		Sn_7, Ca_3, As_3, Mg_1, Al_1, Si_1, Fe_1
D	clear lacquer		25	
E	clear lacquer		15	
12.	Heavily damaged lid or lower part of a box, China, Tang, *c.* eighth century (Linden-Museum, Stuttgart OA 20.744L; Brandt 1986)			
T		wooden core		
(2?)A	black	ground	440	Ca_5, Si_4, Fe_4, Mg_2, Al_2, Ag_2, Hg_1
B	brown	base	60	Ca_4, Ag_2, Hg_2
C	red	ornament	50	Ca_4, Ag_4, Hg_2
13.	Small petal-edged bowl with flat bottom, China, Song dynasty, *c.* twelfth-thirteenth century (Linden-Museum, Stuttgart OA 20.746L)			
T				
A	black	ground	*	
B	clear lacquer		25	Ca_8, Mg_4, P_4, Mn_3, Fe_3, Al_1, Si_1
C	red	ornament	30	

Chinese carved lacquerware

14.	Bowl on a stand, China, Yuan dynasty, probably fourteenth century (Linden-Museum, Stuttgart OA 20.825a+bL; Brandt 1982, 1986)			
T				
A	black	ground	*	Ca_4
4B	yellow	yellow ground	200	Ca_4, As_3, Mg_2
7C	black		230	Ca_4, Mg_2
3D	red		150	Ca_4, Fe_4, Mg_2
4E	black		260	Ca_4, Mg_2
2F	red		200	Ca_4, Fe_4, Mg_2
6G	black		250	Ca_1
15.	Plate, China, Ming dynasty, late fourteenth-early fifteenth century (Linden-Museum, Stuttgart OA 20.824L)			
T				
A	black	ground	*	Si_4, Ca_4, Mn_4, Al_3, Ti_3, Fe_3, Mg_1
B	yellow	yellow ground	*	
4C	yellow	yellow ground	360	As_3, Hg_3, Ca_2
7D E	red black		390	90%D, 10%E: Hg_5, Fe_3, Ca_3, Mg_1, Al_1, Si_1, As_1
7F G	red black		290	
7H I	red black		170	Hg_5, Ca_3, Fe_2, Mg_1, Al_1, Si_1
7J K	red black		230	
7L M	red black		210	
10N	red		380	
16.	Box, China, Ming dynasty, Yongle era (1403–1424; Linden-Museum, Stuttgart OA 20.753L; Brandt 1982, 1986)			
	red layer			Hg_7, Ca_5, Fe_2, Mg_1, Si_1

17.	Large plate, China, Ming dynasty, second half of fifteenth century (Linden-Museum, Stuttgart OA 20.754L); this could also be considerably later and of Japanese origin			
	black layer			Hg_5, Ca_3, Mg_1 (contaminated with red layer by 30%)
	numerous red layers			Hg_6, Ca_3, Mg_1
	aggregate of pigment in red layer			Hg_6, Ca_3, Mg_1
18.	Bowl, China, Ming dynasty, Jiajing era (1522–1566; Linden-Museum, Stuttgart OA 20.796L)			
A	black	ground	*	
B	yellow-brown		<160	
C	black		650	Hg_5, Ca_4, Mg_3, Mn_3, Si_1, Fe_1
D	red		110	Hg_5, Ca_4, Mn_4, Mg_3, Cu_2, Si_1, Fe_1
19.	Large chest, China, Ming dynasty, Jiajing era (1522–1566; Linden-Museum, Stuttgart OA 20.761L; Brandt 1986)			
A	black	ground	*	Ca_4, Hg_3
B	black	ground	*	
C	brown		80	Ca_4, Fe_4, Hg_4, Mn_3, Mg_2, Si_1, As_1
D	black		240	Ca_4, Hg_4, As_3, Mn_3, Mg_1, Si_1, Fe_1
13E	red		1040	Hg_6, Ca_4, Mg_2, Mn_2, Fe_1, As_1 Hg_6, Ca_3, Mg_2, Mn_2, Si_1, Fe_1, As_1 Hg_5, Ba_4, Ca_3, Mg_1, Mn_1, Fe_1
5F	brown		320	Fe_6, Hg_3, Al_3, Si_3, Ca_3, Mn_3, Ti_2, Mg_1
2G	dark red		60	Fe_6, Hg_4, Si_4, Ca_4, Al_3, Mn_3, Ti_2, Mg_1
20.	Large round box with multicolored carved lacquer, China, Ming dynasty, Wanli era, dated 1586 (Linden-Museum, Stuttgart OA 20.807L; Brandt, 1982, 1986)			
T				
A	black	ground	*	Si_4, Ca_4, Fe_4, Mg_3, Ti_3, Mn_3, Al_2, Hg_2
B	red		*	10%A, 90%B: Ca_4, Hg_4, Mg_3, Si_3, Mn_3, Fe_3
2C	red		120	Ca_3, Hg_3, Mg_1, Si_1, Fe_1
D			10	
21.	Large plate with seven dragon medallions, China, Ming dynasty, Wanli era, dated 1592 (Linden-Museum, Stuttgart OA 20.756L; Brandt, 1982, 1986)			
A	black	ground	*	Ca_5, Mg_4, Si_4, Fe_4, Ti_3, Mn_3, Hg_3, Al_2, As_1
4B	yellow	yellow ground	140	Ca_5, Mg_4, Fe_4, As_4, Si_3, Mn_3, Hg_3, Al_1
7C	red		>430	Hg_6, Ca_5, Mg_4, Si_3, Fe_3, Mn_2, Al_1, As_1 Hg_6, Ca_5, Mg_4, Si_4, As_4, Fe_3, Ti_2, Mn_2, Al_1 Ca_5, Hg_4, Mg_2, Si_1, Fe_1, As_1 Ca_5, Hg_4, Mg_2, Si_1, Mn_1, Fe_1, As_1
22.	Octagonal box, China, Ming dynasty, second half of sixteenth century; Linden-Museum, Stuttgart OA 20.763L; Brandt 1986)			
A	black	ground	*	Si_5, Ca_4, Fe_4, Mg_3, Al_3, Ti_3, Mn_2, Hg_1
B	yellow	yellow ground	*	Ca_5, As_4, Mg_2, Si_1, Fe_1, Hg_1
2C	yellow	yellow ground	80	
D	black guide line	black	80	80%D, 20%E: Ca_5, Fe_5, Mg_2, As_2, Si_1, Mn_1
E	dark red		90	Ca_5, Fe_5, Hg_3, Mg_2, Mn_2, Si_1, As_1
39F	red		1550	Hg_5, Ca_3, Ba_3, Fe_2, Mg_1, As_1 Hg_4, Ca_3, Fe_3, Ba_3, Mg_1, Cr_1, Pb_1 Ba_3, Ca_2, Fe_2, Hg_2
23.	Rectangular lidded box, China, Ming dynasty, Wanli era (1573–1619; Linden-Museum, Stuttgart OA 20.819L; Brandt 1986)			
A	black	ground	*	
B	yellow	yellow ground	*	
3C	yellow	yellow ground	190	As_4, Ca_2, Mg_1
5D	black		250	Fe_3, As_2, Ca_2, Hg_2, Mg_1
E	red		100	Hg_5, Fe_4, Ca_3, As_2, Mg_1, Si_1
6F	black		340	
3G	red		65	
8H	black		250	
2I	red		85	
4J	dark red		110	

24.	Rectangular box, China, probably seventeenth century (Linden-Museum, Stuttgart OA 20.765L)			
2B′	black	interior	60	Ca_3, Si_1
A′	black	ground	>850	Ca_8, Ti_5, Mg_4, Mn_4, Fe_4, Al_3, Si_3
T		wooden core		
A	black	ground	*	Si_4, Ca_4, Ti_4, Fe_4, Mg_3, Al_3, Mn_3
B 2C	black black		50 180	50%B, 50%C: Si_4, Ca_4, Fe_4, Mg_3, Al_3, Ti_3, Mn_2
9D	red-brown		425	Ca_3, Hg_3
2E	light red		85	Hg_4, Ca_3, Mg_1
F	black		65	30%E, 70%F: Hg_5, Ca_3, Mg_2, Si_2, Fe_2 20%E, 80%F: Hg_5, Ca_2, Si_1, Fe_1
G H	light brown brown		65 20	70%G, 30%H: Ca_2, Fe_2, Hg_2, Si_1
11I	brown-red		620	Ca_3, Hg_3, Mg_1
J	clear lacquer	exterior	30	30%I, 70%J: Fe_5, Cr_4, Hg_4, Ca_3, Ba_3, Pb_3, Mn_2
25.	Rectangular box, probably China, probably end of Ming dynasty, sixteenth-seventeenth century (Linden-Museum, Stuttgart OA 20.817L)			
A	black	ground	*	
B	yellow	yellow ground	*	
3C	yellow	yellow ground	150	As_4, Ca_2, Hg_2, Mg_1, Si_1, Fe_1
5D E	black red		} 210	Si_2, Ca_2, Hg_2, Mg_1, Fe_1, As_1 Hg_4, Ca_2
5F G	black red		} 290	Ca_1, Hg_1 Hg_4, Ca_2, Mg_1, As_1
5H I	black red		} 250	Hg_3, Ca_1, As_1
5J K	black red		} 235	Ca_4, Hg_3
5L M	black red		} 290	
7N	black		290	
26.	Large table with carved lacquerwork, late seventeenth-early eighteenth century (Linden-Museum, Stuttgart OA 20.748L); samples taken from the upper surface of the table, small particles without ground.			
	numerous red layers alternating			red: Fe_4, As_3, Hg_3, Ca_2, Mg_1
	with red brown layers			red-brown: Fe_5, Ca_2, Hg_2, Mg_1, Mn_1
27.	Square vase in the form of a "ku," China, Qing dynasty, bearing the mark of the Quianlong era (1736–1795) (Linden-Museum, Stuttgart OA 20.055); sample taken from the ornament in the foot.			
	numerous ground layers			Ca_5, Mg_3, As_3, Si_2, Mn_1, Al_1, Fe_1
	numerous red layers			Hg_6, As_3, Ca_2, Mg_1
28.	Twelve-cornered box, China, Qing dynasty, dated 1775 (Linden-Museum, Stuttgart OA 20.787L)			
A	black	ground	*	Ca_4
>20B	black	yellow ground	>1150	Ca_4
8C	red		495	Ca_3, Hg_3

Notes

1. In Garner's estate there is a series of handwritten technical notes dated 6.7.62, 26.7.62, 28.7.62, 1.8.62, 11.2.63, and 25.11.63.

2. The results were communicated to Garner by E.W.S. Press, Director of Chemical Inspection, War Office, London, dated 4.12.61 and 30.10.62.

3. Experiments by A.E. Werner, Research Laboratory, British Museum, London, dated 27.11.62.

4. For example, in a letter dated 27.11.63, the Swedish King Gustav Adolf VI suggested a further examination of a piece from his collection.

5. R.J. Gettens, Freer Gallery of Art, Washington, 7.5.63. A schematic drawing of the construction by E.W. FitzHugh was included.

6. A.R. Sollars, Department of Materials, College of Aeronautics, Cranfield, Bletchley, Buckinghamshire, 17.6.63. It is clear from the letter that Sollars must have examined several samples for Garner. It is possible that the discussion was about one of Garner's cross sections (1963) without it being referred to specifically. There is a photograph in that article (Fig. 4, "magnification about x8") which bears a reference to A.R. Sollars of Cranfield ("magnification x10").

7. For example, H. Trubner of the Royal Ontario Museum, University of Toronto, sent him a sample of a pre-Han or Loyang piece.

8. P.H. Plesch, University of Keele, 29.9.65.

9. On 2.3.64 from Inv. 53.8 of the Freer Gallery of Art, Washington, D.C.

10. On 13.1.66 by A.E. Werner, British Museum, London.

11. V.A. Moss requests an answer to this in his letter of 18.4.67.

12. In a letter to J. Needham on 28.12.66 and in correspondence with his nephew V.A. Moss on 15.10.69 and 20.10.72.

13. In letters to J.S. Mills, National Gallery, London (answered on 11.7.70) and V.A. Moss (answered on 6.5.71) and in debate with R.C.A. Rottländer of the Labor fur Bernsteinforschung, Bonn, on suggested procedures using infrared spectroscopy and field ionization mass spectroscopy.

14. In response to a suggestion from his nephew V.A. Moss, Garner writes on 20.10.72: "I will alter 'urushic acid' to 'urushiol.' Should I not also alter 'acids'. . . . . to 'components'?"

15. Thus on 11.12.69 Garner receives from Sir G. Taylor, Director of the Royal Botanical Gardens, Kew, Richmond, Surrey, the information that "*R(hus) vernicifera* DC is a synonym of *R(hus) verniciflua* Stokes; the latter is the correct name for this species."

16. In many letters and in an updated outline of the second chapter of his book (Garner 1979).

17. In a letter to R.J. Gettens of 8.5.68, Garner writes: "Among the known (Han) lacquer pieces, I would say that less than 5% have fabric." However, Gettens had informed him long before that (2.3.64) that he had found no fabric supports in the pieces studied from the Freer Gallery.

18. The cross section bears the note "15th C. strong black layer," which may refer to the heavily applied "black guide line" (Garner 1963). Unfortunately, the origin of the sample and the name of the preparer are unknown. An additional cross section of a "16th C." piece should also be mentioned here.

19. On 24.4.68 R.J. Gettens sends a "Summary of an examination made of Han lacquer fragments supplied by Sir Harry Garner . . ." written by W.T. Chase.

20. One reason for this may be that urushi is not used as a binding agent in the West, yet the analysis of binding agents has been developed largely in the West.

21. J.S. Mills to Garner, 11.7.70.

22. These results may rest on the observation that earlier lacquers differ radically from later ones, an observation that has never been satisfactorily explained (Burmester 1983b).

23. High-resolution mass spectrometry was carried out on the samples discussed here, nos. 1, 2, 4, 5, 7–9, 11, 12, 14–16, 18–24, 28 on two pieces of Chinese carved lacquerware dated 1586 (OA 20.813L) and second half, eighteenth century (OA 20.804bL), on a Japanese *sake* bowl of the second half of the nineteenth century (9.645B) and on three recent examples from Japan and Thailand. The three additional pieces were also from the collection of the Linden-Museum in Stuttgart. The high-resolution mass spectra of these samples usually showed high intensities for the following masses (in order of their calculated $C_xH_yO_z$ compositions): $C_5H_9O_2$, C_8H_9, C_7H_5O, $C_4H_9O_3$, C_8H_{11}, C_7H_7O, C_7H_8O, C_8H_{13}, C_7H_9O, $C_7H_{11}O$, C_9H_7, C_9H_9, C_9H_{11}, C_8H_9O, $C_7H_7O_2$, $C_7H_8O_2$, $C_6H_6O_3$, $C_{10}H_8$, $C_{10}H_9$, C_9H_7O, $C_9H_{11}O$, C_8,H_8O_2, $C_8H_9O_2$, $C_{10}H_9O$, $C_9H_9O_2$, $C_9H_{15}O_3$.

24. Depending on their date, the pieces examined[23] demonstrated the following high-intensity peaks: $C_6H_5O_2$ for Yuan (Y) and Ming (M); C_8H_{14} for Zhou (Z) and Qing (Q); $C_7H_{10}O$ for M and Q; C_8H_{15} for Z, Q and modern lacquers (mo); $C_7H_{12}O$ M, Q; $C_6H_8O_2$ M, Q; $C_5H_6O_3$ Y, M, Q; $C_7H_9O_2$ late M, Q and mo; $C_8H_{13}O$ Han (H) to mo; $C_6H_8O_3$ Y, M, and early Q; $C_6H_{10}O_3$ Q, mo; C_9H_9O Q, mo; $C_8H_{10}O_2$ late Y to mo; $C_{10}H_{18}$ Z, H; $C_9H_{12}O_2$ Y to Q; $C_9H_{15}O_2$ Z to Q; $C_{11}H_{15}O$ late M and Q; $C_{12}H_{19}$ Z; $C_{13}H_9$ Q and mo; $C_{12}H_{21}$ Z; $C_{12}H_{13}O$ Q and mo; $C_{11}H_{10}O_2$ M to mo; $C_{10}H_9O_3$ Z and mo; $C_{12}H_{11}O_2$ mo; $C_{12}H_{12}O_2$ mo; $C_{15}H_{28}$ Z and $C_{14}H_{28}O_3$ mo. It should be made clear that this note and the previous one[23] are intended merely to convey experimental results, not to deliver broad conclusions.

25. The equipment and parameters used in the experiments conducted here were: Laser Microspectral Analyzer IMA 10 (Jenoptik Jena) with Quartz Spectrograph Q24, Q-switch-cuvette 2, diaphram 1(2), magnification 40(16), lamp voltage 0.9kV, battery I, spark voltage 4kV, induction 125μH, delay time 400μsec., electrode gap 1.6 mm, electrode-to-sample gap 1.2 mm; Q24: width of slit 15μ, slit height 1 mm, aperture 1:15.

26. Zeiss photomicroscope II with dark field illuminator. Film: Kodak Ektachrome, artificial light, 160 ASA.

27. Philips PW 1700 automatic powder diffraction system, Cu-fine-focus tube 40kV/30mA.

28. The high proportion of amorphous binding agent urushi proved to be a problem here, in that it produced a very high background.

29. Lacquer bands, according to Garner's definition (1963) are made up of several lacquer layers which have the same pigmentation and therefore the same color.

30. An analytical review of the corresponding layers of no. 2 shows low intensities of $\mathrm{Ca_3}$ and $\mathrm{Mg_1}$ (for notation see Appendix); of no. 4, cinnabar (HgS, see below) contaminated from the ornamental layer[31], $\mathrm{Ca_3}$, $\mathrm{Hg_3}$, and $\mathrm{Mg_1}$; and, contaminated in the same fashion, in no. 10, $\mathrm{Ca_4}$, $\mathrm{Mg_3}$, $\mathrm{Fe_3}$, $\mathrm{Ba_3}$, $\mathrm{Hg_3}$ and $\mathrm{Si_1}$. As explained in detail below, all elements except manganese can be traced back to the application of natural cinnabar, originating (as in no. 4) in the ornamental zone.

31. In individual cases it was extremely difficult to direct the laser beam onto the selected lacquer layer because the microscope used for selection did not permit color observation.

32. Mänchen-Helfen (1937b) brings to our attention a section of the *Cho-keng lu* of Tao Tsung-yi (A.D. 1366 which is concerned with the ground:

"The lacquer workers buy the objects to be decorated from the turners. They are made of thin, soft pieces of wood bound by glue. First the joints are smoothed and any gaps filled with a mixture of natural lacquer and glue, called *shao-tang*. . . . They cover the object with a layer of lacquer. On top of this they place a linen cloth. The first layer of ash is then applied. In all, three layers of ash are used, mixtures of ash, powdered brick and lacquer and ever finer ash. Each layer, as soon as it is dry, is polished with sand leather [like our glass paper]. For cheaper products a mixture of pig's blood and rice paste is used and, instead of linen, hemp. On top of the last layer of ash, there is another layer of lacquer—*tsao ch'i* . . . and in this way the object has been well primed. When red lacquer is used, there is no *tsao ch'i*."

The question must be raised, both here and for other sources, as to whether the translation is accurate (e.g. linen and hemp). Garner spent years pondering this issue as can be seen by the bibliography of his book (1979).

33. The increased thickness of the layer naturally extends the drying period. In reference to this, Garner (1979) quotes from the *Cho-keng lu* of Tao Tsung-yi (A.D. 1366):

". . . referring to the meticulous care with which the ground work was done and each layer polished before the next was applied. It mentions that several months elapsed between the preparation of the base and the application of the first coloured layer."

In comparison, the drying time for a lacquer layer in the ornamental zone is between four and five days.

34. Mänchen-Helfen (1937b) translates from the *Ko-ku-yao-lun* of Ts'ao Chao of A.D. 1388 (edited 1459):

". . . the objects produced lately under Yüan in Yang-hui and Si-t'ang, Kia-hing-fu have many layers and are carved in strong relief, but the fat content of the lacquer is too low. There are many well-worked pieces but the lacquer flakes off very easily on those with yellow ground."

This is further evidence that the adhesion is weakened between the yellow ground and the ornamental zone. This concept is supported by a comment about a Chinese *guri* lacquer bowl (fifteenth-sixteenth century) by Gabbert (1978):

"the yellow lacquer layers sometimes seem to have the effect of raising the ornamented sections."

However, this raising of the ornamental sections was due not to the yellow layers but to the movement[38] of the wood with respect to the stability of the ornamental zone.

35. In no. 11 (Fig. 18) as ($\mathrm{Sn_5}$), $\mathrm{Ca_4}$, As_3, $\mathrm{Mg_1}$, $\mathrm{Si_1}$, $\mathrm{Fe_1}$ and ($\mathrm{Sn_7}$), $\mathrm{Ca_3}$, As_3, $\mathrm{Mg_1}$, $\mathrm{Al_1}$, $\mathrm{Si_1}$, $\mathrm{Fe_1}$; no. 14 (Fig. 20, 21) as $\mathrm{Ca_4}$, As_3, $\mathrm{Mg_2}$; no. 15 (Fig. 23, 24) as As_3, $\mathrm{Hg_3}$, $\mathrm{Ca_2}$; no. 21 (Fig. 29) as $\mathrm{Ca_5}$, $\mathrm{Mg_4}$, $\mathrm{Fe_4}$, As_4, $\mathrm{Si_3}$, $\mathrm{Mn_3}$, $\mathrm{Hg_3}$, $\mathrm{Al_1}$; no. 22 (Fig. 31, 32) as $\mathrm{Ca_5}$, As_4, $\mathrm{Mg_2}$, $\mathrm{Si_1}$, $\mathrm{Fe_1}$, $\mathrm{Hg_1}$; no. 23 (Fig. 34) as As_4, $\mathrm{Ca_2}$, $\mathrm{Mg_1}$ and in no. 25 (Fig. 39) as As_4, $\mathrm{Ca_2}$, $\mathrm{Mg_1}$, $\mathrm{Si_1}$, $\mathrm{Fe_1}$. Here and in notes 35, 38–40 and 42–45 only those results are given in which there was no possibility of contamination from other layers.

36. Garner (1963) explains evidence of mercury in the yellow ground as contamination from the ornamental zone above, but this interpretation can now be excluded.

37. However, it can be assumed that orpiment, like other pigments (especially cinnabar), was available from other Chinese provinces (Yonezawa 1956) and was imported from abroad.

38. Movement of wood refers to dimensional change arising from a variation in moisture content.

39. On no. 4 (Fig. 9): Hg_1 (low intensity spectrum); no. 6: $\mathrm{Ba_4}$, $\mathrm{Ca_3}$, $\mathrm{Cr_3}$, Hg_3; no. 9 (Figs. 12, 13): $\mathrm{Ca_4}$, Hg_4, $\mathrm{Fe_3}$, $\mathrm{Mg_2}$, $\mathrm{Mn_2}$; no. 10 (Fig. 15): $\mathrm{Ca_4}$, $\mathrm{Ba_3}$ Hg_3, $\mathrm{Fe_2}$, $\mathrm{Mg_1}$, $\mathrm{Si_1}$, $\mathrm{Mn_1}$; no. 12: $\mathrm{Ca_4}$, $\mathrm{Ag_4}$, Hg_2.

40. No. 15 (Fig. 23): Hg_5, $\mathrm{Ca_3}$, $\mathrm{Fe_2}$, $\mathrm{Mg_1}$, $\mathrm{Al_1}$, $\mathrm{Si_1}$; no. 16: Hg_7, $\mathrm{Ca_5}$, $\mathrm{Fe_2}$, $\mathrm{Mg_1}$, $\mathrm{Si_1}$; no. 17: twice Hg_6, $\mathrm{Ca_3}$, $\mathrm{Mg_1}$; no. 18: Hg_5, $\mathrm{Ca_4}$, $\mathrm{Mg_3}$, $\mathrm{Mn_3}$, $\mathrm{Si_1}$, $\mathrm{Fe_1}$ and Hg_5, $\mathrm{Ca_4}$, $\mathrm{Mn_4}$, $\mathrm{Mg_3}$, $\mathrm{Cu_2}$, $\mathrm{Si_1}$, $\mathrm{Fe_1}$; no. 19 (Figs. 26, 27): Hg_6, $\mathrm{Ca_4}$, $\mathrm{Mg_2}$, $\mathrm{Mn_2}$, $\mathrm{Fe_1}$, $\mathrm{As_1}$ and Hg_6, $\mathrm{Ca_3}$, $\mathrm{Mg_2}$, $\mathrm{Mn_2}$, $\mathrm{Si_1}$, $\mathrm{Fe_1}$, $\mathrm{As_1}$ and Hg_5, $\mathrm{Ba_4}$, $\mathrm{Ca_3}$, $\mathrm{Mg_1}$, $\mathrm{Mn_1}$, $\mathrm{Fe_1}$ and $\mathrm{Fe_6}$, Hg_3, $\mathrm{Al_3}$, $\mathrm{Si_3}$, $\mathrm{Ca_3}$, $\mathrm{Mn_3}$, $\mathrm{Ti_2}$, $\mathrm{Mg_1}$ and finally $\mathrm{Fe_6}$, Hg_4, $\mathrm{Si_4}$, $\mathrm{Ca_4}$, $\mathrm{Al_3}$, $\mathrm{Mn_3}$, $\mathrm{Ti_2}$, $\mathrm{Mg_1}$; no. 20: $\mathrm{Ca_3}$, Hg_3, $\mathrm{Mg_1}$, $\mathrm{Si_1}$, $\mathrm{Fe_1}$; no. 21 (Fig. 29): Hg_6, $\mathrm{Ca_5}$, $\mathrm{Mg_4}$, $\mathrm{Si_3}$, $\mathrm{Fe_3}$, $\mathrm{Mn_2}$, $\mathrm{Al_1}$, $\mathrm{As_1}$ and Hg_6, $\mathrm{Ca_5}$, $\mathrm{Mg_4}$, $\mathrm{Si_4}$, $\mathrm{As_4}$, $\mathrm{Fe_3}$, $\mathrm{Ti_2}$, $\mathrm{Mn_2}$, $\mathrm{Al_1}$ and $\mathrm{Ca_5}$, Hg_4, $\mathrm{Mg_2}$, $\mathrm{Si_1}$, $\mathrm{Fe_1}$, $\mathrm{As_1}$ and $\mathrm{Ca_5}$, Hg_4, $\mathrm{Mg_2}$, $\mathrm{Si_1}$, $\mathrm{Mn_1}$, $\mathrm{Fe_1}$, $\mathrm{As_1}$; no. 22 (Figs. 31, 32): Hg_5, $\mathrm{Ca_3}$, $\mathrm{Ba_3}$, $\mathrm{Fe_2}$, $\mathrm{Mg_1}$, $\mathrm{As_1}$ and Hg_4, $\mathrm{Ca_3}$, $\mathrm{Fe_3}$, $\mathrm{Ba_3}$, $\mathrm{Mg_1}$, $\mathrm{Cr_1}$, $\mathrm{Pb_1}$ and $\mathrm{Ba_3}$, $\mathrm{Ca_2}$, $\mathrm{Fe_2}$, Hg_2; no. 23 (Fig. 34) Hg_5, $\mathrm{Fe_4}$, $\mathrm{Ca_3}$, $\mathrm{Mg_1}$, $\mathrm{Si_1}$; no. 24 (Fig. 36): $\mathrm{Ca_3}$, Hg_3, and Hg_4, $\mathrm{Ca_3}$, $\mathrm{Mg_1}$ as well as $\mathrm{Ca_3}$, Hg_3, $\mathrm{Mg_1}$; no. 25 (Fig. 39): Hg_4, $\mathrm{Ca_2}$, and Hg_4, $\mathrm{Ca_2}$, $\mathrm{Mg_1}$, $\mathrm{As_1}$; no. 26: $\mathrm{Fe_4}$, $\mathrm{As_3}$, Hg_3, $\mathrm{Ca_2}$, $\mathrm{Mg_1}$; no. 27: Hg_6, $\mathrm{As_3}$, $\mathrm{Ca_2}$, $\mathrm{Mg_1}$; and no. 28 (Figs. 41, 42): $\mathrm{Ca_3}$, Hg_3.

41. Because of the low contamination in the red of a carved lacquerware object from the fifteenth century, Garner (1963) concluded that vermilion had been used. In view of the experience gained here and a study of the original analysis,[2] these results should not be interpreted in this way; rather they are a clear indication of the presence of natural cinnabar. Garner's reference is to a source not available to the author and which refers to the use of dry-process vermilion since the fourth century B.C. (Read 1936).

42. In no. 14 (Figs. 19–21) with $\mathrm{Ca_4}$, Fe_4, $\mathrm{Mg_2}$; in no. 19 (Figs. 26, 27) with Fe_6, $\mathrm{Hg_3}$, $\mathrm{Al_3}$, $\mathrm{Si_3}$, $\mathrm{Ca_3}$, $\mathrm{Mn_3}$, $\mathrm{Ti_2}$, $\mathrm{Mg_1}$; and in no. 26 with Fe_5, $\mathrm{Ca_2}$, $\mathrm{Hg_2}$, $\mathrm{Mg_1}$, $\mathrm{Mn_1}$.

43. No. 22 (Figs. 31, 32) $\mathrm{Ca_5}$, Fe_5, $\mathrm{Mg_2}$, $\mathrm{As_2}$, $\mathrm{Si_1}$, $\mathrm{Mn_1}$; no. 23 (Fig. 34): Fe_3, $\mathrm{As_2}$, $\mathrm{Ca_2}$, $\mathrm{Hg_2}$, $\mathrm{Mg_1}$.

44. No. 1 (Fig. 5) $\mathrm{Si_5}$, $\mathrm{Fe_5}$, Hg_4, $\mathrm{Mg_4}$, $\mathrm{Al_4}$, $\mathrm{Ca_4}$, $\mathrm{Ti_4}$, $\mathrm{Mn_2}$; no. 8 (Fig. 11): Hg_4, $\mathrm{Si_3}$, $\mathrm{Ca_3}$, $\mathrm{Mn_3}$, $\mathrm{Fe_3}$, $\mathrm{Ba_3}$, $\mathrm{Mg_2}$, $\mathrm{Al_1}$, $\mathrm{Ti_1}$, $\mathrm{Cu_1}$; no. 12: $\mathrm{Ca_4}$, $\mathrm{Ag_2}$, Hg_2; and no. 19 (Figs. 26, 27): $\mathrm{Ca_4}$, Hg_4, $\mathrm{As_3}$, $\mathrm{Mn_3}$, $\mathrm{Mg_1}$, $\mathrm{Si_1}$, $\mathrm{Fe_1}$.

45. No. 14 (Figs. 20, 21) $\mathrm{Ca_4}$, $\mathrm{Mg_2}$ and $\mathrm{Ca_4}$, $\mathrm{Mg_2}$ as well as $\mathrm{Ca_1}$; no. 24 (Fig. 36) $\mathrm{Ca_3}$, $\mathrm{Si_1}$ and $\mathrm{Hg_5}$, $\mathrm{Ca_3}$, $\mathrm{Mg_2}$, $\mathrm{Si_2}$, $\mathrm{Fe_2}$ (contaminated); no. 25 (Fig. 39) $\mathrm{Si_2}$, $\mathrm{Ca_2}$, $\mathrm{Hg_2}$, $\mathrm{Mg_1}$, $\mathrm{Fe_1}$, $\mathrm{As_1}$ and $\mathrm{Ca_1}$, $\mathrm{Hg_1}$ as well as $\mathrm{Hg_3}$, $\mathrm{Ca_1}$, $\mathrm{As_1}$ and $\mathrm{Ca_4}$, $\mathrm{Hg_3}$.

46. The list of evidence of mercury in black layers can be expanded to include a piece of the Jiajing era (Linden-Museum, Stuttgart, OA 20.796L; Burmester, Brandt 1982) which was not included in the series of lacquerware described in detail here.

Bibliography

Brandt, K.J.

1982 "Ostasien." In *Ferne Völker–Frühe Zeiten, Kunstwerke aus dem Linden-Museum*. Catalog. Stuttgart:229–248.

1986 *Lacques chinois du Linden-Museum de Stuttgart, Musée Guimet*. Catalog. Paris.

Burmester, A.

1982 Chemometrische Bewertung komplexer analytischer Befunde–Pyrolyse-Massenspektrometrie an ostasiatischen Lacken. Lecture. Berlin.

1983 "Historical coating materials–East Asian lacquer." In *Proceedings of the 22nd Symposium on Archaeometry 1982*. A. Aspinall and S.E. Warren (eds). Bradford: 184-193.

1983 "Far Eastern lacquer: Classification by pyrolysis mass spectrometry." *Archaeometry* 25:45–58.

1983 "Far Eastern lacquer: No end in sight." In *23rd Symposium on Archaeometry*. Naples.

1985 A description and testing of Oriental lacquer for museum purposes. Lecture at the Symposium of the Urushi Study Group, Tokyo.

Burmester, A. and K. J. Brandt

1982 "Beitrag zur Archäometrie organischer Materialien–Ostasiatischer Lack." *Berliner Beiträge zur Archäometrie* 7:216–251.

d'Incarville, Père L.

1760 "Mémoire sur le vernis de la Chine." *Mémoires de l'Academie Royal des Sciences* III:177.

Doelter, C. and H. Leitmeier

1926 *Handbuch der Mineralchemie,* Vol. IV, Part 1. Dresden/Leipzig.

FitzHugh, E. W.

1979 "A pigment census of Japanese *ukiyo-e* paintings in the Freer Gallery of Art." *Ars Orientalis* XI:27–38.

Gabbert, G.

1978 *Ostasiatische Lackkunst*. Catalog, Frankfurt/Main.

Garner, Sir Harry

1963 "Technical studies of Oriental lacquer." *Studies in Conservation* 8:84–98.

1979 *Chinese Lacquer*. London: Faber and Faber.

Gettens, R.J.

1938–39 "Pigments in a wall painting from central China." *Technical Studies in the Field of the Fine Arts* VII: 99–105.

1976 "Japanese paintings: Technical studies at the Freer Gallery of Art." In *Conservation and Restoration of Pictorial Art*. N.S. Brommelle and P. Smith (eds). London: Butterworths.

Gettens, R.J., R.L. Feller, and W.T. Chase

1972 "Vermilion and cinnabar." *Studies in Conservation* 17:45–69.

Gettens, R.J., and G.L. Stout

1966 *Painting Materials. New York:* Dover Publications.

Goffer, Z.

1980 *Archaeological Chemistry: A Source Book on the Applications of Chemistry to Archaeology*. New York: John Wiley and Sons.

Hora, B.

1981 *The Oxford Encyclopaedia of Trees of the World*. Oxford: Oxford University Press.

Hornblower, A.P.

1962 "Some examples of work with the electron probe microanalyser." *Archaeometry* 5:37–42.

Kenjo, T.

1976 "Effect of pH on the hardening of Japanese lacquer films." *Science for Conservation* 15:1–8.

1978 "Studies on the analysis of lacquer (Part 2). Infra-red spectrometry of lacquer films." *Scient. Antiq.* 23:32–39.

Kumanotani, J.

1979 "Urushi—Lernen von überlieferten natürlichen Materialen." *Kobunshi* (High polymers) 28:871–873.

1983 "Japanese lacquer—a super durable coating (proposed structure and expanded application)." In *Polymer Applications of Renewable Resource Materials*. C.E. Carraher and L. H. Sperling (eds), Plenum Publishing Corporation, pp. 225–248.

1983 "Urushi." *Hifu* (Skin research) 3:416–421.

Kuwayama, G.

1982 *Far Eastern Lacquer*. Catalog. Los Angeles: Los Angeles County Museum of Art.

Lee, Yu-kuan

1972 *Oriental Lacquer Art*. New York/Tokyo.

Majima, R.

1909 "Über den Hauptbestandteil des Japanlackes (Mitteilung): Über Urushiol und Urushioldimethyläther."Berichte der Deutshen Chemishen Gesellschaft 42:1418–1423.

Mänchen-Helfen, O.

1937 "Chinesisches Lackgerät." *Ostasiatische Zeitschrift N.F.* 13:1–5.

1937 "Materialien zur Geschichte des chinesischen Lacks." *Ostasiatische Zeitschrift N.F.* 13:215–222.

Meuzelaar, H.L.C., J. Haverkamp, and F.D. Hileman

1982 *Pyrolysis Mass Spectrometry of Recent and Fossil Biomaterials*. Amsterdam.

Miyama, K.

1906 In an official Japanese report (No. 7000) dated 27 October.

1908 "On Japanese lacquer." *Journal of College of Engineering* (Tokyo) 4(3):89–110.

Moenke, H. and L. Moencke-Blankenburg

1968 *Einführung in die Laser-Mikro-Emissionsspektralanalyse*. Leipzig.

Moss, V.A.

1967 Chemical constitution of lacquer. Unpublished manuscript.

Needham, J.

1976 *Science and Civilization in China*. Vols.III and V. Cambridge.

Occolowitz, J.L.

1964 "Mass spectrometry of naturally occurring alkenyl phenols and their derivatives." *Analytical Chemistry* 36:2177–2181.

Oguchi, H.

1969 "Scientific investigation on color materials in Japanese painting." *Bulletin of the Faculty of Fine Arts* (Tokyo University of Arts) 5:27–81.

Oshima, Ryuichi, Yoshio Yamauchi, Chuichi Watanabe, and Ju Kumanotani

1985 "Enzymic oxidative coupling of urushiol in sap of the lac tree, *Rhus vernicifera." Journal of Organic Chemistry* 50:2613.

Read, Bernard E. and C. Pak

1936 *A Compendium of Minerals and Stones in Chinese Medicine, from the Pen Ts'ao Kang Mu.*

Rees Jones, S.

1975 "Paintings and optical absorption: x-ray absorption." *Institut Royal du Patrimoine Artistique Bulletin* XV:326–336.

Reinhammer, B.

1970 "Purification and properties of laccase and stellacyanin from *Rhus vernicifera."Biochimica et Acta* 205:35–47.

Roy, A.

1979 "The laser microspectral analysis of paint." *National Gallery Technical Bulletin* 3:43–50.

Schafer, E.H.

1955 "Orpiment and realgar in Chinese technology and tradition." *Journal of the American Oriental Society* 75(2):73–89.

Schröcke, H. and K.-L. Weiner

1981 *Mineralogie*. Berlin.

Schrön, W., P. Lange, and W. Kramer

1983 "Einsatzmöglichkeiten des LMA 10 in der Geologie." *Jenaer Rundschau* 3:123–128.

Schulze, G.M.

1966 *Internationaler Code der Botanischen Nomenklatur.* Berlin.

Office of the Shosoin

1975 *Lacquer Works in the Shosoin.* Tokyo.

Tschirch, A. and A.B. Stevens

1905 "Über den Japanlack (Ki-Urushi)." *Archiv für Pharmazie* 243:504–553.

Vaultier, Roger

1957 "The history of china ink." *Papetier* 11(7):44–46.

Wallert, A.

1984 "Orpiment und realgar." *Maltechnik Restauro* 90 (4):45–57.

Winter, J.

1974 "Preliminary investigations on Chinese ink in Far Eastern paintings." In *Archaeological Chemistry,* C.W. Beck (ed). American Chemical Society. Washington, D.C., pp. 207–225.

1984 "Pigments in China–a preliminary bibliography of identifications." In *ICOM Committee for Conservation 7th Triennial Meeting.* Copenhagen. Vol. II:84.19.11–13.

Yamasaki, K. and Y. Emoto

1979 "Pigments used on Japanese paintings from the protohistoric period through the 19th century." *Ars Orientalis* XI:1–14.

Yonezawa, Y.

1956 "On the production of mineral pigments in ancient China." *Tokyo Bunka Kenkyushu Kiyo* (Memoirs of the Institute for Oriental Culture) 11:375–395.

Yoshida, H.

1883 "On urushi-lacquer." *Journal of the Chemical Society*:472.

Zander, R.

1972 *Handwörterbuch der Pflanzennamen.* Stuttgart.

The Production and Use of Chinese Raw Urushi and the Present State of Research

Yumin Du
Wuhan University

Raw (or natural) urushi is the viscous sap of the lacquer tree, a deciduous plant six to ten meters high. The outer bark of the tree is whitish, the inside yellowish. Production of raw urushi in the People's Republic of China has developed steadily in the past ten years and has doubled as a natural resource. Two thousand one hundred seventy tons of raw urushi were produced in 1978 and production has continued to increase.

Distribution of the lacquer tree and collection of sap

Both the climate and soil of the People's Republic of China are favorable for the growth of the lacquer tree. It grows in mountainous districts between latitude 19–42° and longitude 97–126° where the annual rainfall exceeds 0.6 m and the annual mean temperature is 8–20°C. Lacquer trees grow in twenty-three provinces throughout the country, notably Shanxi, Hubei, Sichuan, Yunnan, and Guizhou. There are 161 major urushi-producing districts. The People's Republic of China leads the world in both the production and the export of urushi.

When the bark of the tree is stripped and a cut made, the milky white sap oozes out. The season for tapping varies according to the region, due to differences in climate. In areas where the temperature is slightly higher, tapping starts ten days before the summer solstice and continues until the frosts begin, a period of about 120 days. In areas where the temperature is slightly lower, it begins around the first week in July and continues until the first week of October, a period of only about ninety days. The actual period for the collection of the sap is usually determined by the growth of the lacquer tree (Local Product Bureau 1980).

Chinese urushi farmers have a long history of urushi production and have developed various methods of making the cuts, based upon their experiences. In Hubei and Sichuan we find "willow leaf" and "eyebrow" styles, in Shanxi, Yunnan, and Guizhou "Chinese character," "cattle nose," and "scissor" style cuts. For example, with the "cattle nose" style of cut, the sap is collected in a shell placed under a cut shaped like the nose of a cow. Sap is collected from a tree every two to three years, and only four or five times from the same tree.

Composition and constituents of the sap

The composition of the sap of the lacquer tree varies depending on the species of the tree and also, in the case of trees of the same species, on the place and season of collection. Table 1 shows the composition of sap collected from *Rhus verniciflua* in Maoba, Hubei province, People's Republic of China (Du 1986).

The main constituent of the sap is urushiol. The subconstituents of urushiol were investigated by Majima (1922), Dawson (Sunthankar et al. 1954), Kumanotani (Yamauchi 1982), and Tyman (1982). In recent years, high performance liquid chromatography (HPLC) and high resolution capillary gas chromatography have made such great progress that several new subconstituents of urushiol have been successfully isolated and identified. Under the guidance of Professor Kumanotani and with the cooperation of Dr. Oshima, I found that urushiol in Chinese raw urushi comprises not only those compounds having the formula in Figure 1, but also those in Figures 2 and 3; he also clarified the structure of their side chain (Du 1984, unpublished results). The results obtained are summarized in Figure 4.

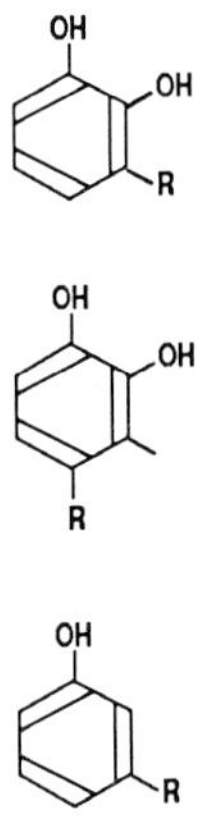

Figures 1–3. Compounds of urushiol in Chinese raw urushi.

Chinese, Korean, and Japanese lacquer trees are all the same species (*Rhus verniciflua*) and their saps have exactly the same urushiol composition (very similar to that of American poison ivy except that the American plant contains fewer triene subconstituents; Elsohly et al. 1982).

Recently, I explored the chemical structure and proportion of the subconstituents of the main components of Taiwanese and Vietnamese lacquer trees (*Rhus succedanea*), laccol, and that of Thai and Burmese lacquer trees (*Melanorrhoea usitata*), thitsiol (1986). The results obtained are shown in Figures 5 and 6 in which are listed some new compounds found.

Table 1

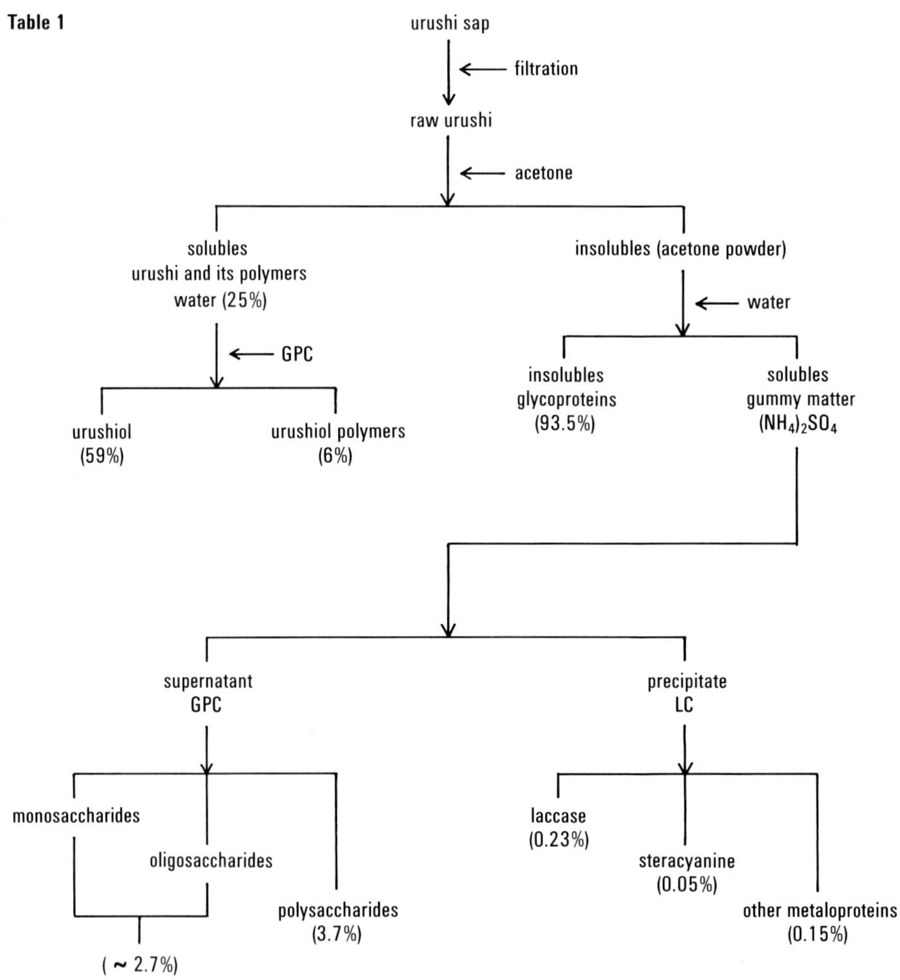

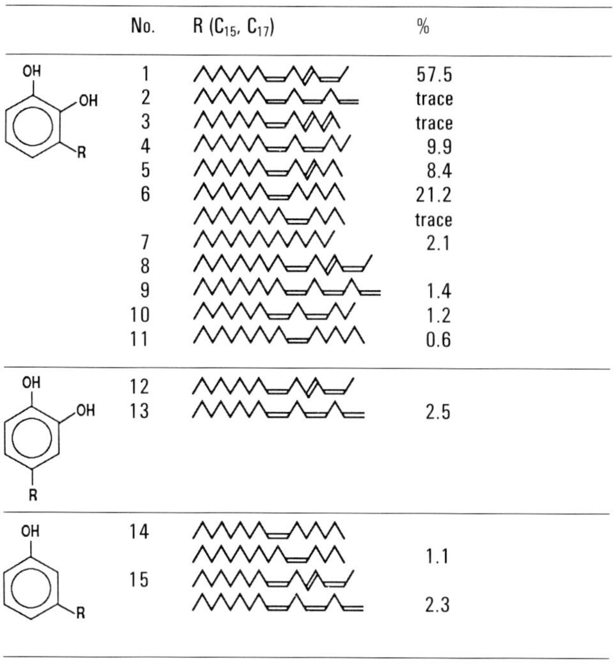

Figure 4. Urushiol constituents of the Chinese and Japanese tree, Rhus verniciflua.

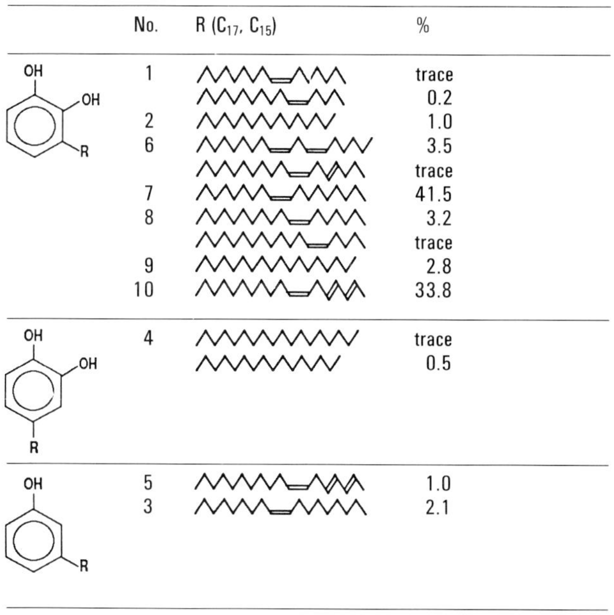

Figure 5. Laccol constituents of the Taiwanese and Vietnamese lacquer tree (Rhus succedanea).

Figure 6. Thitsiol constituents of the Thai lacquer tree (Melanorrhoea usitata).

		R	%
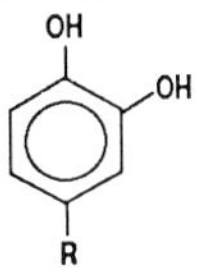	1a	$C_{17}H_{29}$ (all *cis*)	0.7
	1b	$C_{17}H_{29}$ (includes at least one *trans* double bond)	
	1c	$C_{17}H_{31}$ (all *cis*)	20.5
	1d	$C_{17}H_{33}$ (*trans*)	
	1e	$C_{15}H_{29}$ (*cis*)	
	1f	$C_{15}H_{31}$	0.7
	3a	$-(CH_2)_{10}C_6H_5$	1.1
	3b	$-(CH_2)_{12}C_6H_5$	3.6
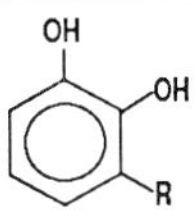	2a	$C_{17}H_{29}$ (all *cis*)	
	2b	$C_{17}H_{29}$ (includes at least one *trans* double bond)	
	2c	$C_{17}H_{31}$ (all *cis*)	19.7
	2d	$C_{17}H_{33}$ (*cis*)	
	2e	$C_{17}H_{33}$ (possibly *trans*)	
	2f	$C_{15}H_{29}$ (*cis*)	
	2g	$C_{15}H_{31}$	3.9
	5a	$-(CH_2)_{10}C_6H_5$	7.5
	5b	$-(CH_2)_{12}C_6H_5$	36.
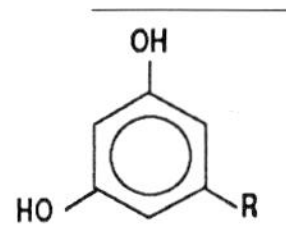	4a	$-(CH_2)_{10}C_6H_5$	0.7
	4b	$-(CH_2)_{12}C_6H_5$	2.1
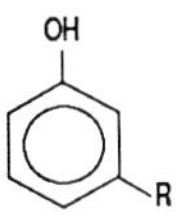	6a	$C_{17}H_{31}$	
	6b	$-CH_2CH(OH)C_{17}H_{31}$	
	6c	$-CH_2CH(OH)C_{17}H_{29}$	
	6d	$-(CH_2)_{10}C_6H_5$	1.4
	6e	$-(CH_2)_{12}C_6H_5$	
	6f	$-CH_2CO(CH_2)_{10}C_6H_5$	0.4
	6g	$-CH_2CO(CH_2)_{12}C_6H_5$	
	6h	$-(CH_2)CH(OH)(CH_2)_{12}C_6H_5$	

Figure 7. Chromatograms of urushiol and laccol from the sap of (a) Rhus verniciflua *(Shensi, The People's Republic of China) and (b)* Rhus succedane a *(Taiwan) in the underivatized form. Conditions: column, methylsilicone, 12.5 m x 0.2 mm i.d., df = 0.33 m; column programmed from 23C (maintained for 8 min) to 28C at 5C/min; injection-port and detector temperature 28C; carrier gas (helium) linear velocity 40 cm/sec; splitting ratio 50:1. Numbers on peaks represent compound numbers; the peak marked with an arrow was not identified.*

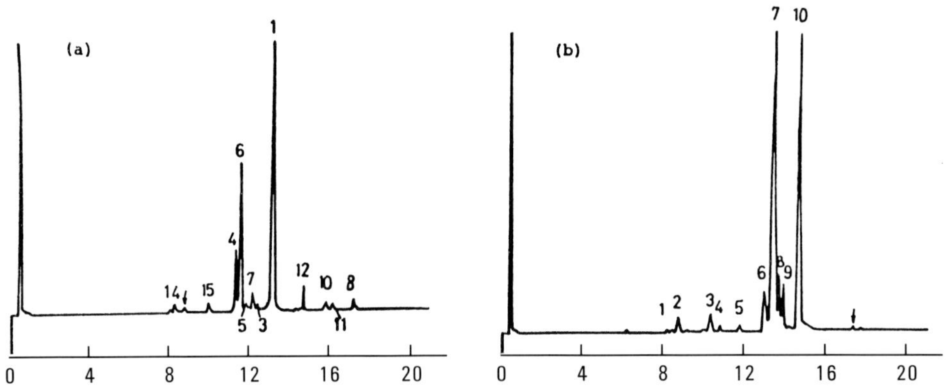

Table 2

*Compound**	*China*						*Japan*		
	Hupei	*Shenst*	*Huxian*	*Fengju*	*Songya*	*Kiangs*	*Yamagata*	*Naruko*	*Ibaraki*
1	64.13	53.72	53.84	57.37	59.37	38.49	51.30	60.75	57.24
3	–	0.32	–	–	–	0.42	0.19	0.23	0.35
4	2.86	9.66	24.80	12.88	13.08	2.82	14.49	10.75	5.21
5	0.85	1.47	0.50	0.65	–	2.80	1.03	1.23	1.50
6	23.01	25.51	12.61	19.08	12.54	47.18	25.03	19.33	25.82
7	3.06	3.55	2.02	2.45	1.94	3.90	3.13	3.25	4.35
8	0.59	0.47	2.11	1.79	6.51	0.29	0.57	0.56	1.00
10	1.32	0.72	0.85	1.37	1.99	0.28	0.71	0.33	0.27
11	0.18	0.39	0.64	0.52	0.78	–	0.27	–	1.25
12	1.19	2.05	1.77	2.29	1.94	0.51	2.45	1.90	0.90
Monomer content (%)	88.6	77.9	83.8	75.8	80.8	–	90.9	92.6	–

*Compounds 1, 8 and 12 contain compounds 2, 9 and 13 respectively, as trace contaminants.

Table 3

Number	*Analysis*					*Average ± S.D.*
	1	*2*	*3*	*4*	*5*	
1	0.165	0.147	0.153	0.161	0.172	0.160±0.009
2	0.990	0.982	0.963	0.935	0.969	0.968±0.019
3	2.143	2.186	2.206	2.192	2.163	2.178±0.022
4	0.465	0.425	0.473	0.481	0.492	0.467±0.023
5	0.984	0.961	0.978	0.954	0.963	0.986±0.011
6	3.531	3.602	3.556	3.541	3.582	3.562±0.026
7	41.499	41.508	41.704	41.829	41.756	41.672±0.151
8	3.163	3.212	3.186	3.201	3.154	3.183±0.022
9	2.772	2.706	2.689	2.690	2.714	2.714±0.030
10	33.771	33.051	33.647	33.893	33.465	33.571±0.295

So far, separation and analysis of mixed phenolic compounds in natural urushi have been conducted by gas chromatography (GC) or liquid chromatography (LC) after being converted into their methylated (Craig et al. 1978; Baer et al. 1980) or acetylated (Yamauchi et al. 1980) derivatives. I used HPLC and capillary gas chromatography for quantitative analysis of the subconstituents of urushiol without modifying them and found that phenolic subconstituents could be separated directly from natural urushi. This direct method is expected to play a great part in the analysis of phenolic subconstituents of urushiol, its uses, the synthesis of derivatives, tests for its sensitivity to the human body and immunotherapy.

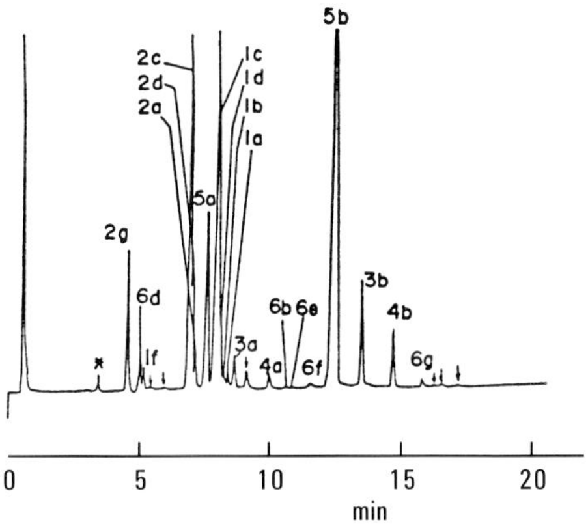

Figure 8. Chromatograms of the oily component of the sap from the Thai lacquer tree (Melanorrhoea usitata) *in the underivatized form. Conditions: column, methylsilicone, 12.5 m x 0.2 mm i.d., df = 0.33 m; column programmed from 25°C (maintained for 1 min) to 30°C at 5°C/min (d); injection-port and detector temperature 300°C; carrier gas (helium) linear velocity 36 cm/sec; splitting ratio 50:1. Numbers represent compounds; peaks with an arrow were not identified, and that with an asterisk is due to phthalate from a plastic container.*

Table 4

Number	*Compound*	*Analysis*						*Average ± S.D.*
		1	2	3	4	5	6	
1	2g	3.773	4.101	3.983	3.968	3.863	3.768	3.909±0.132
2	6d	1.120	1.638	1.124	1.364	1.511	1.493	1.375±0.214
3	1f	0.620	0.879	0.632	0.663	0.759	0.841	0.734±0.110
4	2c	21.051	20.585	20.688	17.699	17.701	20.550	19.712±1.568
5	5a	7.173	7.178	8.292	7.467	7.863	7.212	7.531±0.458
6	1c	20.111	20.162	20.441	20.560	21.210	20.259	20.457±0.406
7	1a	0.638	0.616	0.899	0.647	0.969	0.600	0.728±0.162
8	3a	1.062	1.076	1.197	1.130	1.244	1.059	1.128±0.078
9	4a	0.622	0.705	0.675	0.729	0.705	0.708	0.691±0.004
10	6f	0.346	0.361	0.342	0.400	0.326	0.334	0.352±0.003
11	5b	36.305	35.025	34.795	37.777	36.181	35.689	35.962±1.075
12	3b	3.562	3.638	3.531	3.733	3.684	3.633	3.630±0.007
13	4b	2.035	2.124	2.066	2.178	2.151	2.110	2.104±0.005

Assessment of the quality of raw urushi

According to Kumanotani (1983), when it is allowed to stand in the air, raw urushi that contains urushiol with a readily oxidizable catechol nucleus and olefinic side-chain is oxidized enzymatically in its polymerizable urushi phase to produce oxidized films having a lower density of cross-linkages. It is believed, therefore, that the quality of a given raw urushi depends mainly on the content of urushiol triene subconstituents and the amount and activity of enzyme. The composition of raw urushi has been scientifically verified by conventional methods in combination with the chromatographic technique that we have developed.

As mentioned above, the urushiol in Chinese urushi does not differ from that in Japanese urushi as to subconstituents and composition. Both have a large content of triene subconstituents (Du 1984) and are readily dried (Fig. 7a and Table 2).

Laccol in Taiwanese and Vietnamese urushi unexpectedly contains much monoene subconstituent (Du n.d.), though it also contains triene constituents. These urushi are said to be poor in drying, which is partly attributable to their insufficient triene content (Fig. 7b and Table 3).

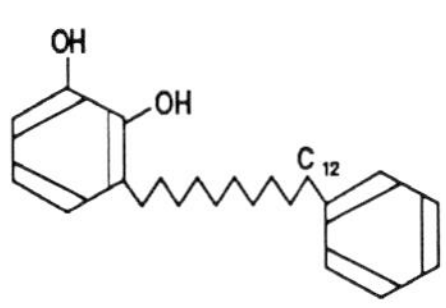

Figure 9. Thai urushi compound.

Thai urushi contains a compound having the interesting structure in Figure 9 in an amount of about 38% (Du 1985). Since the compound has no triene-b bearing side-chain, Thai urushi is very difficult to dry. However, it has been found that the compounds of Thai urushi can be used as a new material for the chemical industry by isolating them from the sap (Fig. 8 and Table 4).

Recently, I also successfully developed a method for the rapid analysis of metaloprotein enzyme in natural urushi by HPLC, resulting in good reproducibility and accuracy (Du n.d.). Samples of raw urushi were taken to Japan from the People's Republic of China for the 1985 Urushi Study Group and analyzed there in detail. Both the quantity and the activity-level of oxidized enzyme in raw urushi were measured and it was found that Chinese urushi contained more oxidized enzyme, and that it was also more active. Therefore it may be said that Chinese urushi is better.

Table 5

Place of production	*China*[a]						*Taiwan*	*Japan*[b]		
Peak areas (mg)	*Hupei*	*Shenei*				*Kiamgei*		*Ibaraki*	*Naruko*	*Slowly dried urushi*
		Pingli	*Langquan*	*Songya*	*Huxian*					
Peak area (mg)	13.04	12.9	19.3	15.9	14.3	5.2	17.4	13.1	8.8	9.1
Activity	54.68	45.2	67.3	50.0	26.4	7.4	21.0	12.2	6.7	10.0
Specific activity (activity/peak area)	4.19	3.5	3.49	3.15	1.85	1.42	1.21	0.93	0.76	1.10

[a] Urushi obtained in China in 1982 and Taiwan in 1979.
[b] Urushi obtained in Japan in 1979.

Present uses of urushi

Raw urushi has been used in China for more than six thousand years. It has been employed as a coating for objects in everyday use, works of art, and industrial equipment. It has also been found that urushi on ancient items unearthed from tombs thousands of years old still retains its vivid luster and incomparable durability.

Raw urushi and lacquer coating materials are characteristically hygienic, water repellent, hard wearing, and good insulators. Factories processing raw urushi in the People's Republic of China produce three types of urushi: traditional, purified, and improved. Traditional urushi is used for wood furniture, laboratory benches, and flooring. Purified and improved urushi are used as industrial coating materials. At present, more than twenty kinds of these urushi are produced and extensively used (Training Course 1976).

Raw urushi is also used in the spinning, printing, chemical, and petroleum industries, in gravel production, mining, production of Chinese lacquerware, as a coating for the bottoms of ships and boats, and for the repair of ancient buildings. For example, pipes used for extracting crude oil formerly required cleaning every day to remove deposits on their inner walls, but a single application of improved urushi coating material gives such excellent results that now the walls can be kept clean for at least three months. Marine growth on the hulls of ships immersed in seawater for prolonged periods, results in corrosion. But it has been found that application of improved urushi prevents such growths, thus maintaining the sailing speed and prolonging the life of the ship. It is also a good coating material for undersea cables.

Recently, a great many Chinese researchers have been engaged in developing new applications for both raw urushi and improved urushi. Urushi has been repeatedly examined for toxicity and verified nontoxic. Consequently, it can be used as a coating material for the inner walls of water tanks and cans to store fresh water, soy sauce, sake, vinegar, and so on. Hardened films of improved urushi were applied to the inner walls of a chimney two hundred meters high and six meters in diameter; they exhibited no change upon testing for thermal stability at 15°C and they resisted 5% solutions of nitric acid, hydrochloric acid, and potassium hydroxide, and a 20% solution of sulfuric acid.

Future urushi research in the People's Republic of China

In the People's Republic of China today, scientific research into urushi has made many advances in addition to the development of processes for and the increase in the quantity of the production of raw urushi. A special urushi research program has been organized and a Chinese Society of Urushi Chemistry has been established. National meetings and cooperation on urushi research have also been arranged. Study groups have been organized, with Wuhan University playing a leading part on the academic side and the Xian Raw Urushi Institute representing the urushi industry (Kenjo 1982). There are also some local institutes. In the past, raw urushi in was investigated mainly by the lacquer industry, but now many researchers from various specializations—chemistry, macromolecular science, physics, enzymology, botany, medicine, pharmacology, and archaeology—are taking part. Research is being conducted in various aspects, including the growth, breeding, and biological morphology of lacquer trees; the chemistry of urushi; the scientific examination of urushi works of art; various applications of raw urushi; and the prevention and treatment of hypersensitivity to urushi.

Although synthetic coating materials are developing all the time, they have not replaced urushi and its unique qualities. It is believed that raw urushi will be further developed as a renewable coating material, whose many colors, nontoxicity, and quick-drying properties are equivalent to those of synthetic coating materials.

At the present time, saturated urushiol-30-crown-10 which is synthesized from urushiol, is being used as a potassium ion selective electrode of the highest quality (Yu et al. 1982). In addition, saturated urushiol polymers have relatively high thermal stability, with a thermal decomposition temperature reaching 30°C, (Huang et al. 1984), and can be used as a stationary phase in chromatography and as ion-selective membranes for chromatography.

Hardened urushi is believed to have anticarcinogenic properties. "P235 pinxiaopin" produced in the People's Republic of China are herbal tablets with dried urushi as the active principle. The tablets have been shown to reduce tumors in patients with cancers of the lung, stomach, and esophagus, inhibiting the growth of cancer cells and thus prolonging the lives of the patients. This medicine has won great popularity in the People's Republic of China and elsewhere.

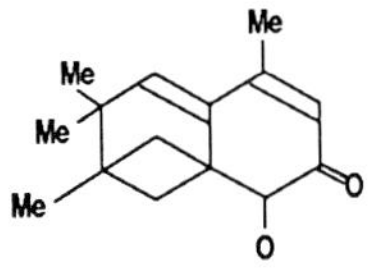

Figure 10. An urushiol subconstituent: six-membered, unsaturated-lactone.

Recent research into allergens (for allergy to Chinese urushi) has shown that an urushiol subconstituent, six-membered, unsaturated-lactone (see Fig. 10) can also be an allergen that may possibly play a part in the treatment and prevention of hypersensitivity to raw urushi. It is believed that there will be splendid results from urushi research in the near future.

Finally, I hope that raw urushi will attract the attention of many researchers to its complex and versatile properties, and that it will continue to be widely developed both for general use and for specialized processes.

Bibliography

Baer, H., M. Hooton, H. Fales, A. Wu, and F. Shaub

1980 "Catecholic and other constituents of the leaves of toxicodendron radicans and variation of urushiol concentrations within one plant." *Photochemistry* 19:799–802.

Craig, J.C., C.M. Walker, S. Billets, and M.A. Elsohly

1978 "New GLC analysis of urushiol congeners in different plant parts of poison ivy, toxicodendron radicans." *Journal of Pharmaceutical Sciences*. 64:483–485.

Du, Y.M.

1986 "Recent advance in study of Chinese lacquer." *Chemistry Bulletin* 1:1–9.

Du, Y.M.

n.d. Direct separation and analysis of urushiol congeners in Chinese lacquer by GPC-GC/MS. Unpublished manuscript.

Du, Y.M.

n.d. Chemical composition of laccol essential oil from the Chinese Taiwan lacquer. Unpublished manuscript.

Du, Y.M.

n.d. *Rhus vernicifera* Chinese lacquer laccase and stellacyanin separation and analysis accelerated oxidation. Unpublished manuscript.

Du, Y.M., T. Miyakoshi, and J. Kumanotani

n.d. The structural identification of the laccol in the sap of the Taiwan and Vietnam lac tree, *Rhus succedanea*. Unpublished manuscript.

Du, Y.M. and R. Oshima

1985 "Analysis of long-chain phenols in the sap of the Burmese lac tree, *Melanorrhoea usitate* by capillary gas-liquid chromatography." *Journal of Chromatography* 318:378–383.

Du, Y.M., R. Oshima, H. Iwatsuki, and J. Kumanotani

1984 "High-resolution gas-liquid chromatographic analysis of urushiol of the lac tree, *Rhus vernifi-cifera*, without dervatization." *Journal of Chromatography* 295:179–186.

Du Y.M., R. Oshima, and J. Kumanotani

1984 "Reversed-phase liquid chromatographic separation and identification of constituents of urushiol in the sap of the lac tree, *Rhus vernicifera*." *Journal of Chromatography* 284:463–473.

Du, Y.M., R. Oshima, Y. Yamauchi, and J. Kumanotani

1985 "3-(W-phenylalkyl)-catechols: novel phenolic lipids found in sap of the Burmese lac tree, *Melanorrhoea usitata*." *Journal of the Chemical Society, Chemical Communications* 15 (May):630–631.

Du, Y.M., R. Oshima, Y. Yamauchi, J. Kumanotani, and T. Miyakoshi

1986 "Long chain phenols from the Burmese lac tree, *Melanorrhoea usitata*." *Phytochemistry*. 25:2211–2218.

Elsohly, M.A., P.D. Adawadkar, C.Y. Ma, and C.E. Turner

1982 "Separation and characterization of poison ivy and poison oak urushiol components." *Journal of Natural Products* 45:532–538.

Huang, Z.F., J.Y. Shu, and Z.Y. Yu

1984 "Studies on the saturated urushiol crown ether polymers." *Polymer Communications* (The People's Republic of China) 3:220–222.

Kenjo,T.

1982 "The information about Chinese lacquer." *Bunkacho Geppo* (Japan) 162:16–17.

Kumanotani, J.

1983 "Japanese lacquer-a super durable coating. "*Polymer Applications of Renewable-Resource Materials* C.E. Carraher and L.H. Sperling (eds). 225–248.

Local Product Bureau of Supply and Marketing Cooperation of The People's Republic of China

1980 *Lacquer Tree and Lacquer*. Agricultural Press 17–38.

Majima, R.

1922 "Stellung der Doppelbindungen in der Seitenkette des Urushiols und Beweisfiihrung, dass das Urushiol eine Mischung ist." *Chemische Berichte* 55B:172–174.

Sunthankar, S.V. and C.R. Dawson

1954 "The structural identification of the olefinic components of Japanese lac urushiol." *Journal of the American Chemical Society* 76:5070–5074.

Training Course on Painting Technique of The People's Republic of China

1970 *Paint Technique (I)*. Petrochemical Press:203–211.

Tyman, J.H.P. and A.J. Matshews

1982 "Compositional studies on Japanese lac (Rhus vernicifera) by chromatography and mass spectrometry." *Journal of Chromatography* 235:149–164.

Wen, Y.Y., C.X. Wang, and C.X. Hu

1984 "Studies on the allergic principle of lacquer," *ZhiWu Xuebao* (The People's Republic of China) 26:523–526.

Yamauchi, Y., R. Oshima, and J. Kumanotani

1980 "Separation of Japanese lac urushiol diacetate on silver nitrate-coated silica gel columns by high-performance liquid chromatography." *Journal of Chromatography* 198:49–56.

1982 "Separation and identification of components of dimethylurushiol by means of reductive ozonolysis and high-performance liquid chromatography." *Journal of Chromatography* 243:71–84.

Yu, Z.Y., Z.F. Huang, M.Y. Zhang, and X.X. Zhou

1982 "Studies of saturated urushiol crown ether type PVC membrane potassium electrode." *Acta Chimica Sinica* (The People's Republic of China) 40:1076-1080, 40(11):1076–1080.

X-ray Radiography in the Study of Oriental Lacquerware Substructures

Billie Milam and Hélène Gillette[1]

The J. Paul Getty Museum and the Los Angeles County Museum of Art

1. Hélène Gillette was an intern from the State University College at Buffalo, New York, working in 1985-86 under Ms. Milam (1975-85 Los Angeles County Museum of Art, 1986 to present J.Paul Getty Museum) on this project at both the Los Angeles County Museum of Art and the Getty, but Ms. Gillette was not part of theUrushi Study Group.

Interest in the substructural characteristics of urushi developed during the 1985 meeting of the Urushi Study Group in Japan. There is little or no information on the subject of urushi substructures published in English. During visits to the Shosoin in Nara, the Tokyo National Museum, and the Tokyo National Research Institute of Cultural Properties, the Study Group participants examined x-ray radiographs of urushi objects and discussed the substructural characteristics of the objects, but arrived at no definitive conclusions.

In 1983 the Los Angeles County Museum of Art (LACMA) Conservation Center began a comprehensive interdepartmental project to study and examine the pieces in LACMA's extensive urushi collection. We recorded our visual and microscopic findings with photographs and written documentation. X-ray radiography was used to examine the substructures of the LACMA pieces and pieces from the Detroit Institute of Arts (DIA); the Freer Gallery of Art, Smithsonian Institution, Washington, D.C. (FGA); and the Asian Art Museum of San Francisco, Avery Brundage Collection (AAM). We are especially indebted to the staff of these institutions for sharing their information with us.

Oriental lacquerwares are made from the sap of the tree, *Rhus verniciflua*, refined and colored with various inorganic and some organic pigments. This refined sap or lacquer is applied in thin coats on a substructure. There are many core materials used for the substructure: wood, leather, cloth, paper, shell, ceramic, horn, fish skin, etc. There is very little published about these substructures, how they are manufactured, and what materials are used.

X-ray radiography is a tool that facilitates a visual reading of substructural construction as well as later additions, alterations, and repairs. Specific construction methods are repeatedly seen in the radiographs, but we do not yet have enough information to determine if these patterns are representative of a particular period or culture. In this report, we have catalogued the characteristic details seen in over one hundred radiographs of Oriental lacquerwares with wood substructures, as follows:

1. The entire substructure can be fabricated from a single piece of wood (LACMA M.53.12). A variation of this would be a single piece of wood forming the central section with added handles, feet, sides, and/or foot ring (LACMA L.78.25.6).

2. The substructure of the central base can be fabricated from the following types of jointed wood sections (Shosoin 1975:154): (a) parallel sections of wood with

with rub joints (Fig. 1a),[2] (b) parallel sections of wood with rub joints and a single perpendicular butt joint (Fig. 1b), (c) parallel sections of wood with rub joints and two perpendicular butt joints (Fig. 1c).

2. A rub or rubbed joint is used to join two pieces of wood edge to edge. The edges are planed true, coated with adhesive and rubbed together to make a firm joint.

These types of construction for the central base can be mechanically joined by dowels, cleats, split bamboo canes, fabric saturated with urushi, and/or with urushi used as an adhesive.

The single butt style of joinery may have been used to control distortion in Oriental lacquerware due to changes in relative humidity (RH) and/or poorly seasoned wood. Strengthening one side with a single perpendicular butt joint (LACMA M.78.121.1) would not necessarily control distortion; however, this would be achieved by introducing a second perpendicular butt joint on each side (AAM BL.77.M.19a.b and LACMA M.78.71.la.b).

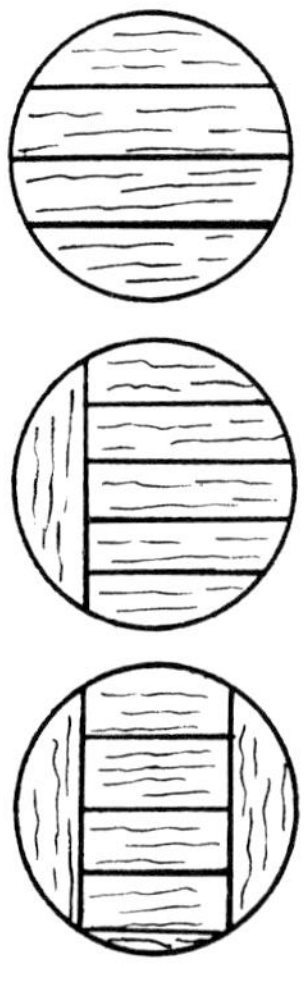

Figure 1 a-c. Construction of central base.

Visual and x-ray radiographic examination were used in determining the above variations in style of joinery. Visual examination generally shows that the urushi coating buckles or cracks perpendicular to the grain of the wood substructure. Major cracks are observed to be parallel to the rub joints between the wood sections and splits in the wood.

3. Coiled wood strips can be used to form the flared sides of Oriental lacquerware (Fig. 2). The sides are formed by a series of bent wood or bamboo strips laminated or tied together. Once the wood strips are tightly fitted together, they are carved or sanded to form the sides. This makes it easy to construct round, flaring forms and also helps to counteract warpage (Shosoin 1975:148-154; AAM BL.77.M.19a.b and LACMA L.78.25.6).

4. Basket weave of reeds, grasses, or bamboo can be used to form the sides and sometimes the entire substructure of an object. This is a variation on the wood substructure. The basket weave was also used as a design element or texture on Oriental lacquerwares. Judging from the objects we have observed, this is not as durable as a wood substructure (LACMA M.78.121.9).

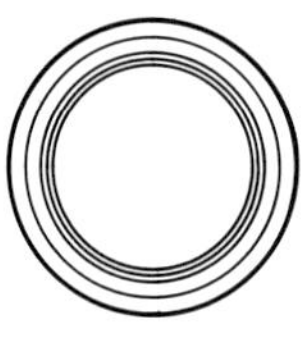

Figure 2. Coiled wood strips.

5. Reinforcements such as wood dowels, cleats, and split bamboo canes were used for joining and strengthening the wood substructure of the base in urushi pieces. Wood dowels and cleats are seen in some radiographs. Split bamboo canes are used as exterior supports for the wood substructures. They are adhered with urushi and/or fabric (LACMA M.78.121.9).

6. Score lines that appear to be associated with the wood substructure are radiographically opaque. There is no visible evidence of these lines on the urushi surface (LACMA L.78.19.10 and L.78.25.6) and their purpose is not known. Woodworkers score wood either to align sections during gluing or to position a pattern for tracing a design. However, many of the radiographs of Oriental lacquerware show score lines totally out of alignment or following no visible pattern (FGA 65.25).

7. The centering point was made by a sharp instrument or resulted from turning the bowl on a lathe. The centering point usually appears opaque in the radiograph. This opacity could be due to a buildup of *sabi* (urushi mixed with pulverized ceramic, clay, or rice flour) in the indentation (LACMA M.78.121.1). Occasionally the centering point appears as a dark spot in the radiograph. This could be due to the type of *sabi* material used or even to the lack of a *sabi* layer.

8. Various metal strips, sheets, or wires were used to reinforce the rim, foot, or edges of urushi (AAM BL.77.M.40a.b). Wire was also used to tie together the coiled wood strips and bundles of reed, grass, or bamboo in basket weave pieces (LACMA M.78.121.9, M.79.182.6, and M.80.96.2).

9. Grouting, probably *sabi*, was used as a filling material in the wood substructure (AAM BL.77.M.19a.b).

10. Fabric, probably a bast fiber textile, was used in the dry lacquer technique as a substructure (FGA 68.67). In wood substructures, fabric was sometimes

used to hold the wood sections together and to reinforce the rim and the foot ring. In addition, fabric was used to give tooth or texture to the subsequent urushi layers (LACMA M.78.25.7 and M.80.96.2; AAM BL.77.M.19a.b).

11. Restorations in urushi are visible as dark areas or occasionally as light areas in the radiographs. These areas differ from the original urushi because of the different pigments and resins used in the repair (LACMA M.78.121.1).

X-ray radiography was used to examine the substructures of lacquerwares with wooden substructures. Radiography produces a shadowgraphic image created by the interaction of x-rays with an object. X-rays can be absorbed, transmitted, or scattered, depending on the materials present in the object. The equipment used at LACMA was a General Electric M01/CE5 x-ray tube that emits fairly low energy x-rays. The radiographs were done by direct exposure onto Kodak Industrex M film, a very sensitive, high contrast film. Extensive testing was done to determine the kilovoltage (kV) and milliamperage (mA) required to show various structural details.

During our study, we found that Oriental lacquerwares pose some unique problems in achieving sharp, high contrast x-ray radiographic images. Organic materials commonly used in Oriental lacquerware, such as urushi itself, wood, carbon black, and fabric are very x-ray transparent. These wares require low kVs of 20–35 and mAs around 5, with short time-exposures of about one minute. However, most red urushi and some multicolored urushi have very x-ray opaque pigments incorporated in the urushi. These pigments include litharge (lead oxide), orpiment (yellow arsenous sulfide), realgar (orange arsenic sulfide), ochers (iron compounds), and vermilion (mercuric sulfide; e.g. LACMA M.53.12). Lacquerwares containing these x-ray opaque pigments require higher kVs around 40–45 and longer time-exposures of 2–5 minutes, with the mA remaining at 5 (5 is the highest mA setting on the LACMA Radiation Source or x-ray tube). Note that lead screens are not used with settings below 200 kV.

Carved Oriental lacquerwares present other problems. They can be either red, solid black, or alternating layers of red and black, or three or more colored layers of urushi carved to varying depths. These urushi pieces are very high contrast subjects, with x-ray transparent and x-ray opaque areas. Since each of these areas would require different energy levels or the use of filters to achieve detail in all areas, multiple x-ray radiographs or multiple x-ray exposures on a single film sheet can give satisfactory results.[3]

Once the quality of contrast and density in the x-ray radiographic image is clear and legible, the various components of the lacquerware substructure can be observed. The dark areas in the radiograph are formed by x-ray transparent materials in the urushi lacquerware. The lighter areas in the radiograph are formed by the more x-ray opaque materials.

Reading of the radiograph can be complicated by the fact that light or opaque areas can sometimes be the result of overlapping or differing thicknesses of components within the object. The shadowgraphic depiction of carved urushi is a good example (LACMA L.78.19.10 and M.80.96.1; FGA 65.25). X-ray opaque images such as score lines, centering points, grouting, and worm holes can be caused by a build-up of fill material (*sabi*). Because of the pulverized ceramic contained in the *sabi*, it is more x-ray opaque than wood or textile.[4]

3. "To insure proper density for multiple kV exposures, the total amount of exposure (mAs) required for the most satisfactory single kV radiograph is proportionally divided among the different kVs according to the empirically derived formula $mAs=K/(kV)^2$." (Kushel 1984:5): see also Rawlins 1937:38-39.

4. The x-ray radiograph when viewed in stereo may reveal on which side of the urushi substructure the centering point and/or score lines are inscribed (see for example, FGA 67.14, Fig.6).

Radiographs reveal areas of additions, alterations and repairs as they generally have different densities due to the different resins and/or pigments used in these restorations. Also, they will sometimes reveal inscriptions or design motifs that are covered by later layers of urushi, if x-ray opaque pigments were used in these motifs.

Metal reinforcements can also be distinguished in an x-ray radiograph. Metals are among the most x-ray opaque materials, therefore they appear as light areas in the radiographic image.

The following lacquerware x-ray radiographs were chosen to illustrate structural characteristics. It should be noted that in these examples the substructural characteristics are from various periods and cultures.

Figure 3. Wine cup with winged handles, Chinese, Eastern Zhou dynasty; height 3.2 cm, width 16.5 cm, depth 15 cm; Los Angeles County Museum of Art M.83.147, gift of Donald and Iris Blackmore. Cups of this type are found in tombs excavated around Changsha, Hunan province.

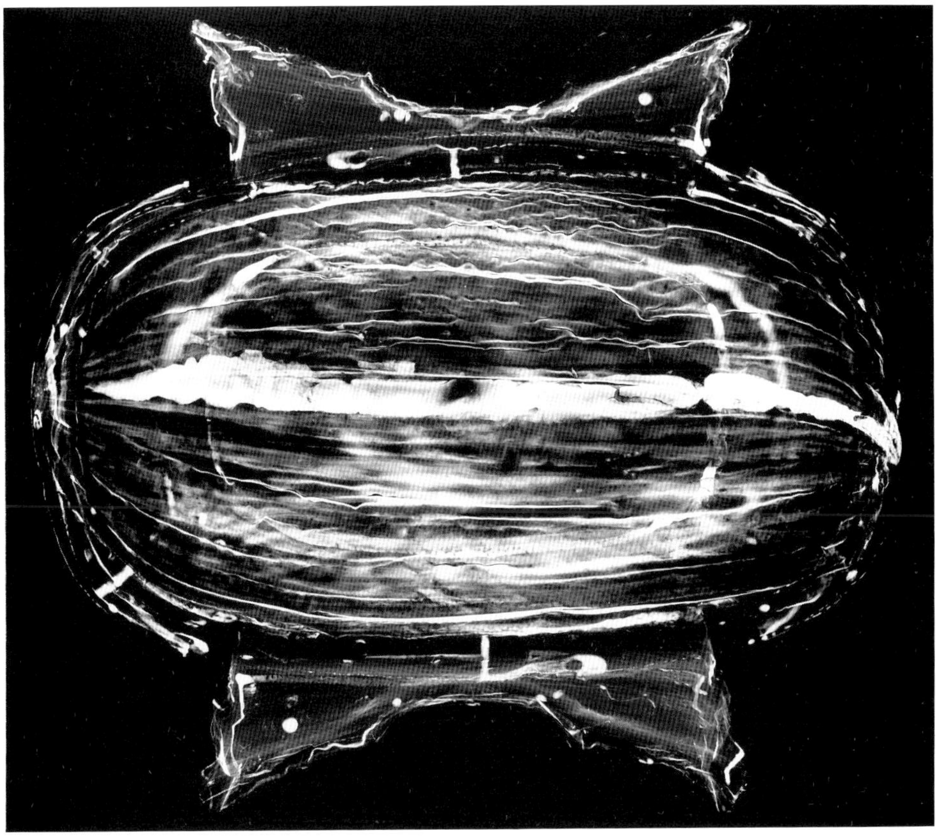

The black urushi was applied over a wood substructure, then decorated with designs in red urushi along the mouth, rim, and handles. Rapid change in relative humidity after excavation deformed the handles and caused the wood substructure to shrink. Many surviving examples of these cups are known to have been excavated from a waterlogged environment. When the objects are moved to an environment with lower relative humidity, the wood undergoes cellular collapse and shrinks across the grain.

Technical data: 35 kV, 5 mA, 3 minutes, distance from tube 90 cm. The light area in the center is a repair. The painted geometric design in cinnabar urushi is also light in the radiograph.

Another example of a similar substructure with the wood grain running side to side as opposed to lengthwise is FGA 47.24.

Figure 4. Shallow bowl, Chinese, Han dynasty (early second century B.C.); height 3.5 cm, width 21.9 cm; Los Angeles County Museum of Art M.53.12, Mr. and Mrs. Allan C. Balch Fund.

The wood substructure is fabricated from a single piece of wood and is covered with a thin base coat of dark ocher-colored urushi. Motifs of whorls, "C" shapes, and spirals in thin red urushi decorate the bowl. There is no *sabi* layer between the wood substructure and the fragile urushi layers. The urushi surface has buckled and cleaved from the substructure, particularly along the wood grain and the edges of the bowl.

Technical data: 40 kV, 5 mA, 1 minute, distance from tube 90 cm. The designs in red urushi are visible in the radiograph. Repairs are visible as light areas along the rim of the bowl.

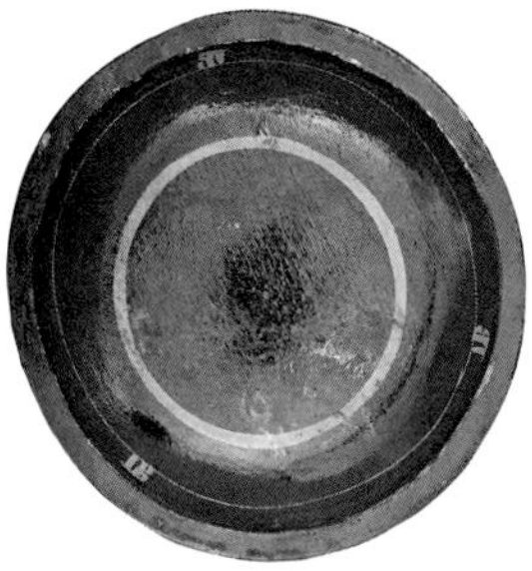

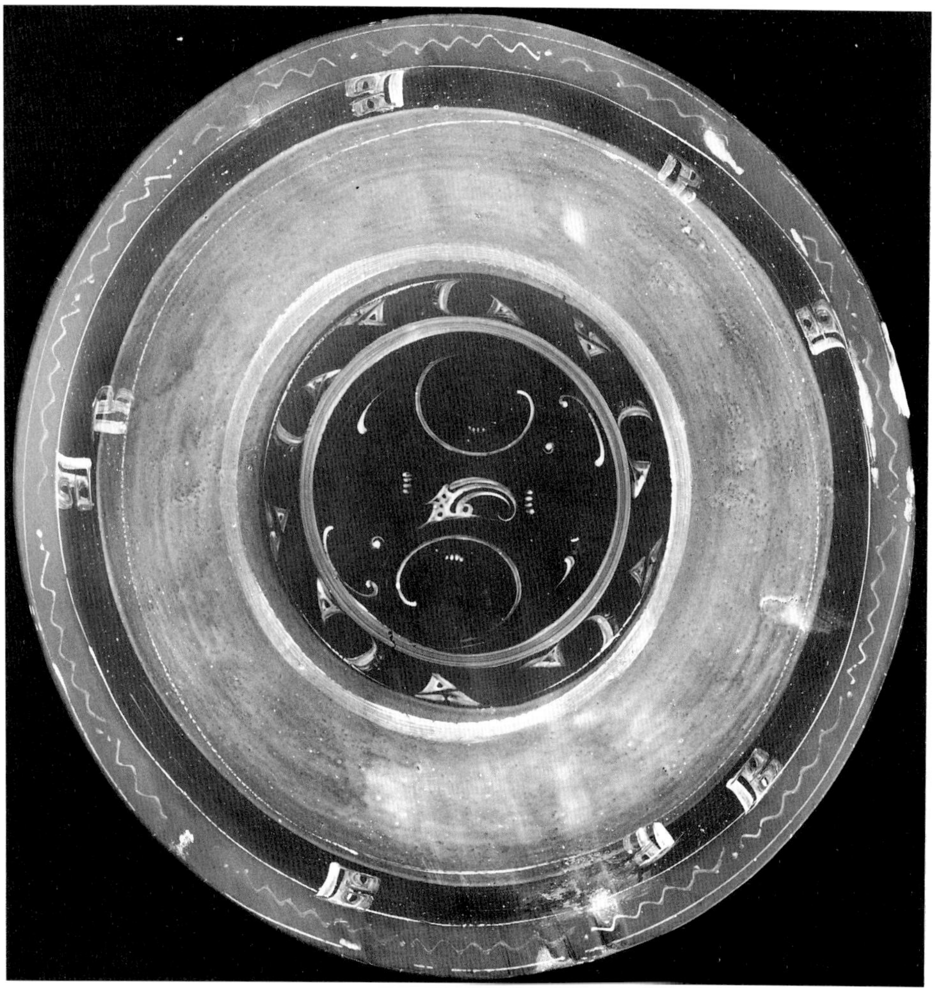

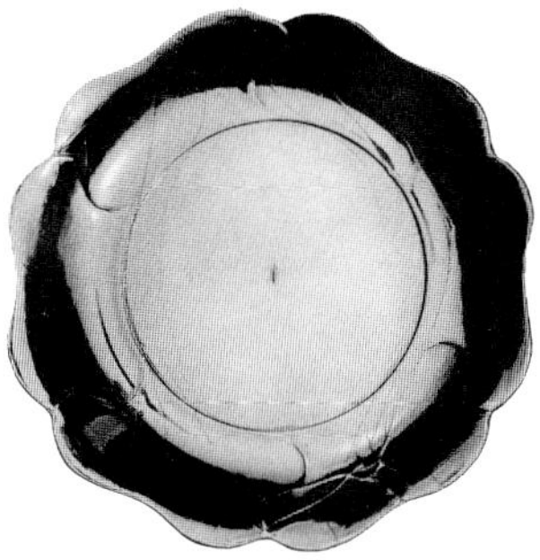

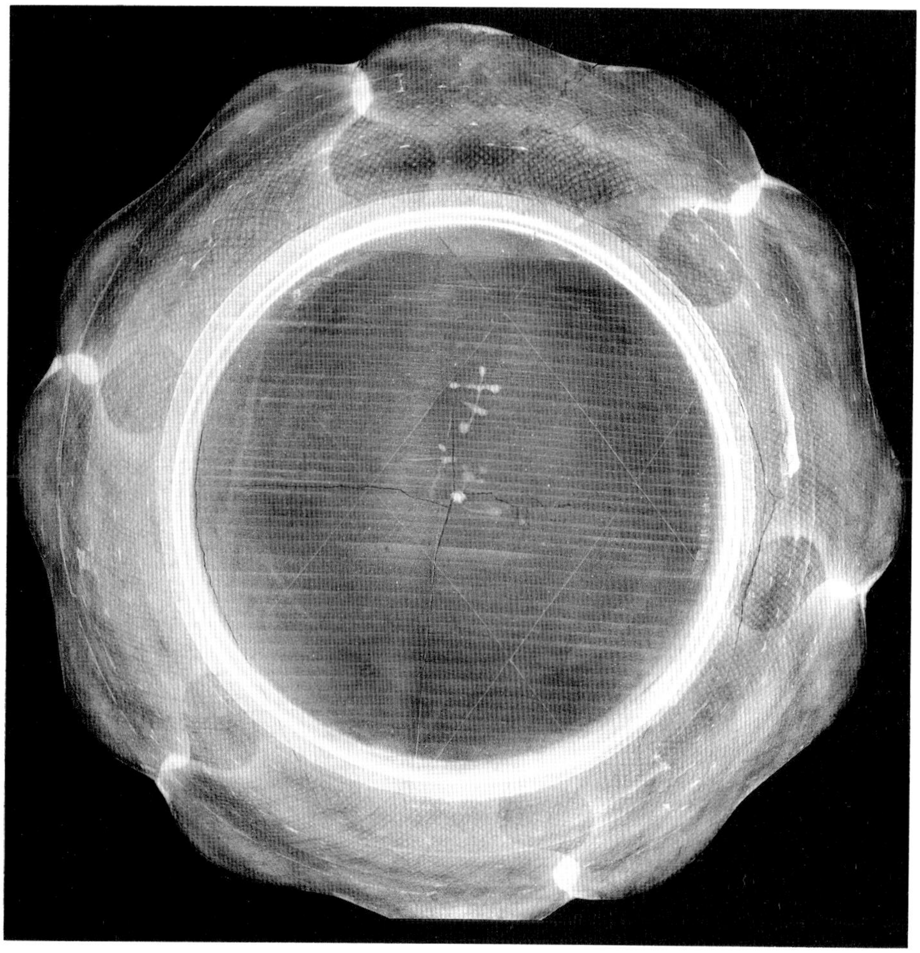

Figure 5. Six-lobed dish in the shape of a mallow flower, Chinese, Southern Song dynasty; height 3.8 cm, width 15.9 cm; Los Angeles County Museum of Art L.78.25.6, collection of Mr. and Mrs. John H. Nessley.

The six overlapping carved petals that form the side have a raised undulating edge. The wood substructure is coated with transparent and opaque brown urushi. The inscription in red urushi is painted within the circular foot ring and is visible as a light area in the radiograph. The urushi surface is damaged along the rim. There are cracks and small losses along the sides. Within the circular foot ring, the urushi surface has major cracks running in a cross pattern.

Technical data: 30 kV, 5 mA, 1 minute, distance from tube 90 cm. The central base appears to be constructed from a single wood section. The side is constructed of coiled wood strips. A coarse fabric is applied over the wood substructure. There is a centering point and an open crosshatched pattern of fine score lines. The foot ring is off center.

Other examples of Oriental laquerware with similar substructures are LACMA M.83.148.2, chrysanthemum dish, Chinese, Ming dynasty (sixteenth century); DIA 79.148, chrysanthemum dish, Chinese, Song/Yuan dynasty (thirteenth century); LACMA M.80.212, tray with multilobed rim (chrysanthemum), Chinese, Ming dynasty.

Figure 6.Circular tray, attributed to Chinese, Song dynasty (c. A.D. *960–1279); height 4.1 cm, width 31.8 cm; Los Angeles County Museum of Art M.78.121.1, gift of Mr. and Mrs. Eric Gilberg.*

The wood substructure is covered with *sabi*, fabric, and several layers of urushi. The interior of the tray is red urushi. The rim is accented with black urushi, and the underside is decorated with strips of bamboo that radiate out toward the rim. The urushi has parallel and perpendicular cracks. Separation of the wood sections caused the parallel cracks, while the swelling and contraction of the wood sections caused the perpendicular cracks. The outer rim has several radial cracks; the sides of the tray have horizontal cracks that parallel the underlying coiled wood strip construction. The urushi surface is cleaving away from the *sabi* layer along the major cracks.

Technical data: 45 kV, 5 mA, 1 minute, distance from tube 90 cm. The substructure is composed of parallel wood sections with a single perpendicular butt joint. Coiled wood strips from the sides. Faintly visible is the fabric reinforcement of the foot and the rim. There is a centering point and four faint scorelines in the central section. Repairs are visible as dark areas in the center and the rim and light areas along the side.

A similar example of this wood substructure is LACMA M.82.135, circular platter, Chinese, Yuan dynasty.

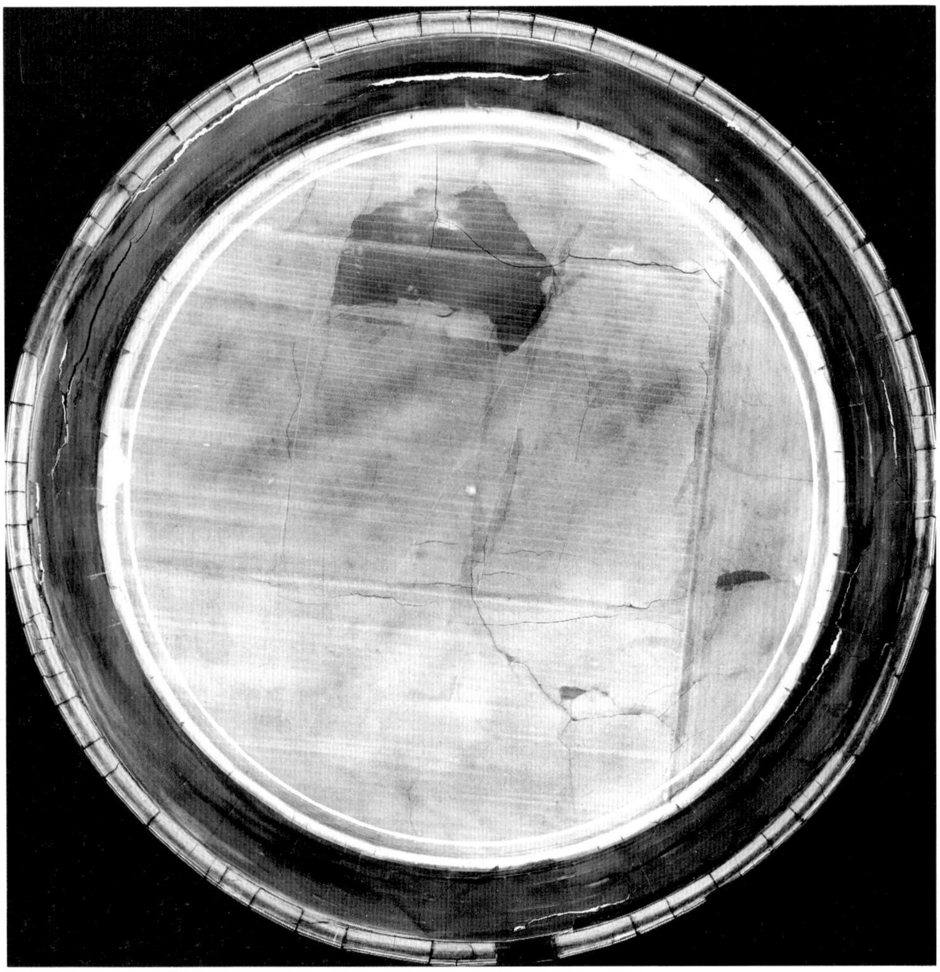

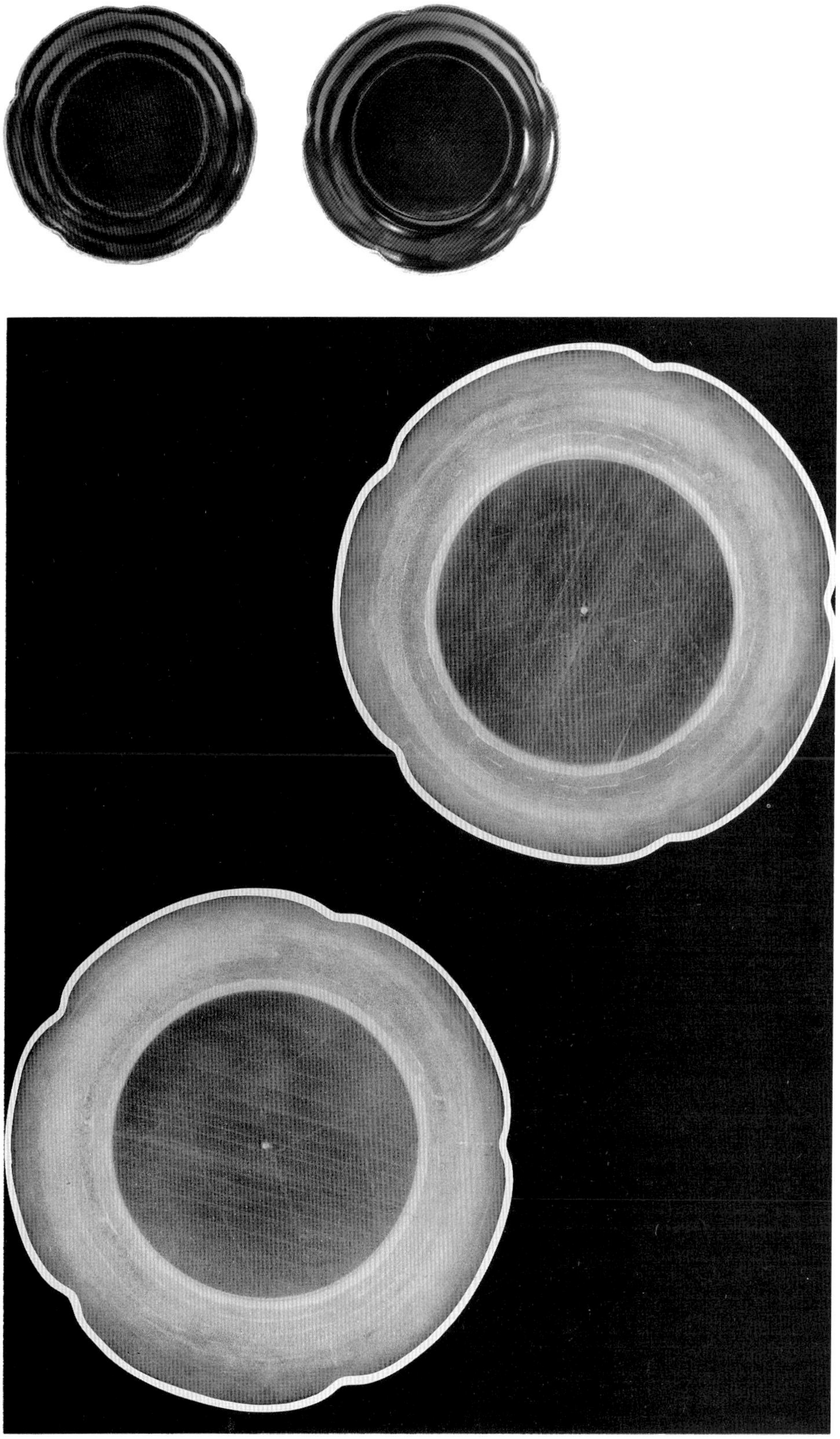

Figure 7. Pair of lobed dishes, Chinese, Song dynasty (c. A.D. 960–1279); height 1.9 cm, width 12.7 cm; Asian Art Museum of San Fransisco, Avery Brundage Collection, BL.77 M.40a and b, formerly in the collection of Jean-Pierre Dubosc, lent to the Asian Art Museum by the Christensen Fund.

Technical data: 40 kV, 7 mA, 30 seconds, distance from tube 65 cm. The central base of each dish appears to be constructed of a single wood section. The sides of both dishes are constructed of coiled wood strips. Both dishes have x-ray opaque centering points and score lines. There is a double set of score lines inscribed in the upper dish. The rim of both dishes is reinforced with a metal strip.

Other examples of metal strip used as a rim reinforcement are FGA 67.14, lobed dish, Chinese, Song dynasty, and DIA 79.145 and 79.146, two red urushi dishes in the shape of a mallow flower, Chinese, Southern Song dynasty.

Figure 8. Dish with five-lobed base, cavetto, rim, and foot, Chinese, Song dynasty; height 4.3 cm, width 19.5 cm; courtesy of the Freer Gallery of Art, Smithsonian Institution, Washington, D.C. 67.14.

Technical data: 20 kV, 3 mA, 3 minutes, distance from tube 90 cm. Five planks of wood form the central base. Score lines are present in the base in an open crosshatched pattern. Coiled wood strips form the five-lobed side. A metal strip reinforces the rim and foot ring. The centering point and the areas between the coiled wood strips contain x-ray dense *sabi* material.

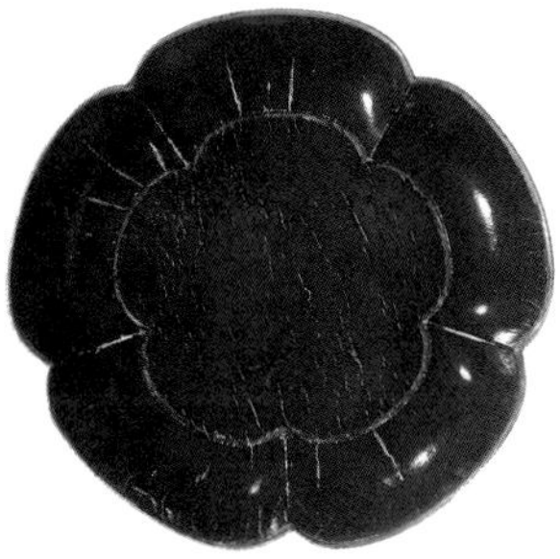

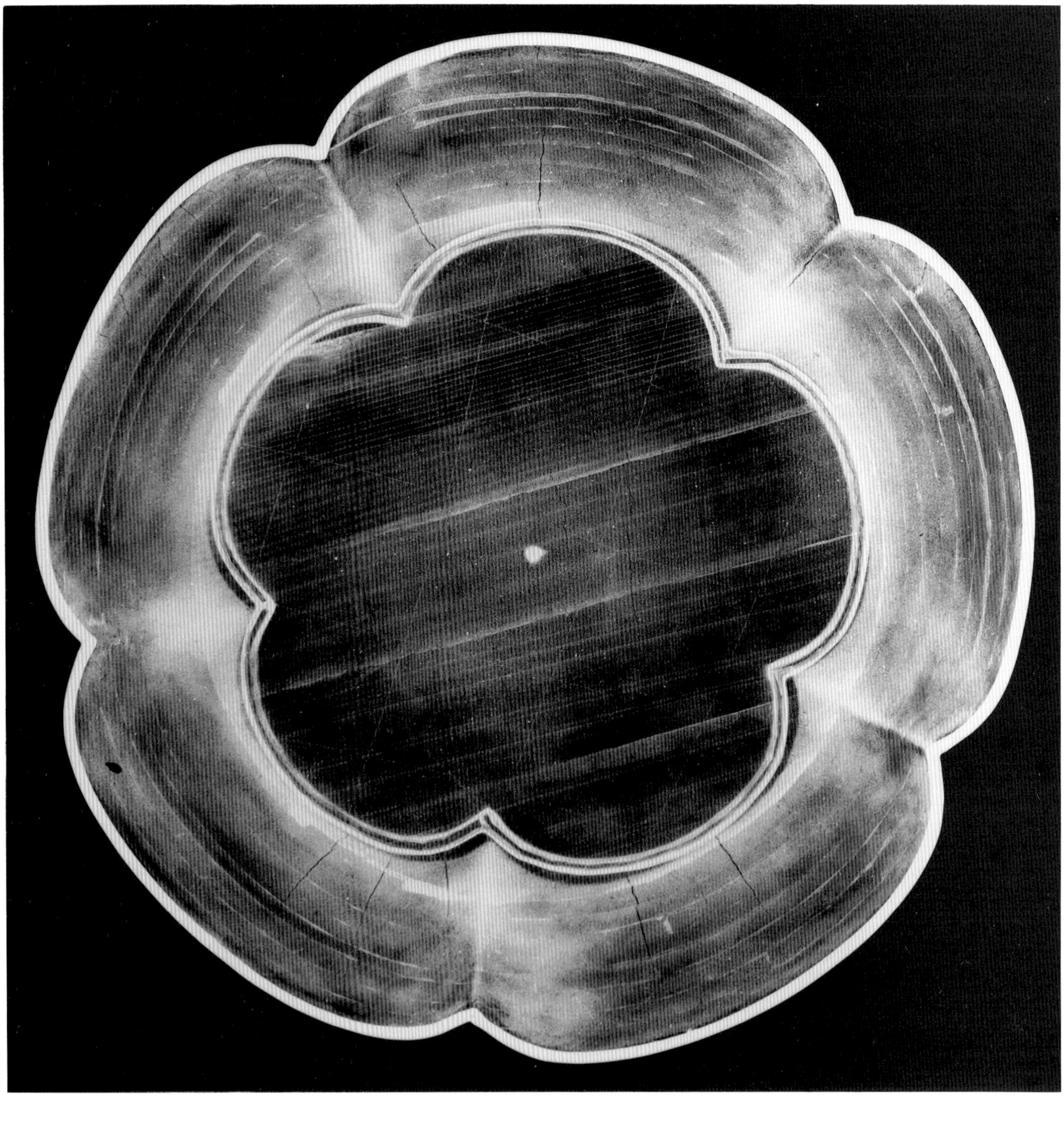

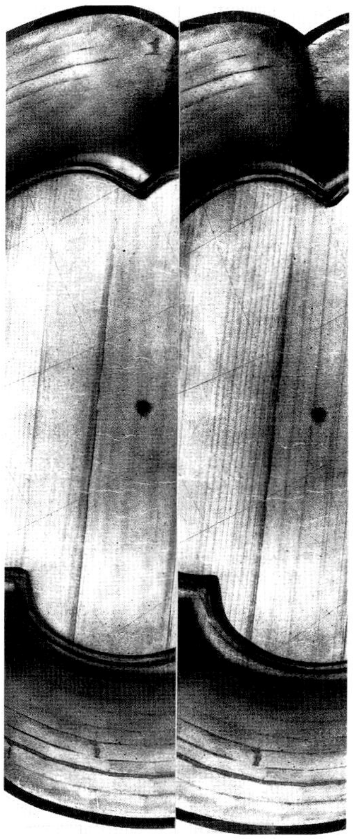

Figure 9 (see Fig. 8).
The details illustrated here are an enhanced stereo pair of x-ray radiographs of the central area of the Freer Gallery of Art lobed dish 67.14. Since the details were made from the radiographs by contact printing, the values are reversed from normal radiographs. The x-ray opaque areas such as the centering hole and score lines appear dark here. Cracks show up as white lines.

If a stereo viewer is not available, the stereo image can still be seen by placing a 75 mm x 125 mm file card or suitable piece of cardboard vertically on one's nose at a right-angle to the face; the left eye will see only the left image and the right eye will see only the right image. The muscles of the eyes must be relaxed to make the eyes converge by prolonged staring at one fixed spot (such as the centering point or a crossing of the score lines), or by raising the eyes to a distant object and then gazing back at the paper without refocusing.

When stereo fusion is achieved, a three-dimensional structure will be perceived. The score lines can be seen to lie below the cracks, in the uppermost layer of the wood. The urushi layer on the bottom of the piece (inside the foot ring) can be distinguished because it contains small, x-ray opaque dots, probably *sabi*. The centering point appears to be below the surface. Since the radiograph was taken with the bowl base down on the x-ray film, the centering point and score lines are on the underside of the plate in this example.

Technical data: 30 kV, 5 mA, 3 minutes, distance from tube 100 cm, 900 mAs; Film Kodak Type M, ready pack, normal development. Enhanced stereo pair taken with the tube 15° off vertical to the left and right.

Figure 10. Circular platter, Chinese, Yuan dynasty (fourteenth century); height 3.4 cm, width 27.6 cm; Los Angeles County Museum of Art M.82.135, gift of the Sammy Yu-Kuan family in honor of his eightieth birthday.

The black urushi is carved in a pattern of sword-pommel scrolls. A symmetrical arrangement of spirals and arcs is carved in the center of the platter. The underside of the rim is carved in a pattern of shorn sword-pommel scrolls. The base within the foot ring has been coated with layers of black urushi. Major cracks and small losses run along the crevices in the carved urushi design. The *craquelure* within the foot ring of the base is not perpendicular to the grain of the wood substructure. This condition could have resulted from a later reinforcement of the base with fabric and urushi.

Technical data: 40 kV, 5 mA, 1 1/2 minutes, distance from tube 90 cm. The base is built of wood sections that have a single perpendicular butt joint. The rim is constructed from coiled wood strips. The wood substructure is covered with fabric. There are three small restorations visible as light areas near the 10, 11 and 1 o'clock positions. There is a centering point.

Another example of this substructure is LACMA M.78.121.1, circular tray, Song dynasty (c. A.D.960–1279).

Figure 11. Tray with cranes and chrysanthemems, Yuan dynasty (fourteenth century); height 3.5 cm, width 29.2 cm; Detroit Institute of Arts 80.25, Founders Society purchase, Stoddard Fund for Asian Art.

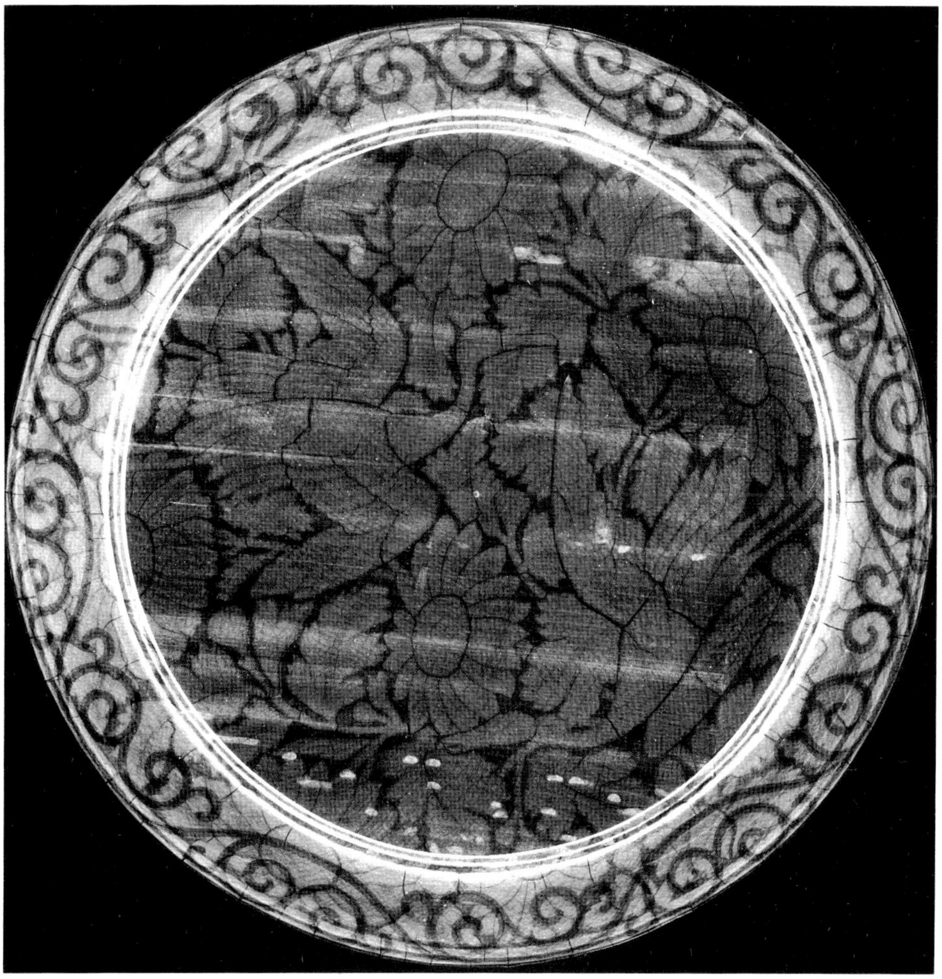

Black urushi tray carved with cranes and chrysanthemums, with a *xiangcao* border of arabesque scrolls.

Technical data: 45 kV, 5 mA, 4 minutes 10 seconds, distance from tube 90 cm. The base is constructed of parallel wood sections with a single perpendicular butt joint. The sides are constructed from coiled wood strips. The entire substructure is reinforced by coarsely woven fabric. There is a centering point and other dense splotches (possibly insect or worm holes filled with *sabi*).

Another example of a similar wood substructure is LACMA M.81.92, tray with two flying peafowl among peonies, Chinese, Yuan dynasty.

Figure 12. Dish with flattened foliated rim, Chinese, Yuan dynasty (first half of fourteenth century); height 2.8 cm, width 21.6 cm; courtesy of the Freer Gallery of Art, Smithsonian Institution, Washington, D.C., 68.67.

Technical data: 20 kV, 3 mA, 3 minutes, distance from tube 60 cm. The substructure of the dish appears to be only fabric, with no visible wood elements. The dish was manufactured in the dry lacquer technique, where fabric saturated with urushi is stretched over a form. The cracks in the urushi appear to be random.

The dry lacquer technique was used to form the sides of many urushi bowls. An example of this is LACMA M.83.148.2, chrysanthemum dish, Chinese, Ming dynasty (sixteenth century).

Figure 13. Black lobed box with painted decoration, Chinese, Yuan or early Ming dynasty (fourteenth-fifteenth century); height 7.6 cm, width 15.6 cm; Asian Art Museum of San Fransisco, Avery Brundage Collection BL 77 M.19a and b.

Technical data: 40 kV, 7 mA, 30 seconds, distance from tube 65 cm. The central sections of both the lid (Fig. 12a) and the base (Fig. 12b) are constructed from parallel wood sections with two perpendicular butt joints. There is fabric reinforcement extending over the entire wood substructure. Several bent wood strips are laminated together to form the eight-lobed sides. *Sabi* used as grouting is visible as light areas between the wood strips.

Examples of this lobed side construction are: DIA 80.49, red and black urushi dish with a design of prunus-blossom, Chinese, Yuan dynasty; LACMA M.80.154, carved red urushi five-lobed dish, Chinese, Yuan dynasty; LACMA L.82.19 collection of Mr. Irwin Jaeger, five-lobed dish, Chinese, Ming dynasty; LACMA M.78.70, lobed food-box, Chinese Qing dynasty (eighteenth century).

Figure 14. Chrysanthemum dish, Ming period (sixteenth century); height 3.2 cm, width 27.9 cm; Los Angeles County Museum of Art M.83.148.2, gift of Mr. and Mrs. John H. Nessley.

The dish is red urushi with a painted urushi vine motif over the fluted border. The back of the dish is black. A major crack in the urushi surface runs along the joint of the base and the rim. There are fine parallel cracks on the base of the dish.

Technical data: 40 kV, 5 mA, 1 minute 15 seconds, distance from tube 90 cm. The base is constructed of parallel wood sections. Fabric starts at the edge of the central wood substructure and continues to the edge of the fluted rim. There is no evidence of coiled wood strip construction. The type of damage along the rim and the lack of a coiled wood substructure suggests that it could be a dry lacquer construction. There are faint score lines through the central substructure, and a visible centering point.

Another example of this substructure is DIA 79.148, chrysanthemum dish, Chinese, Song/Yuan dynasty (thirteenth century).

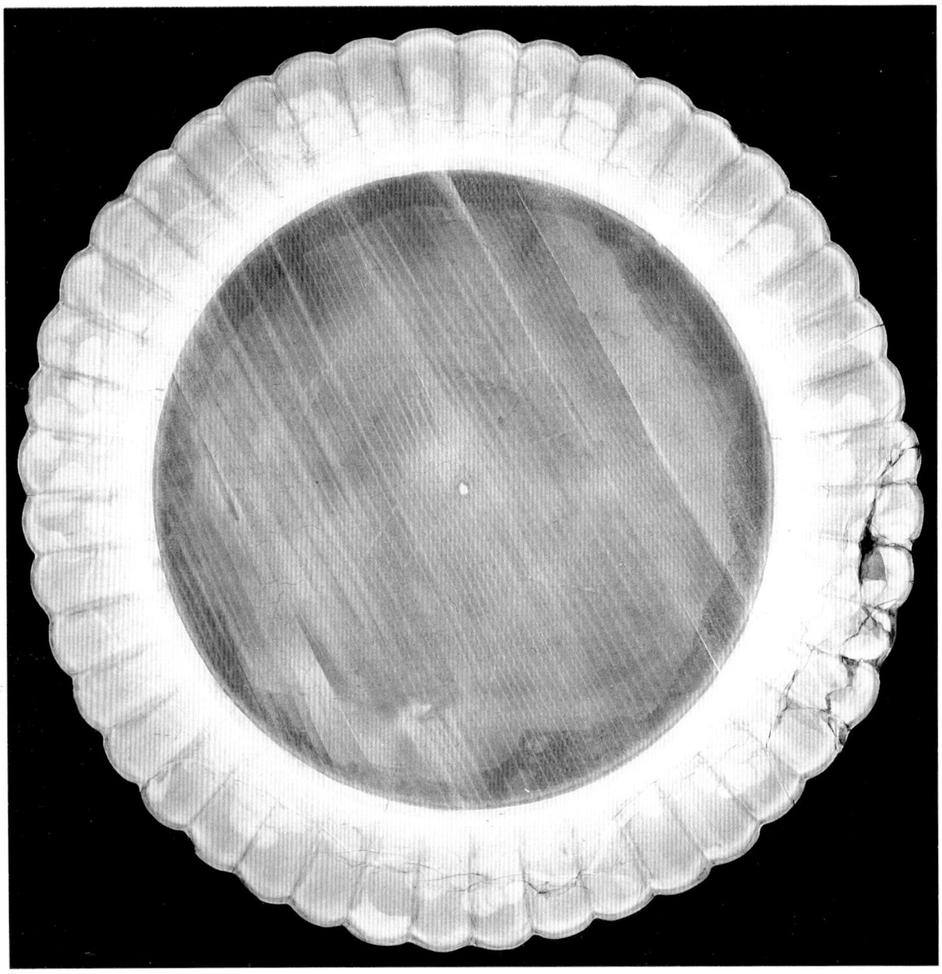

Figure 15. Octagonal box, Chinese, Ming dynasty (fifteenth century); height 12.1 cm, width 26.7 cm; Los Angeles County Museum of Art M.80.96.2 a and b, gift of Dr. and Mrs. Michael Robert Terk.

Black urushi with *raden* (mother-of-pearl inlay) design. The copper reinforcing wire is visible through the urushi. The thinly applied urushi layer has a distinct *craquelure* that is perpendicular to the grain of the wood substructure.

Technical data: 40 kV, 5 mA, 2 minutes 15 seconds, distance from tube 90 cm and 100 cm. The central area of both the lid (Fig. 14a) and the base (Fig. 14b) is constructed with parallel wood sections. Several wood strips are laminated to form the eight sides of the lid and the base. The wood substructure is covered with fabric. A two-ply twisted copper wire is used to reinforce the edges as a decorative motif.

Other examples of this side construction and use of wire reinforcement are LACMA M.82.207, octagonal multi-tiered box, Chinese, Yuan dynasty, and AAM BL 77 M.9, octagonal box, Chinese, Yuan to early Ming dynasty (fourteenth-fifteenth century).

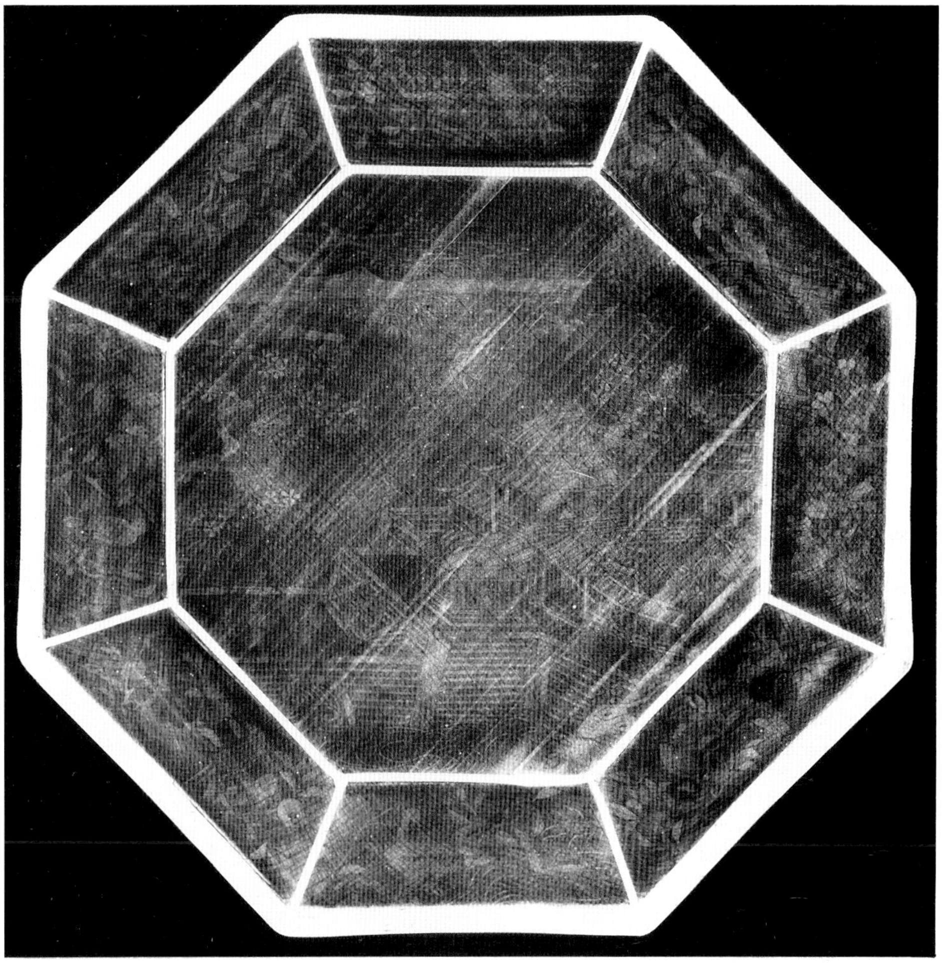

Figure 16. Tray with multi-lobed rim (chrysanthemum blossom), Ming dynasty (sixteenth century); height 6 cm, width 38.4 cm; Los Angeles County Museum of Art M.80.212, gift of Mr. and Mrs. Richard M. Baker.

The side of the tray is carved with concave grooves. The red urushi field is painted with a green and gold design of four lions around a central green disk. The surface of the tray is abraded. The urushi layers have a fine *craquelure*. There are associated losses of urushi down to the *sabi* layer.

Technical data: 40 kV, 5 mA, 1 minute, distance from tube 90 cm. The substructure of the central base has parallel wood sections, with coiled wood strips forming the side. There is a visible centering point and fabric reinforcement. The central circle is distinct and it has a different *craquelure* from the green and gold urushi.

Figure 17. Square dish with two peafowl in flight, Chinese, Ming dynasty; height 3.8 cm, width 27.9 cm; Los Angeles County Museum of Art L.78.19.10, collection of Mr. H.K. Lee.

The deep carving of the black urushi reveals a ground of ocher-colored urushi. The underside is carved in a *xiangcao* pattern of arabesque scrolls. Two Chinese characters are painted with red urushi on a highly polished black background within the foot ring. The urushi layer is in good condition with a pattern of fine cracks parallel to the underlying wood sections. There are major cracks following the carved motif on the obverse of the dish.

Technical data: 45 kV, 5 mA, 2 minutes, distance from tube 90 cm. Two exposures revealed different characteristics of fabrication used in this dish. The less exposed radiograph illustrated here shows that the central base is constructed from parallel wood sections with a single perpendicular butt joint. Also visible is an open crosshatched pattern of score lines and a faint image of fabric. The x-ray radiograph clearly illustrates the Chinese characters painted on the urushi surface. The denser radiograph more clearly illustrates the construction of the four-lobed sides, where several wood strips are laminated. Fabric is used to reinforce the sides. Also note the shadowgraphic image in the radiograph caused by the different thicknesses of the carved urushi layer.

Figure 18. Circular food-box, Chinese, Ming dynasty (1635); height 21.6 cm, diameter 41.9 cm; Los Angeles County Museum of Art M.78.121.9 a and b, gift of Mr. and Mrs. Eric Gilberg.

The lid of the box is decorated with a spray of peonies painted in red and green urushi outlined in gold on a black background. The sides are basket weave covered with transparent urushi. The urushi surface has a *craquelure* that is perpendicular to the underlying wood substructure. Major cracks in the urushi run parallel to the joints of the wood sections.

Technical data: 40 kV, 5 mA, 21/2 minutes, distance from tube 100 cm. The central areas of the lid (Fig. 17a) and the base (Fig. 17b) are constructed of two wood sections that are internally doweled, and externally bridged with split bamboo canes. Centering points are cut into the bamboo strips. Fabric covers both the bamboo strips and the wood sections. Basket weave with fabric reinforcement is attached to the circumference of the wood substructure of both the lid and the base. Pieces of wire are used to bunch wood strips in both the foot ring, the rim of the base, and the rim of the lid.

Other examples of mechanical reinforcements such as dowels, cleats, bamboo, and wire to join and strengthen the wood substructures used in basket weave pieces are: LACMA M.78.121.3, oblate box, Chinese, Southern Song or Yuan dynasty, gift of Mr. and Mrs. Eric Gilberg; LACMA M.78.121.14, circular tray, Chinese, Qing dynasty (dated 1754).

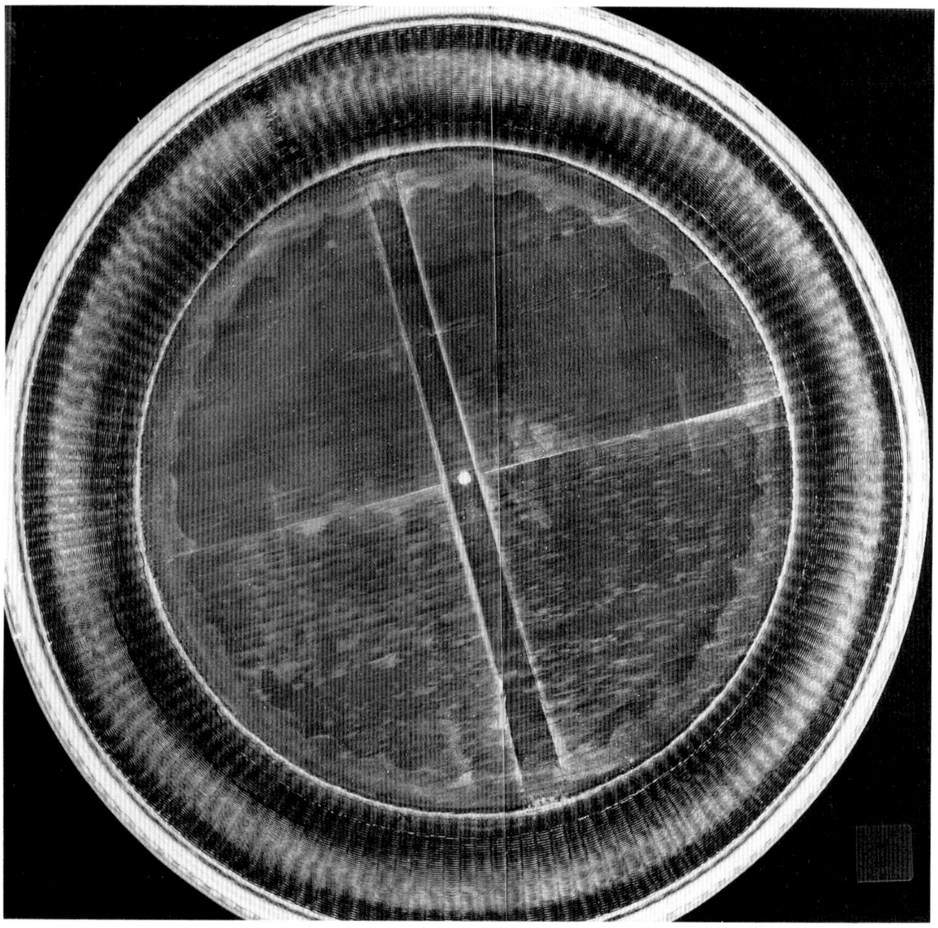

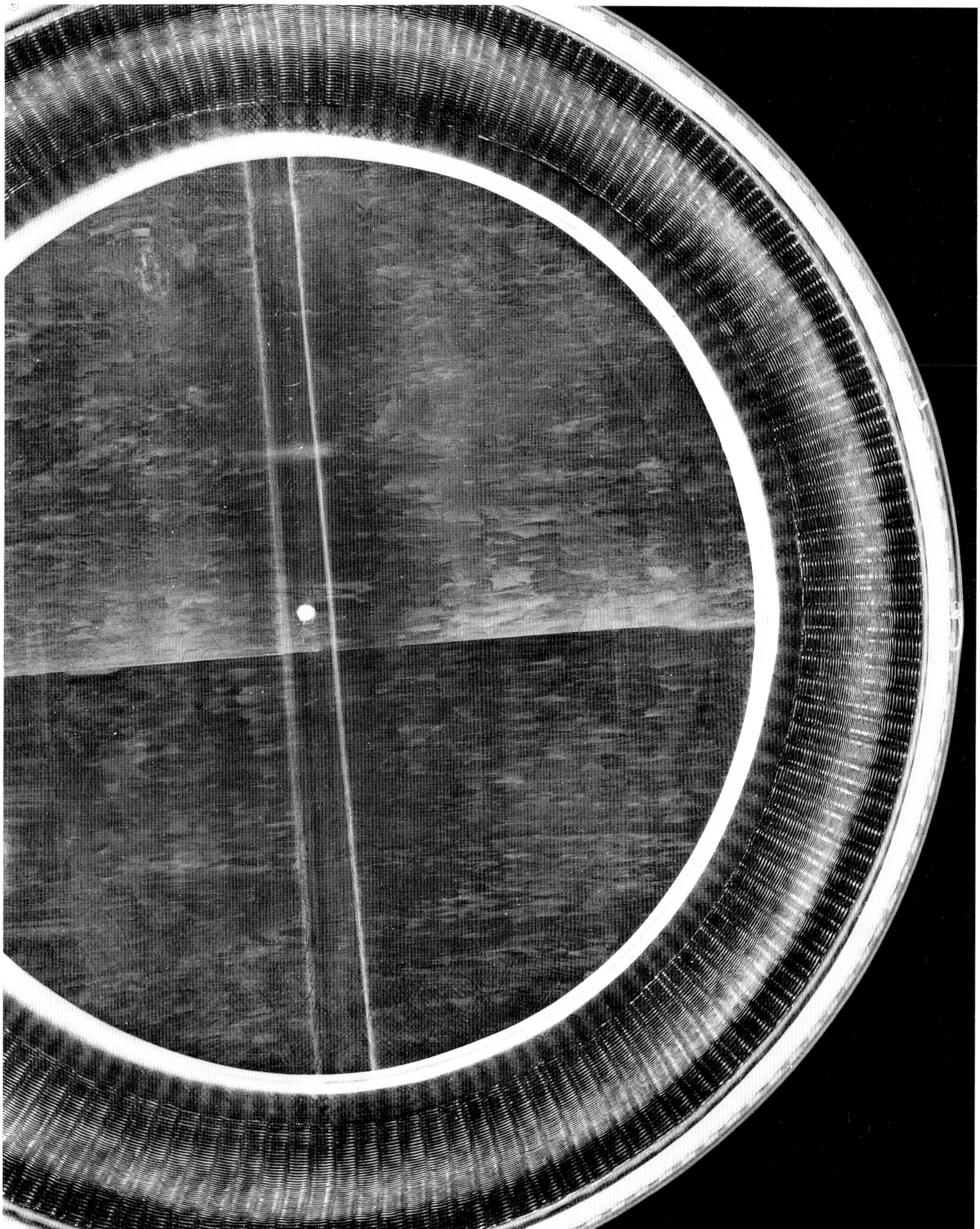

Figure 19. Round covered box, Chinese, Ming dynasty (fifteenth century); height 3.8 cm, width 11.0 cm; courtesy of the Freer Gallery of Art, Smithsonian Institution, Washington, D.C., 65.25.

Technical data: 30 kV, 5 mA, 7 minutes, distance from tube 120 cm. The central base is constructed of parallel wood sections. There are discontinuous score lines through the wood sections, which could have been scrap pieces left over from another object. The shadow image of the carved urushi illustrates the different thicknesses and densities of the urushi layers.

Another example of discontinuous score lines is AAM BL 77 M.25 a and b (Avery Brundage Collection), round box with carved floral decoration, Chinese, late Yuan dynasty (A.D. 1279–1368).

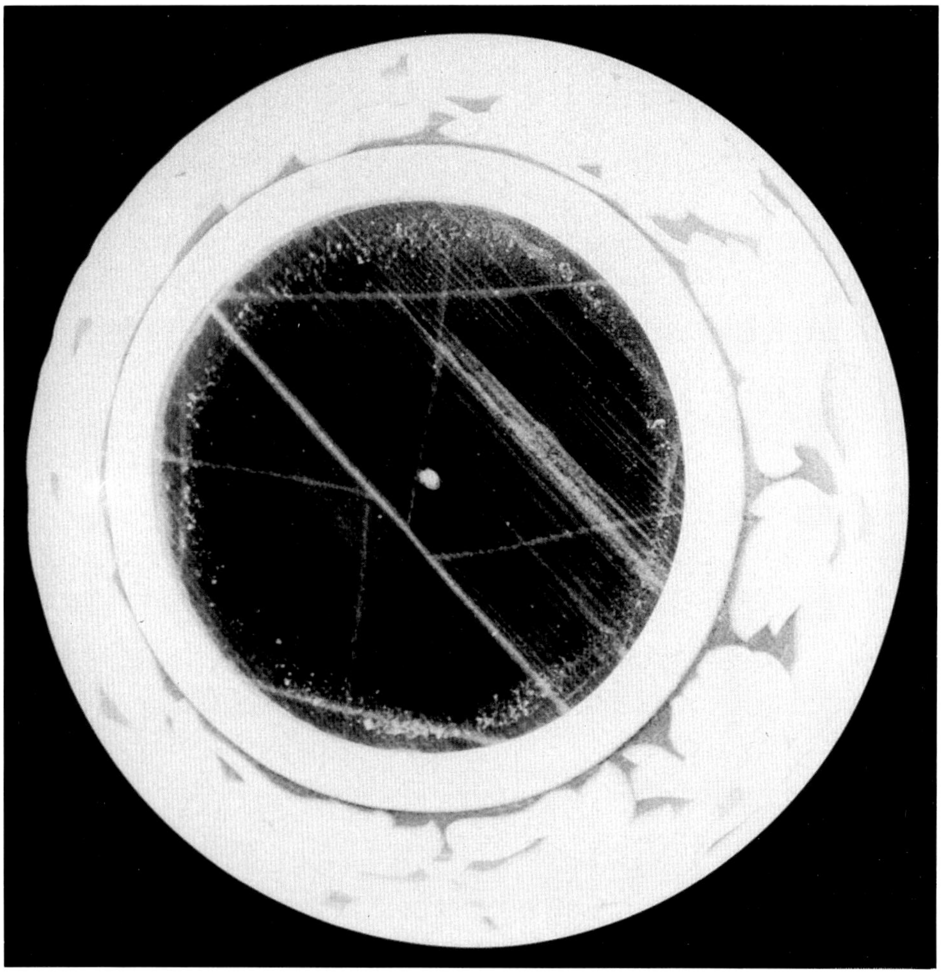

Figure 20. Circular footed tray, Ryukyu (eighteenth century); height 4.5 cm, width 22.6 cm; Los Angeles County Museum of Art M.80.109.2, gift of Mr. and Mrs. Leo Krashen.

Red urushi tray decorated with qiangjin (incised gilt lines). The thin layer of urushi is cracked along the edge of the central section.

Technical data: 40 kV, 5 mA, 1 minute, distance from tube 90 cm. The central base is constructed of three parallel wood sections, with a single perpendicular butt joint. The side is constructed from coiled wood strips. A centering point is visible. There is wire reinforcement along the join of the foot ring.

Chinese examples of similar wood substructures are: LACMA M.81.92, tray with two flying peafowl among peonies, Yuan dynasty; DIA 80.25, black urushi tray carved with cranes, chrysanthemums and a *xiangcao* pattern of arabesque scrolls, Yuan dynasty (fourteenth century).

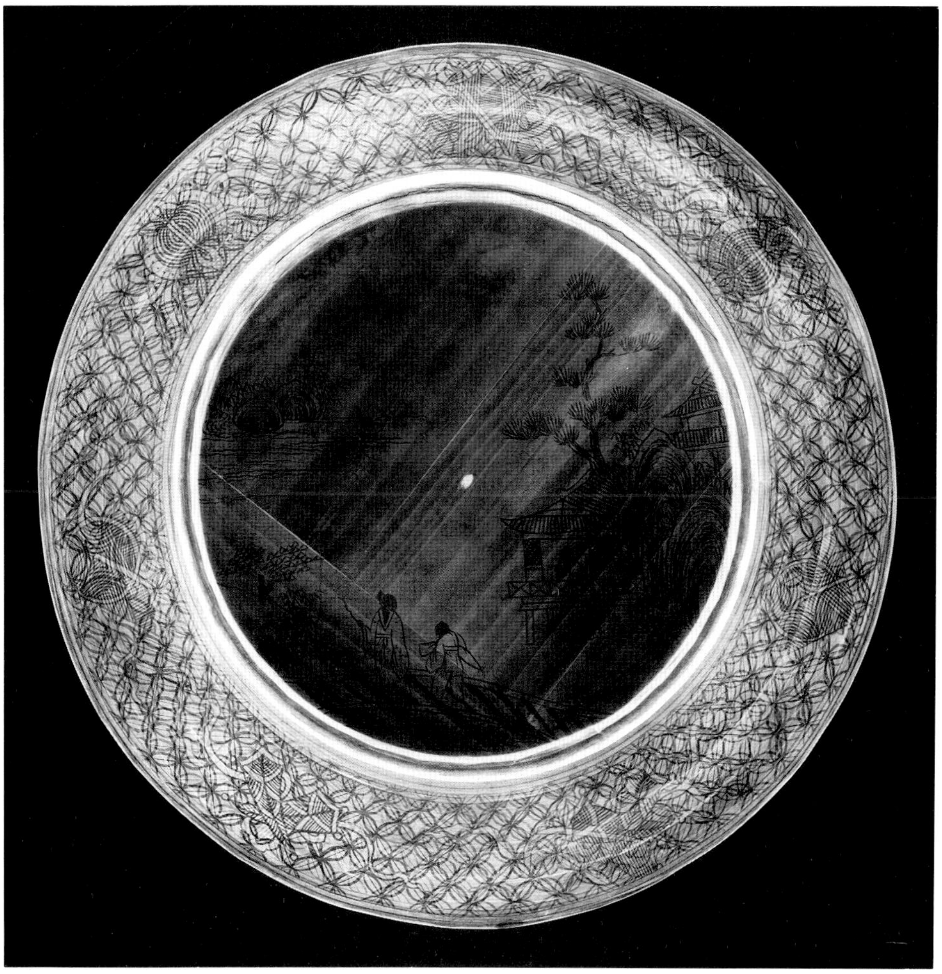

Figure 21. Circular food box with two pheasants and peonies, attributed to Japanese, Muromachi period; height 5.1 cm, diameter 19.0 cm; Los Angeles County Museum of Art M.80.96.1, gift of Dr. and Mrs. Michael Robert Terk.

The decorative motifs are deeply carved into the urushi layers.

Technical data: 45 kV, 5 mA, 1 minute, distance from tube 90 cm. The central areas of the lid and base were fabricated from parallel wood sections. Coiled wood strips form the side. A centering point is visible on both the lid and the base of the box. There are repairs on the foot ring and two floral motifs on the lid.

A Chinese example of a similar substructure is LACMA L.82.26.9 (collection of Mr H.K. Lee), carved red urushi circular box with peonies and flowers, Ming dynasty.

Figure 22. Incense box, Japanese, Edo period (seventeenth century); height 2.5 cm, width 6.7 cm; Los Angeles County Museum of Art M.79.182.6, gift of Mr. and Mrs. M.E. Wright.

The box is red and black urushi carved in both low and high relief.

Technical data: 45 kV, 5 mA, 2 minutes, distance from tube 90 cm. Both the lid and the base of the box contain a metal disk in the substructure, possibly used for weighting the object and/or as a support.

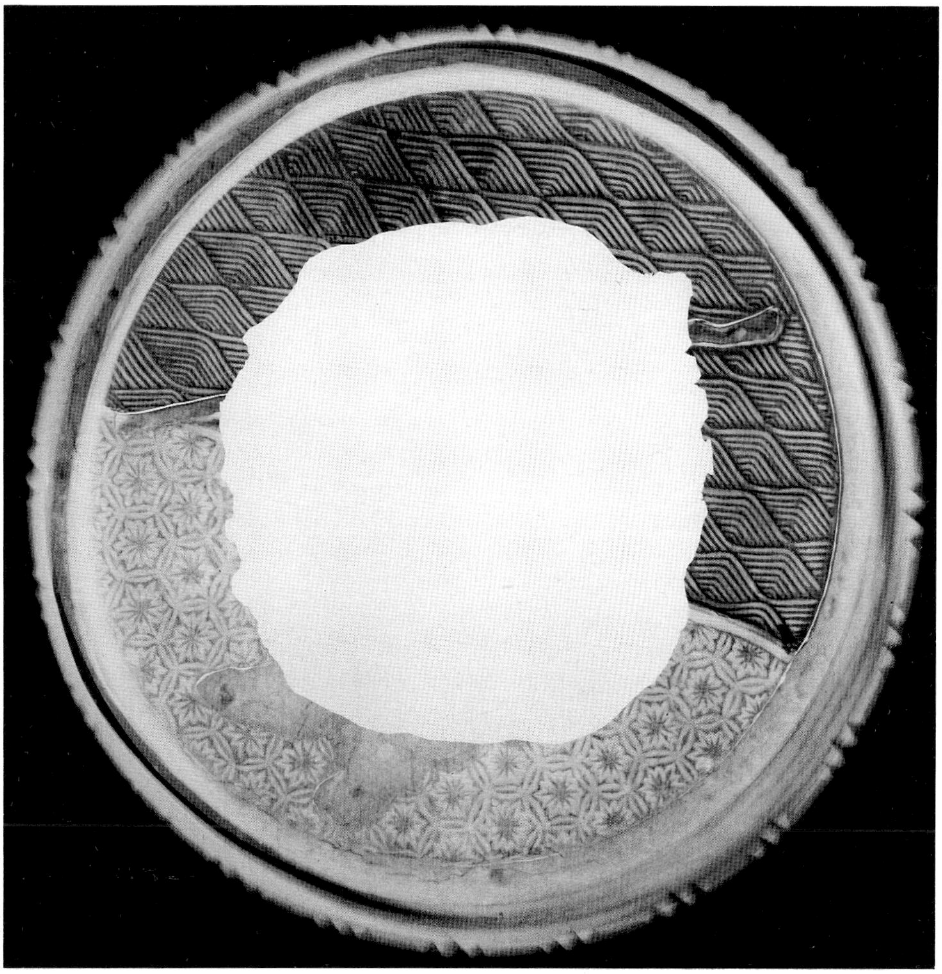

Conclusion

Compiled in this paper are the structural characteristics found in a sample of Oriental lacquerware substructures. At this point all the questions raised about urushi substructures and their technology cannot be answered, though we are able to note certain differences.

We think that the use of parallel sections of wood with a single perpendicular member, and parallel sections with two perpendicular members, was developed to prevent the object from warping over time. The coiled wood strip construction used to form the sides of urushi lacquerware may have been a means of preventing warpage or simply a method of forming a circular body without a lathe.

The variety of materials and methods used to reinforce the wood substructure are practical and in some cases decorative. Metal used to protect the rim and the foot was also a decorative element. The use of reinforcing materials such as fabric could be the result of the technological advance to thinner wood substructures, creating a need to give added strength to the delicate wood support. Fabric could also have been introduced to prime or give tooth to the wood substructure for the subsequent layers of urushi.

Several questions still need to be answered. We do not know the purpose of the score lines in the central section of the wood substructure, nor do we know how urushi technology developed, nor how the technology migrated from one culture to another. The examination of urushi substructures by means of radiography gives information that, when collated, will aid the art historian, the conservator and the conservation scientist. By compiling a large body of urushi radiographs, art historians will perhaps be able to determine the provenance of pieces of doubtful origin. Knowing both the surface and the substructural characteristics of Oriental lacquerware, improvements can be made in the conservation, storage, environment and special packing and handling requirements. X-ray radiography will prove invaluable when used in a comparative manner and in conjunction with other analytical and historical data.

Acknowledgments

The information compiled in this paper would not have been possible without the participation of the following individuals and institutions: for x-ray radiography Hélène Gillette, Don Menveg and Dr. Pieter Meyers of the Los Angeles County Museum of Art; Richard L. Barden and Alexis Pencovic of the Asian Art Museum of San Fransisco, The Avery Brundage Collection; Dr. Gary Carriveau of the National Gallery of Art, Washington, D.C.; and the Freer Gallery Technical Laboratory, courtesy of the Freer Gallery of Art, Smithsonian Institution, Washington, D.C. For photography, I would like to thank John Gebhardt of the Los Angeles County Museum of Art, and Thomas Moon of the J. Paul Getty Museum (Selections from Freer Gallery, Los Angeles County Museum of Art, and Asian Art Museum of San Fransisco); for stereo x-ray radiography, the Freer Gallery, the Smithsonian Institution, Washington, D.C.

In addition, we are indebted to the following persons for the information which came out of numerous discussions: Dr. Pieter Meyers of the Los Angeles County Museum of Art; W. Thomas Chase, Dr. John Winter, and Paul Jett of the Freer Gallery of Art, Smithsonian Institution, Washington, D.C.; Dr. Gary Carriveau of the National Gallery of Art in Washington, D.C.; Suzanne Mitchell of the Detroit Institute of Arts; and Mr Toshikatsu Nakasato of the Tokyo National Research Institute of Cultural Properties.

Bibliography

Anonymous

1896 "Lacquer—Japanese and Chinese." *Workshop Receipts* (third series). London: C.G. Warnford Lock.

1983 "New technique exposes fake lacquerware." *New Scientist*, April 21:149.

Booth, L.

n.d. "Chemical composition of urushi (Japanese lacquer)." Student Project, Winterthur University of Delaware Art Conservation Program (unpublished manuscript).

Bourne, J.

1984 *Lacquer: An International History and Collector's Guide*. Marlborough: The Crowood Press (in association with Phoebe Phillips Editions).

Burmester, Andreas

1983 "Far Eastern lacquer: Classification by pyrolysis mass spectrometry." *Archaeometry 25: 45–58*

Carriveau, G. and S. Mitchell

1984 "The scientific analysis of lacquer ware at the Detroit Institute of Arts: A preliminary report." *Shikkoshi* (History of Lacquer Art) 7:4–9.

Eastman Kodak Company

1980 *The Fundamentals of Radiography*. Rochester, New York.

1980 *Radiography in Modern Industry*. Rochester, New York.

Emoto, Y.

1966 "Study on the materials of ancient cultural properties by x-ray analysis, Part III." *Science for Conservation* 2:27–35.

1968 "X-ray analysis of materials of ancient cultures, Part I." *Science for Conservation* 4:51–60.

1969 "X-ray analysis of materials of ancient cultural properties, Part II." *Science for Conservation* 5:57–67.

1980 "X-ray analysis of materials of ancient cultural properties, Part III." *Science for Conservation* 19:35–41.

1978 **Franke, Anita**

"The use of beeswax and resin in the restoration of east Asian lacquer work." In *Conservation of Wood in Paintings and the Decorative arts*. London: IIC, p. 45–47.

Garner, Sir Henry

1979 *Chinese Lacquer.* London: Faber and Faber.

Gilardoni, A., R. Orsini, and S. Taccani

1977 *X-Rays in Art.* Mandello Lario: Gilardoni SpA.

Graham, D. and T. Eddie

1985 *X-Ray Techniques in Art Galleries and Museums*. Bristol: Adam Hilger, Ltd.

Herberts, K.

1962 *Oriental Lacquer: Art and Technique*. London: Thames and Hudson.

Kushel, D.

1984 "X-radiographic exposures: Suggestions and techniques." Buffalo: State University College of New York, Art Conservation Department (handout).

Kuwayama, George

1982 *Far Eastern Art*. Los Angeles: Los Angeles County Museum of Art.

Lee, Yu-Kuan

1972 *Oriental Lacquer Art*. New York: Wetherhill.

Lefebvre-d'Argence, Réné-Yvon

1980 "Chinese lacquerware of the late medieval period." *Apollo* 113:6–19.

Okada, Barbara Teri

1983 *A Sprinkling of Gold: The Lacquer Box Collection of Elaine Ehrenkranz*. Newark: The Newark Museum.

Office of the Shosoin

1975 *Shosoin-no Shikko* (Lacquer works from the Shosoin). Tokyo: Heibonsha Showa.

Pencovic, Alexis

1986 "Lacquer and the use of x-ray radiography in its study and conservation." In *Marvels of Medieval China: Those Lustrous Song and Yuan Lacquers*. San Francisco: Asian Art Museum of San Fransisco.

Quin, John J.

1882 *Report of Her Majesty's Acting Consul at Hakodate on the Lacquer Industry of Japan*. London: Harrison and Sons.

Rawlins, F.I.G.

1937 "The physics and chemistry of paintings." *Journal of the Royal Society of Arts* 85:(1)933–948, (2)949–968, (3)971–988.

Saratani, T.

n.d. "Some thoughts and facts about the preservation and restoration of urushi." Vienna: Österreichische Museum für Angewandte Kunst (manuscript).

von Ragué, Beatrix

1976 *A History of Japanese Lacquerwork* (Anne Wassermann, trans.).Toronto and Buffalo: University of Toronto Press.

Weintraub, S., K. Tsujimoto, and S. Walters

1978 "Urushi and conservation: The use of Japanese lacquer in the restoration of Japanese art."In *Ars Orientalis* II:39–62.

Yonemura, Ann

1979 *Japanese Lacquer* (catalogue). Washington, D.C.: Smithsonian Institution, Freer Gallery of Art.

FTIR Analysis of Authentic and Simulated Black Lacquer Finishes on Eighteenth Century Furniture

M. Derrick, C. Druzik, and F. Preusser
The Getty Conservation Institute

Background

In eighteenth century Europe, Oriental lacquer was a popular furniture finish. This popularity, combined with the scarcity and expense of lacquerware, resulted in the manufacture of furniture with imitation lacquer finishes. Some furniture pieces were made with imported urushi panels mounted on carcasses painted with imitation urushi finishes. These black or red simulated urushi finishes were typically made with natural pigmented resins that lacked the durability of genuine urushi. Today, determination of whether or not a finish is urushi is often based on historical records, appearance (*craquelure*), and solubility tests. Alternatively, instrument analysis may determine the chemical composition and identity of the finishes. In addition to answering the basic question of whether or not the material is urushi, compositional analysis can provide supplementary information for the conservation of the object and on its historical treatment, or past restorations.

Previous researchers have examined the components of urushi by several methods. Burmester (1982) used pyrolysis mass spectrometry (PMS) as a method for the identification of restored areas and forgeries and for the relative dating of East Asian lacquerware. This technique required 300 g of unrecoverable sample as well as complex data handling methods.

Infrared analysis has been used to provide a unique spectral pattern of organic and inorganic functional groups present in urushi compounds. Urushi, a complex mixture of materials, produces overlapping infrared bands which may be difficult to decipher. Kenjo (1977) and Kumanotani (1976) made studies on the changes that occur in the infrared absorption bands during the urushi drying process. From their results it is apparent that the complexity of the urushi matrix increases as the material hardens. This subsequently obscures many of the infrared absorption bands that were distinct in the liquid. However, the remaining pattern of band positions and intensities for cured and aged urushi remains unique. Using a dispersive infrared spectrophotometer, Kenjo (1978) characterized several infrared bands that are significant in the determination of the age and origin of ancient lacquerware.

FTIR

Our studies have been performed on a Fourier Transform Infrared (FTIR) spectrophotometer that has significantly greater spectral sensitivity and resolution when compared to the wavelength dispersive instruments. The design of the FTIR provides high energy throughout, fast scanning speed, and precise wavelength positioning. The computer, required for the complex mathematical calculations, provides the additional benefits of increased signal-to-noise ratio through multiple scan accumulations as well as extensive data storage capabilities. Digitally-stored spectra can be mathematically matched to a large reference library, based upon band intensity and position. These advantages of FTIR over dispersive IR allow smaller samples to be analyzed and bands otherwise difficult to detect to be elucidated.

Microanalytical techniques

To retain the integrity of an art object, compositional analysis by a nondestructive method is preferred. However, for most organic analysis methods, such as IR, a sample must be taken from the object. In these cases, it is desirable to remove as small a sample as possible, doing as little damage to the object as possible. To analyze these minute samples by FTIR, microanalytical techniques must be applied.

We have taken samples that are barely visible to the naked eye and prepared them for analysis using two microanalytical methods. These methods use beam condensers to focus a high energy infrared beam onto millimeter-size areas. The standard method of IR microanalysis uses a few micrograms of sample diluted with anhydrous KBr and pressed into a transparent pellet. The high pressure diamond cell is another micro-IR method that has been applied successfully to microgram samples from art objects by McCawley (1975) and Laver and Williams (1978). With a diamond cell, any material that flows under pressure may be placed between two diamonds and pressed into a film. This requires minimal sample preparation, and the sample is easily recoverable for further analyses. There will be no side-effects due to sample interaction with the KBr or to the hygroscopic tendencies of the KBr.

There are inherent problems when sample size is restricted and microanalytical methods are applied. Homogeneity of the material at the microscopic level is critical if the sample is to be representative of the object. At the microgram level any source of contamination, even a dust speck, can significantly bias the results. Careful sample selection, handling, and preparation procedures can reduce these problems. As the sample size increases, the relative consequence of micro-variations decreases. These considerations must be weighed against the amount of sample that may be taken from an object.

Objective

The objective in this study was to apply both the high resolution and good sensitivity of FTIR and the minimal sampling requirements of microanalytical methods to the analysis of finishes on oriental-style furniture. The primary question was whether the furniture finishes were urushi or imitation. IR analysis proved successful for the identification of these finishes.

Experimental

IR spectra were obtained at 4 cm^{-1} resolution on a Digilab 15-E FTIR spectrophotometer equipped with a Motorola 3200 computer and a dry nitrogen gas purge. A wide-range, cryogenically cooled mercury-cadmium-telluride (MCT) detector was used for the examination of the mid-IR range from 4000-500 cm^{-1} (2.5-20 μm).

KBr micropellet

Microgram quantities (20–50 μg) of sample were diluted with anhydrous KBr powder, uniformly mixed and pressed under vacuum with a Perkin Elmer die into a 1.5 mm diameter micropellet. This formed a transparent disk which was placed at the focal point of a Harrick Scientific 4x beam condenser for transmissive analysis. An accumulation of two hundred scans was Fourier transformed and ratioed with a background spectrum of nitrogen traversing the same path length. All data were represented as absorbance spectra.

Diamond cell

Microgram quantities (20–50 μg) of sample were placed between two diamonds and pressed with minimal pressure into a transmissive film. The high pressure Diamond Optics diamond cell was then placed at the focal point of a Harrick Scientific 6x beam condenser that was specially modified to accommodate the cell. The beam condenser provides increased energy throughout, which is required since the diamonds only transmit 5–6% of the total beam energy. Diamonds are considered opaque in the region from 1750–2650 cm^{-1}. Five hundred scans were co-added, Fourier transformed, and ratioed against a background spectrum obtained from the empty diamond cell. The diamonds were cleaned between samples with methanol and checked with an optical microscope for cleanliness and alignment.

Sampling

Reference samples of aged urushi were obtained from objects of known provenance from the Far East Asian collection of the Los Angeles County Museum of Art. The reference set contained six selections, two each from Japan, China, and Ryukyu, with either red or black pigmentation. The sample set was composed of furniture pieces from the Decorative Arts collection at the J. Paul Getty Museum (Table 1). Several of these pieces are illustrated in the paper by Westmoreland in this volume. Each piece was of eighteenth century European manufacture in the Oriental style with a black lacquerlike finish. Samples were chosen arbitrarily from obscure areas on the furniture, near or under gilt brass mounts. In each case, the area to be sampled was cleaned with solvent to remove any dust or superficial coating. A scalpel was used to remove small, barely visible amounts of the surface layers of the finish. Since only a few samples of minimal size were taken from each object, the assumption had to be made that the selections were representative.

Table 1. Summary of the sampling and analysis of Oriental lacquer-like finishes on furniture

Acquisition number	*Description of furniture*	*Sampling area*	*Finish analysis interpreted from infrared spectra*
55.DA.2	French commode *c.* 1750 set with panels of black Japanese lacquer	1. Right side 2. Front, right side	Urushi Urushi
65.DA.4	French commode *c.* 1737 set with panels of black Japanese lacquer, surrounding areas painted with *vernis Martin*	1. Front, left door inside panel 2. Right side, inside panel 3. Right side, under panel	Urushi Urushi Non-Urushi
84.DA.74	Chinese black lacquer plate, bowl, and lid	1. Plate—underneath near lip 2. Bowl—inside near handle 3. Bowl—outside near foot 4. Lid—outside under knob 5. Lid—inside next to rim	Urushi Urushi Urushi Urushi Urushi
83.DA.280	French cartonnier *c.* 1745 decorated with black *vernis Martin*	1. Front, left half 2. Top portion, right side in panel	Shellac Shellac
78.DA.119	Pair of corner cupboards, *c.* 1755. Veneered with panels of Japanese lacquer. Set with gilt-bronze mounts.	6. Right cupboard—left front—under mount—black 7. Right cupboard—left front—under mount—yellow varnish 8. Left cupboard—left front—under mount	Copal or sandarac Copal or sandarac Copal or sandarac
71.DA.104	French secretaire *c.* 1785, veneered with panels of Japanese lacquer	1. Front center, in panel 2. Bottom shelf, rear inside 3. Front hole under horse's leg	Shellac Shellac Urushi

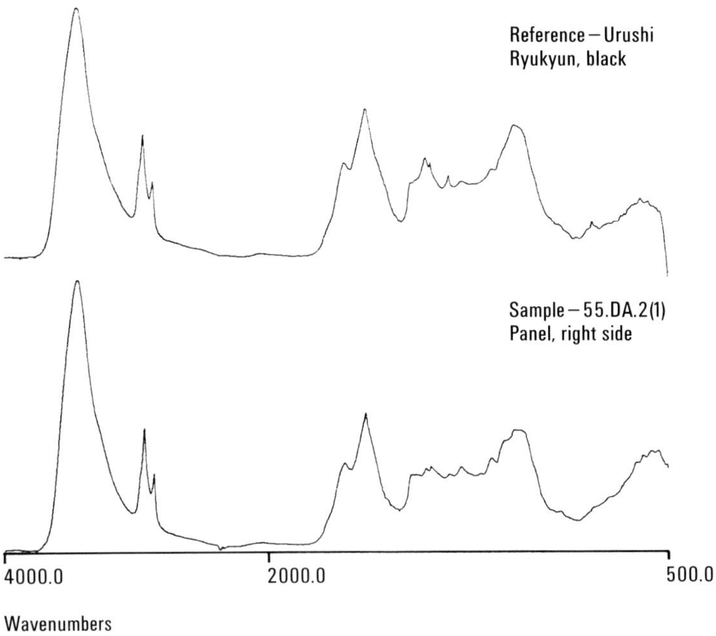

Figure 1. Infrared absorbance spectra of two urushi samples. The sample on top is from a black lacquered Ryukyun tray, and the sample on bottom is from a panel on a black lacquered commode. This visual identification was concurred by computer matching.

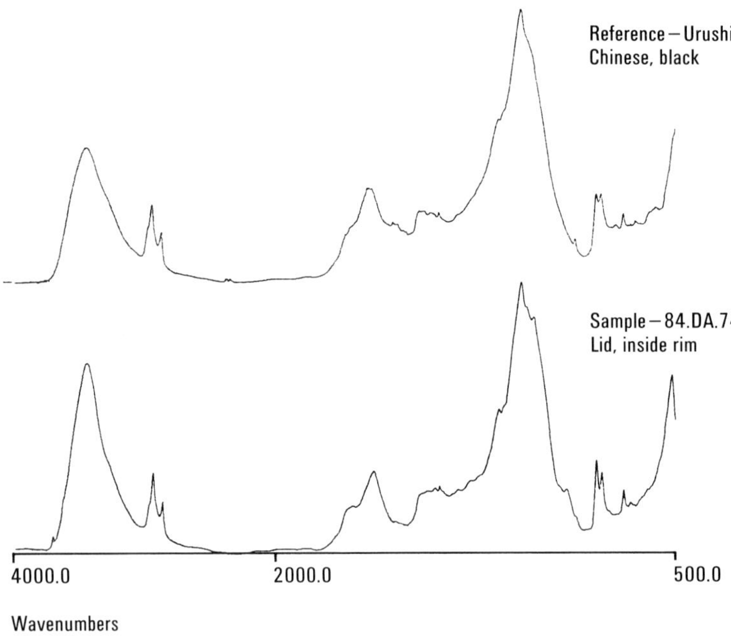

Figure 2. Infrared absorbance spectra of two urushi samples. The sample on top is from a Chinese black lacquered food bowl, and the sample on bottom is from the lid of a black lacquered bowl set thought to be of Japanese origin. This visual identification was concurred by computer matching.

Results

Reference spectra

Two distinct patterns were observed in the infrared spectra of the urushi reference sample. Spectra representative of these two variations are shown in Figures 1 and 2. The differences do not preclude the identification of the urushi complex. Bands previously identified by Kenjo (1978) as source-independent are observed in both patterns. These and additional absorbance bands consistently found in our set of reference spectra are listed in Table 2. The reference set contained urushi of differing manufacture, but the limited number of samples prohibited any conclusions regarding the source-dependent spectral variations.

The major difference in the two patterns consisted in relatively large absorbances due to siliceous material (Fig. 2). The strong band at 1070-1080 cm^{-1}, along with the absorbances at 796 and 779 cm^{-1}, are due to quartz. Additional absorbances at 1030, 3620 and 3698 cm^{-1} are due to clay. Garner (1963) indicated that the silicon portion of the lacquer in his samples was due to contamination from the ground layer. However, no ground layer was present in these samples. This seems to indicate that the silicon material may have been an additive in the manufacturing process.

Neither the red nor the black lacquer samples in the reference set produced differentiable infrared bands that could be attributed to the pigmentation. This corresponds to Garner's (1963) determination that pigments typically used to color lacquer are vermilion and carbonaceous material for red and black, respectively. These colorants do not produce significant infrared bands in the spectral region examined.

Table 2. Characteristic absorption bands observed in Oriental lacquer

Source-Independent		*Source-Dependent*
Kenjo (cm^{-1})	*This study* (cm^{-1})	*Kenjo* (cm^{-1})
3400	3450	1595–1720
2925	2926	993
2850	2885	
1430–1465	1435–1465	
1065	1315	
	1145–1165	
	1050–1090	

Sample Spectra

Interpretations of the spectral features of the samples from the oriental-style eighteenth century European furniture are given in Table 1. The characteristic band positions and intensities were taken into consideration in the comparison of the sample spectra with the reference urushi and natural resin spectra. In addition to the visual comparison, computer matching was done. A reference library of spectra, which included the six urushi samples and other natural resins, was searched for the best fit to the unknown spectra. The computer matches corresponded to the visual identifications. The matches and their hit values, based on peak position and intensity, are given in Table 3.

Urushi-based resin finishes were found on several pieces. Panels from the French commode, 55.DA.2, have an urushi finish as illustrated by the spectra in Figure 1. Samples from a plate, bowl, and lid, 84.DA.74, were verified as urushi, as illustrated by the spectra in Figure 2. Slight spectral differences found between the pieces in the set may indicate that the bowl has a different source of manufacture from the lid and plate. As seen in Table 3, the computer matched the bowl with the Japanese urushi reference, while matching the lid and plate with the Chinese urushi reference. This is a tentative assignment that may be confirmed by comparison with

Table 3. Results of comparison of finish samples with spectra library

Acquisition number	*Description of furniture*	*Sampling area*	*Hit quality index**	*Computer match reference material*
55.DA.2	French commode *c.* 1750 set with panels of black Japanese lacquer	1. Right side 2. Front, right side	0.07 0.07	Urushi, Ryukyu, black Urushi, Ryukyu, black
65.DA.4	French commode *c.* 1737 set with panels of black Japanese lacquer, surrounding areas painted with *vernis Martin*	1. Front, left door inside panel 2. Right side, inside panel 3. Right side, under panel	0.06 0.10 0.10	Urushi, Ryukyu, black Urushi, Ryukyu, black Shellac
84.DA.74	Chinese black lacquer plate, bowl, and lid	1. Plate—underneath near lip 2. Bowl—inside near handle 3. Bowl—outside near foot 4. Lid—outside under knob 5. Lid—inside next to rim	0.07 0.06 0.08 0.03 0.04	Urushi, China, black Urushi, Japan, black Urushi, Japan, black Urushi, China, black Urushi, China, black
83.DA.280	French cartonnier *c.* 1745 decorated with black *vernis Martin*	1. Front, left half 2. Top portion, right side in panel	0.10 0.07	Shellac Shellac
78.DA.119	Pair of corner cupboards, *c.* 1755. Veneered with panels of Japanese lacquer. Set with gilt-bronze mounts.	6. Right cupboard—left front—under mount—black 7. Right cupboard—left front—under mount—yellow varnish 8. Left cupboard—left front—under mount	0.13 0.07 0.10	Copal Copal Copal
71.DA.104	French secretaire *c.* 1785, veneered with panels of Japanese lacquer	1. Front center, in panel 2. Bottom shelf, rear inside 3. Front hole under horse's leg	0.09 0.08 0.06	Shellac Shellac Urushi, Japan, black

*The hit quality index is a measure of the difference between the unknown spectra and the library spectra. A HQI of 0.0 indicates that there is no difference and higher HQIs indicate an increasing amount of dissimilarity. A HQI below 0.15 indicates nearly identical spectra (Sadtler 1985).

Figure 3. Infrared absorbance spectra of two shellac varnishes. The sample on top is shellac resin, and the sample on bottom is from the surface of a black lacquer-like secretaire. Another sample from below the surface was urushi.

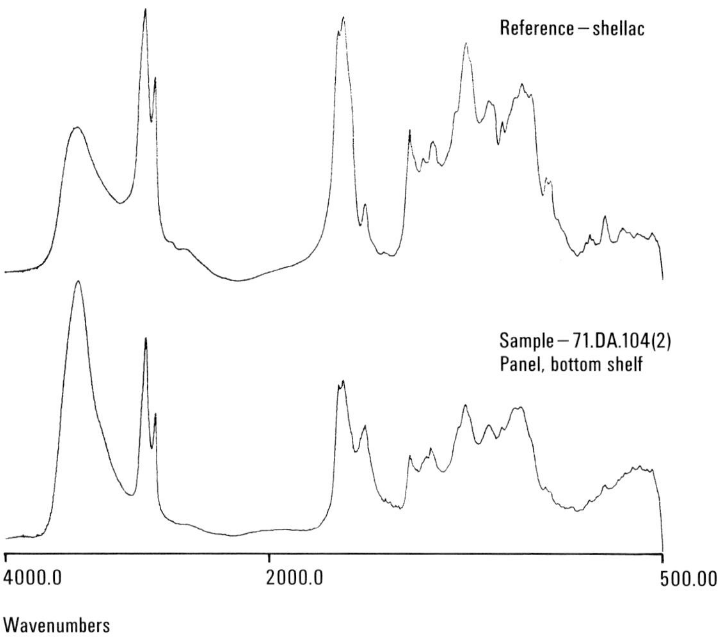

Figure 4. Two infrared absorbance spectra of the finish on the right corner cupboard, 78.DA.119. For the spectrum on top, the sample was prepared as a KBr micropellet, while the spectrum on bottom was analyzed using a diamond cell.

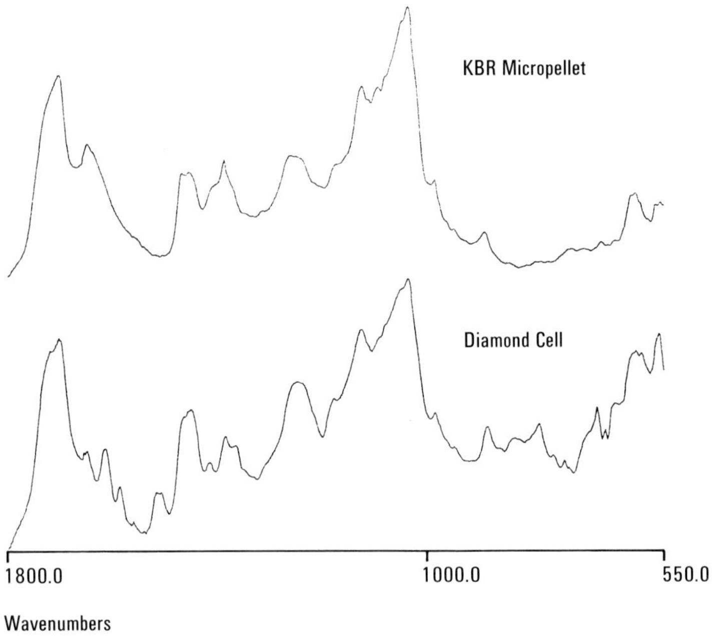

a larger reference base. Since we analyzed only two pieces for each manufacturing source (Japan, China, and Ryukyu) as references, it is not clear whether variations similar to those seen between pieces made in different countries may occur between pieces manufactured within one country.

Some European furniture pieces consisted of custom-made frames for imported Oriental lacquer panels. An example of this is the French commode labeled 65.DA.4. Samples taken from two of its panels were found to be urushi. However, the sample taken from below the panel on the right side of the commode is not urushi, indicating that the framework is painted with a pigmented varnish.

In a third group of pieces, pigmented natural resin finishes were found on the inset panels. The varnish on the French cartonnier, 83.DA.280, is shellac-based, while the varnish on the pair of cupboards, 78.DA.119, is copal or sandarac. This is consistent with the cracked appearance of the finish on these pieces.

The results for the finishes described above corresponded to the curatorial information provided. However, the finish on the French secretaire, 71.DA.104, was inconsistent with the expected results. As shown in Table 1 and Figure 3, samples from the front center section and the center panel on the bottom shelf are shellac. Since the piece was expected to be urushi, another sample was taken from a nick in the bottom shelf. This third sample was found to be urushi, which suggests that the piece was originally urushi but had been coated with shellac. One must be cautious in assuming that microgram quantities of material are representative of the entire piece when interpreting analytical results.

Microanalytical Methods

Each of the samples was analyzed by both KBr micropellets and a diamond cell. The spectra obtained from both techniques are similar (Fig. 4). Slight spectral deviations are attributable to the different methods.

KBr is hygroscopic and easily absorbs trace amounts of organic compounds from the atmosphere. When very small amounts of sample, i.e., 10 μg, are mixed with KBr, any contamination in the KBr is more apparent in proportion to the intensity of the sample absorption bands. Hence, the hydroxy bands (3450, 1630 cm^{-1}) and the C–H bonding vibrations (1400, 1384 cm^{-1}) are more prevalent in the spectra from KBr micropellets than in the corresponding spectra obtained from the diamond cell. Nonetheless, the spectra obtained by KBr micropellet are reproducible and are directly comparable to most published and digitally stored reference spectra.

The diamond cell eliminates the effects of water absorption or sample interaction due to the presence of KBr, and the sample is readily recoverable. However, there is an artificial enhancement of the absorbance bands at lower wave numbers, and diamonds do not transmit the infrared beam from 2650–1750 cm^{-1}. Because of the opaque region and this apparent amplification at low wavenumbers, agreement with reference spectra may not be obvious. Band position information is more reliable than band intensity information obtained from the diamond cell spectra. This limits the usefulness of computer matching techniques which rely on both sources for comparison. Although the two microanalytical techniques, KBr micropellet and diamond cell, are not equivalent, they are comparable and provide complementary information. Each method has its advantages and either may be used for the identification of the microgram quantities of resin analyzed in this study.

Conclusion

This study demonstrates that FTIR can be used to identify urushi versus pigmented natural resins for the characterization of oriental-style European furniture. Microgram quantities of sample, barely visible to the naked eye, were removed from each piece for spectral analysis. The samples were found to fall into three categories: urushi-based resin finish on the entire piece; urushi-based resin on inset panels only; resin other than urushi.

Microanalytical FTIR techniques were used to classify the samples in these categories, based on a reference set of samples of known origin. Attributions to country of origin were not made in this study because of the limited size of the reference set. However, when more pieces of known manufacture have been analyzed to provide a statistically sound reference base, attributions may be possible.

As in any microanalytical technique, caution must be exercised when extrapolating the results of a small sample. This consideration must be weighed against the amount of sample which it is desirable to take from a valuable object. When only limited samples may be taken, the two IR microanalytical techniques, micropellets and diamond cell, may be used for accurate characterization of the sample.

Acknowledgment

We would like to thank George Kuwayama, curator of Far Eastern Art at the Los Angeles County Museum of Art, for generously allowing us to obtain Oriental lacquer samples from artifacts of known provenance and age for our reference spectra.

Bibliography

Burmester, A.

1982 "Historical coating materials—East Asian lacquer." In *Proceedings of the 22nd Symposium on Archaeometry*. Bradford, England.

Garner, Sir Harry

1963 "Technical studies of oriental lacquer." *Studies in Conservation* 8:84–98.

Kenjo, T.

1977 "Effect of humidity on the hardening of lacquer." In *Conservation of Wood*. International Symposium on the Conservation and Restoration of Cultural Property, Tokyo, 1977. Tokyo National Research Institute of Cultural Properties, 151–163.

1978 "Studies on the analysis of lacquer, II. Infrared spectrometry of lacquer films." *Scientific Papers on Japanese Antiques and Art Crafts* 23: 32–39.

Kumanotani, J.

1976 "Chemistry of Japanese lacquer: model of coatings for the future." In *XIII FATIPEC Congress*, Cannes: France, 360–369.

Laver, M. and R.S. Williams

1978 "The use of a diamond cell microsampling device for infrared spectrophotometric analyses of art and archaeological materials." *Journal of IIC-CG* 3(2):34–39.

McCawley, J.C.

1975 "Diamond Cell Infrared Spectroscopy in the Analyses of Paints and Pigments." *ICOM Committee for Conservation 4th Triennial Meeting*, Venice, 75/4/5:1–11.

Sadtler

1985 *Search32 User's Manual, M091-0305*. Cambridge, Massachusetts: Bio-rad Digilab Division.

Solvent Testing Method for Identification of Oriental Lacquer Used in European Furniture

Rosamond Westmoreland
Private Conservator

The collection of the J. Paul Getty Museum in Malibu, California, contains eleven pieces of furniture made in Paris between 1735 and 1785 that incorporate lacquer panels surrounded by *vernis Martin* or veneers and decorated with gilt bronze mounts. Of these eleven pieces, nine are stamped with the name or initials of the *ebeniste*, but these marks provide no information about the creator of the lacquerwork.

It was a practice throughout Europe to employ both Far Eastern lacquer and European imitations in the manufacture of furniture and other objects. Today, this presents problems of identification and classification for museum curators and others. This paper describes an experimental method of using solvents to distinguish between Oriental and European lacquer. The solvent testing was followed by sampling and technical analysis at the Getty Conservation Institute (see Derrick, Druzik, and Preusser, this volume). The results were identical, with a single exception which will be described.

These methods may not be the ultimate answer to the problem of accurately identifying each section of a large piece when modifications, restorations, or other changes have taken place. However, when a museum has no access to analytical facilities or when funds for such investigations are limited, the solvent method may be of use.

Seventeenth and eighteenth century European lacquerware

When the Dutch and British East India companies began importing Asian lacquerware, they created a fashion that was to extend throughout Europe and colonial America. To meet the increasing demand, the Dutch East India Company provided the specifications for various objects, especially cabinets and chests, and had the pieces made in the Far East (Jarry 1981:128). In 1670, English craftsmen were sent to China to teach Chinese artisans how to make objects that would be suitable for sale in Western markets. This practice soon created a threat to the domestic industry, and in England at the beginning of the eighteenth century the master craftsmen complained of the volume of Oriental lacquerware being imported (Jarry 1981:132). John Stalker and George Parker published in 1688 their *Treatise of Japanning and Varnishing*, a booklet of recipes and instructions for varnishing and lacquering in the Oriental style. Thus it was a relatively simple matter for craftsmen in the West to develop their own methods and materials for obtaining the desirable appearance of lacquer. In England and America the term "japanning" came to mean a style of furniture decoration that employed a painted black background with shallow relief ornamentation made with gesso, gilded or painted with metal powder, and finally coated with several coats of varnish to simulate the brilliance of lacquer.

In Italy in 1690, Cosimo III de' Medici commissioned a scientific study of the resin used in the fabrication of Chinese lacquer; however, the only result was that the investigator became ill (Jarry 1981:134). In 1720 the Jesuit Filippo Bonanni published *Trattato sopra la vernice.* He used the studies in Florence between 1690 and 1700 as the basis of a thorough and authoritative account of the origins of the raw material of Chinese lacquer, as well as the difficulties of handling the raw lacquer and its toxic potential. He drew the conclusion that Europeans "should content themselves with substitutes, since Chinese resin would not tolerate transfer from the Far East to Europe and since the tree called Tri would not grow in a Western climate" (Jarry 1981:134).

Bonanni's treatise was translated into French in 1723. By this time lacquered furniture was quite the rage, although the type of piece (commode, secretaire, carton-nier, etc.) had nothing to do with the Orient. The terms applied to these pieces were imprecise: *vernis de la Chine, vernis de lacque, lacq ancien,* and *vernis Martin.* By 1730 it was fashionable to mount Chinese or Japanese lacquer panels directly on the furniture like inlay or pieces of veneer, and certain cabinetmakers specialized in making carcasses to which the lacquer could be attached. Frequently the Oriental lacquer was retouched by a Parisian cabinetmaker in order to make the piece appear as a unified whole. Practices such as this can make it extremely difficult to distinguish imported Asian lacquer from the domestic European kind.

In the middle of the eighteenth century, there was an atelier of extraordinarily skilled craftsmen, the Martin family, who worked through two generations. Much of their work was for the royal court at Versailles. They had numerous imitators and pieces can generally only be attributed to them on account of the excellence of the craftsmanship, since no signed piece exists. Their method of finishing required the surface to be prepared with forty-three layers of chalk white. Five layers of paint were then applied, followed by fifteen layers of varnish (Huth 1971:96). In 1753, the two oldest brothers were granted a royal patent for the invention of a lacquer with a copal resin base. The dissolving of the hard copal resin in alcohol created a thick, hard, dark varnish, and was probably the *vernis Martin* for which they became world renowned (Huth 1971:95). The Martins are also known to have made repairs on Oriental lacquer; the repair of lacquer was an important part of their craft.

Therefore, use of "lacquered" to describe furniture from this period is often no more than a general description of a decorative style. The tests described here may help to identify the technique and materials used. It should be noted that an experienced curator specializing in this field can generally distinguish between European and Oriental material by stylistic judgment. An experienced conservator can also make judgments based on deterioration of the object, as for example when cleavage reveals lower layers.

Tests with solvents

Solvent testing should be carried out where there is good access to the questionable areas. For example, removal of a door may facilitate the removal of a mount, permitting inconspicuous testing. As the tests proceed, strong lighting will make it easier to see areas of overpainting on the dark surfaces. These areas can be eliminated at the outset if there are other areas that seem to have fewer layers to be penetrated.

The location of tests should be precisely recorded. If samples are to be taken later for analysis, the test area will already be clean. Test areas should be original panels and not locations that may have had alterations, such as a door with a lock-plate; these later additions or modifications may have been painted to match the original lacquer. The test spot should be located either under a removable mount or in a part of the design where it will be unnoticed. It is also advisable to have all possible treatment records at hand, as certain types of coating material applied in a past restoration may be insoluble and thus cause uncertainties.

The selection of solvents is based on the solubility of various films and coatings in organic solvents (Gettens and Stout 1966:204–205). The uppermost layers may be furniture wax or polish, and these can be removed with turpentine or xylene, using a small cotton swab and cleaning an area approximately 1 cm^2 or smaller.

Other modern finishes used by restorers may be insoluble in the above solvents and reagents and an expanded range of solvents and solvent mixtures may be required. Urushi appears to be soluble only in nitric acid. Paint strippers based on methylene chloride (dichloromethane) can be tried.

Table 1. Coatings and solvents

Surface coatings		*Solvents and reagents*
Natural resins	Dammar	Acetone
	Mastic	Acetone, diacetone alcohol
	Sandarac	Acetone, diacetone alcohol
	Copal	Acetone, ethoanol, diacetone alcohol, Cellosolve® (2-ethoxyethanol)
	Shellac, seed-lac	Ethanol, diacetone alcohol
Waxes		Turpentine, xylene, toluene
Synthetic resins	Polyvinyl acetate	Xylene, toluene, ethanol
	Acrylic polymer (Paraloid®)	Xylene, toluene
Dried linseed oil		Ethanol, morpholine, dimethylformamide

AU ROY D'ESPAGNE.
Rue de la Monnoie, prés le Pont-neuf, à Paris.
DARNAULT Marchand, vend tout ce qu'il y a de plus beau & de plus nouveau ; Sçavoir, toutes sortes de Miroirs, Glaces de Cheminées, Trumeaux, avec leurs bordures sculptées & dorées ; de toutes grandeurs.
Toutes sortes de Grilles, ou Feux de Cheminées, des Bras de toutes façons, à deux & trois branches ; Ecritoires, Flambeaux, Porte-Mouchettes, Girandoles, & Lustres à six, huit & dix branches, le tout de bronze cizelé, doré d'or moulu, d'or en feuilles, argenté, & en couleur d'or.
Des Lustres & Girandoles de Crystaux, Lustres de Bois doré, toutes sortes de belles Pendules en bronze cizelé & doré d'or moulu, & couleur d'or, avec leur mouvement, tant à répétition qu'autrement, que l'on ne vend qu'avec garantie, toutes sortes de Tables de Marbre à choisir, avec leurs pieds à Consoles sculptées & dorées.
Des Tableaux pour dessus de porte, de toutes grandeurs, avec leurs bordures de bois doré.
Des Toilettes complettes en vernis de toutes couleurs.
Cabarets, Cabinets, Paravents, & Ecrans en vernis de la Chine, & autres.
Des Bureaux pour écrire, Serre-papiers, Commodes de toutes grandeurs avec leurs dessus de marbre, des Sécretaires, Armoires, Bibliothéques, le tout en bois des Indes, de toutes especes, en vernis de la Chine & du Japon, garnis de bronze doré d'or moulu & en couleur d'or.
Et toutes sortes d'autres choses pour meubler les Appartemens ; le tout à juste prix & en conscience.
A PARIS.

Figure 1. Commode (a), attributed to Joseph Baumhauer, French, c.1750; Darnault label underneath (b). Oak veneered with black lacquer panels, gilt bronze mounts. J. Paul Getty Museum, 55.DA.2.

Test area no. 1: top drawer, front, under mount, proper left end. Acetone removed varnish and black overpaint and revealed a hard, brown, insoluble surface. Ethanol produced the same results.

Test area no. 2: proper left side, bottom center of panel. Acetone removed varnish and black overpaint and revealed a hard, black, insoluble surface. Ethanol produced the same results.

We concluded that these were urushi panels with additional European surface coatings or overpaint.

Figure 2. Commode (a) by Bernard van Risenburgh, French, c.1737. Veneered with Japanese black lacquer and vernis Martin, *gilt bronze mounts. J. Paul Getty Museum, 65.DA.4.*

Test area no. 1: keyhole, front, proper left door (b). Acetone removed yellow varnish. Ethanol removed varnish and black paint, exposing white gesso.

Test area no. 2: front, proper left door, bottom of panel (b). Acetone removed yellow varnish and black paint and revealed a black insoluble surface. Ethanol produced the same results.

Test area no. 3: proper left side, bottom left corner (c). Acetone removed varnish and revealed a hard, black, insoluble surface. Test area no. 4: proper left side, center (c). Acetone removed varnish and revealed a hard, black, insoluble surface.

We concluded that this was urushi with additional European surface coating and overpaint.

Figure 3. Secretaire, attributed to Jean-Henri Riesener, French, c.1785. Veneered with Japanese lacquer and ebony, gilt bronze mounts. J. Paul Getty Museum, 71.DA.104.

Test area no. 1: front, lower left side, immediately adjacent to central mount; the mount was difficult to remove and removal might have caused damage. No surface coatings were removed with xylene, acetone or ethanol.

Test area no. 2: circular panel on bottom shelf, proper right top corner. No surface coatings were removed with xylene, acetone or ethanol.

At this stage, test area no. 2 was sampled. The results showed a shellac-based material, giving a contradictory result to the solvent test. The test area was slightly enlarged and the surface was cleaned with dimethylformamide. No residue was visible on the swab. A further sample was taken for analysis, the results of which showed the coating to be urushi. This piece was restored in 1973 on a contract basis for the J. Paul Getty Museum and no treatment report is on file. A modern surface coating may have been used, which might account for the discrepancy in the initial result. This experience indicates that caution should be exercised when interpreting solvent tests.

We concluded that this was urushi with additional European surface coatings and overpaint.

Figure 4. Cartonnier and serre papier *by Bernard van Risenburgh, French, c.1745–49. Oak veneered with* vernis Martin, *enameled and painted metal, gilt bronze mounts. J. Paul Getty Museum, 83.DA.280.*

Test area no. 1: front, proper right side of lower panel. Acetone very slowly removed yellow varnish. Ethanol removed black paint, exposing white ground.

Test area no. 2: proper left side of *serre papier.* Acetone very slowly removed yellow varnish. Ethanol removed black paint, exposing white ground.

We concluded that this was a European shellac-based coating.

Figure 5. Pair of corner cupboards by Jacques Dubois, French, c.1755. Oak veneered with vernis Martin *and set with lacquer panels, gilt bronze mounts. J. Paul Getty Museum, 78.DA.119 a and b.*

Test area no. 1 (a): proper right side of door, under upper mount. Acetone removed yellow varnish followed by black paint. Ethanol quickly removed black paint. Beneath this there was no sign of gesso but, as paint was removed, fine horizontal cracking remained and the area continued to be tacky. Visual examination revealed much prior loss and overpaint.

Test area no. 2 (b): between front legs of proper left-hand horse. The results were the same as for test area no. 1.

We concluded that further testing should be done on this piece. X-radiographs might show other decorated layers beneath the present surface, which is a European coating of shellac or some other resin.

Acknowledgments

I would like to thank the J. Paul Getty Museum for the opportunity to investigate pieces in the collection, Gillian Wilson for her support of the project, Diane Biehl for preparing the transcript, and Barbara Roberts for assisting with solvent testing and the final preparation of the paper. The analytical work (see Fig. 3) was carried out by Michele Derrick.

Bibliography

Gettens, R.J. and G.L. Stout.

1966 *Painting Materials: A Short Encyclopedia*. New York: Dover Publications Inc.

Huth, Hans

1971 *Lacquer of the West: The History of a Craft and an Industry, 1550–1950*. Chicago: University of Chicago Press.

Jarry, Madeleine

1981 *Chinoiserie: Chinese Influence on European Decorative Art—17th and 18th Centuries*. Published for Sotheby Publications by Philip Wilson Publishers Ltd, London.

Stalker, John and George Parker

1688 *Treatise of Japanning and Varnishing*. Reprinted by A. Tiranti, London, 1971.

Watin

1755 *L'Art du Peinture, Doreur et Vernisseur.* Liege: B. de Boubers. Facsimilie addition reprinted in 1975, Paris: Chez Leonce Laget.

The Chemistry of Oriental Lacquer (*Rhus Verniciflua*)

Ju Kumanotani
Ehime University

The art or technique of lacquerware came to Japan from China via Korea with Buddhism in the sixth century. Since then, it has been modified and developed in response to the characteristics of the culture of each period as well as to local natural features, such as weather and products such as clay, wood, etc.

Lacquerware techniques cannot be understood without a knowledge of the chemistry of sap and lacquer as materials, especially in the preservation and restoration of cultural properties.

Properties of lacquer

When compared with synthetic coatings, the most prominent property of Oriental lacquer is its excellent durability, which has been demonstrated by the use of lacquerware since ancient times. In Japan and China, and in some museums in the United States and Europe, there are large numbers of Japanese lacquer objects that have been preserved for thousands of years or excavated in their original condition.

Lacquer layers exposed to open air degrade to a powdery efflorescence without losing the beauty of their appearance.

Lacquer is adhesive and electrically insulating, and it is sometimes used as a structural material. Moreover, lacquerware is beautiful, smooth to the touch, and yet, solid. These characteristics are unique to this material.

Constituents of the sap of the lacquer tree, **Rhus verniciflua**

The sap of the lacquer tree is a water-in-oil emulsion. Analysis of the sap constituents, carried out by the application of gel permeation chromatography (GPC) and high performance liquid chromatography (HPLC) in combination with the Japanese Institute of Standards method K5950, is shown in Table 1.

Table 1. Analysis of sap constituents

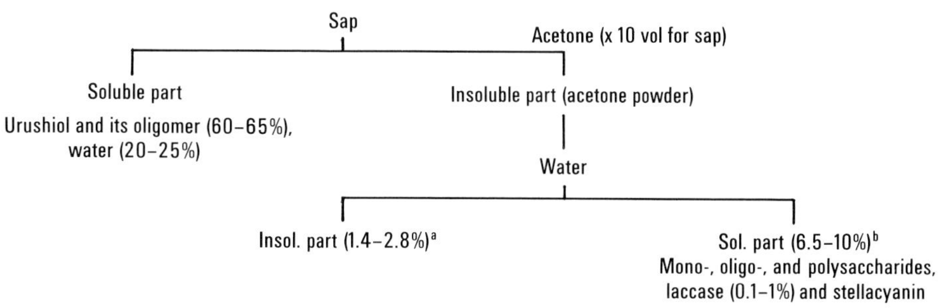

[a]Nitrogenous substances and [b]gummy substances in JIS K5950

*Figure 1. Urushiol isomers in sap (*Rhus verniciflua*).*

		R	%
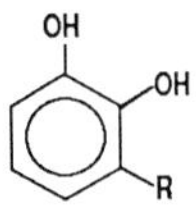	1	15:3 8'cll't13'c	67.0
	2	15:3 8'cll'c14'e	0.1
	3	15:3 8'cll't13't	0.3
	4	15:2 8'cll'c	2.5
	5	15:2 8'cll't	1.2
	6	15:1 8'c	20.6
	7	15:0	3.6
	8	17:3 10'c13't15'c	0.4
	9	17:3 10'c13'c16'e	0.1
	10	17:2 10'c13'c	0.4
	11	17:1 10'c	0.1
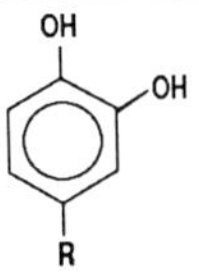	12	15:3 8'cll't13'c	3.0
	13	15:3 8'cll'c14'e	0.1

Figure 2. Structural features of acid from polysaccharides.

```
4-O-Me-ß-GlcpA                                        4-O-Me-ß-GlcpA
      1                                                      1
      ↓                                                      ↓
      6                                                      6
   ß-Galp                  ∝-Galp                         ß-Galp
      1                       1                              1
      ↓                       ↓                              ↓
      6                       6                              6
   ß-Galp                  Galp-(1→3)-ß-Galp-(1→3)-ß-Galp-(1→3)-ß-Galp
      1                                                                1
      ↓                                                                ↓
      6                                                                6
→3)-ß-Galp-(1→3)-ß-Galp-(1→3)-ß-Galp-(1→3)-ß-Galp-(1→3)-ß-Galp-(1→3)-ß-Galp-(1→3)-ß-Galp-(1→
                              6                        6                        6
                              ↓                        ↓                        ↓
                              1                        1                        1
         4-O-Me-ß-GlcpA-(1→3)-ß-Galp                 ß-Galp                   ß-Galp
                              6                        6                        6
                              ↓                        ↓                        ↓
                              1                        1                        1
                  ß-Galp-(1→3)-ß-Galp           4-O-Me-ß-GlcpA               ∝-Galp
                    6
                    ↓
                    1
         4-O-Me-ß-GlcpA
```

Urushiol, the major component of the sap, is a mixture of 3-substituted pyrocatechol derivatives with a saturated or unsaturated chain of fifteen (major) or seventeen (minor) carbon atoms as shown in Figure 1 (Du 1984:463).

Nitrogenous substances, isolated as a water-insoluble part of the acetone powder (see Table 1), are insoluble in most organic solvents, but dispersible or dissolved in urushiol as a stabilizer of the water droplets in the native sap, because the original sap emulsion cannot be made without these nitrogenous substances (Kumanotani 1985:163). The ^{13}C/MAS spectrum, IR spectrum and chemical analysis lead to the conclusion that the nitrogenous substances are glycoproteins (sugar/protein = 1/9 w/w). They become soluble in SDS-water when reduced with 2-mercaptoethanol, and were found to be giant molecules linked with S-S bonds between the fragments, with molecular weights of 8,000 and 17,000 (major) and 25,000 and 47,000 (minor).

Gummy substances are mainly dissolved in the water droplets in the sap and are composed of mono-, oligo-, and polysaccharides (Oshima 1984:43). The polysaccharide was separated by aqueous-phase GPC into two fractions having molecular weights 84,000 and 27,000. The fractions contain D-galactose (65 mol%), 4-O-methyl-D-glucuronic acid (24 mol%), D-glucuronic acid (3 mol%), L-arabinose (4 mol%) and L-rhamnose (3 mol%). Smith degradation of the carboxyl-reduced polysaccharides

Figure 3. SEM photomicrograph of the fibrous polysaccharides.

gives products of halved molecular weight, and these consist of a β-(1-3)-linked galactopyranan main chain and side-chains made up of galactopyranose residues. Peripheral groups, such as α -D-Galp-, α -D-Galp-(1→ 6)-β-D-Galp-,4-O-methyl-β-D-GlcpA- and 4-O-methyl-β-D-GlcpA-(1 →6)-β-D-Galp-, are attached to this interior core through β-(1→ 3)- or β-(1→6)- linkages as shown in Figure 2.

Despite the branched structure, the polysaccharides are able to make a fibrous network (see Fig. 3) that participates in producing durable lacquer films, as will be explained later.

Laccase is a copper glycoprotein, p-quinol-O_2-oxidoreductase with M_n of 120,000 and four atoms of copper per molecule, consisting of common amino acids (55%) and sugars (45%), indicating a reduction potential 415 mV at 25°C and pH 7.4 (hydroquinone as substrate) respectively (Nakamura 1958:538). Stellacyanin is also a copper glycoprotein with M_n of 20,000, which shows a low reduction potential (184 mV, pH 7.0; Lappin 1981:34). Its role in the sap or lacquer is not yet known. Laccase and stellacyanin are likely to be distributed in both the water droplets and the urushiol phase of the sap.

Durability of Oriental lacquer films

From measurements of the dynamechanical property of sap and lacquer films by the torsional pendulum method, it has been demonstrated that sap films undergo a large degree of chemical cross-linking over a period of three years, but the lacquer film keeps almost the same degree of cross-linking (Kuwata 1961:1678). Recently the same conclusion was extended to lacquer and sap films stored over twenty years, from the measured dynamechanical property of the same films (Kumanotani 1981:643; 1983:225). Variation of density is shown in Table 2 for sap and lacquer films stored for nineteen years. This indicates almost no variation of density for lacquer films, but a remarkable change for sap films. These results demonstrate that the lacquer is substantially different from the sap. The lacquer gives highly durable films; the sap gives films oxidizable by air (oxygen). This finding corresponds to the excellent long-term durability of lacquerware and articles buried in the earth remaining in their original condition.

Table 2. Density of Japanese sap and lacquer films at 25°C

Film	J(S,3M)	J(S,19Y)	J(L,19Y)	J(L,1Y)
Density	1.128	1.217	1.185	1.180

S, sap; L, lacquer; M and Y are symbols of month and year

A way of making lacquer from sap

Sap is stirred at room temperature for half an hour, then at 20–45°C for two to four hours in a specially designed open vessel. The conclusion of the stirring is determined by the changing color and viscosity of the sap under treatment. The lacquer film thus made usually contains 2–4% water and has an appropriate viscosity for coating.

The sap is used for the base coat on the substrate, and the lacquer for the middle and top coats. These are dried in a moist air chamber at 70–80% RH for one day.

In this process, the following developments take place: water evaporation, deposition of the polysaccharides from the water droplets in the sap, subdivision of the deposited polysaccharides into fine particles, followed by the dispersion into urushiol, grafting of urushiol onto the glycoproteins and its association with the polysaccharides (making the dispersion of the polysaccharides in urushiol easier), and urushiol polymerization. Thus, lacquer is a mixture of mono- and oligo-urushiol, fine particles of the polysaccharides or their aggregates associated with the glycoproteins or urushiol-g-glycoproteins, laccase, stellacyanin, mono- and oligo-sugars and water.

This account is derived from the results obtained in studying (a) the laccase-catalyzed polymerization of urushiol (Oshima 1985:2613), (b) IR-spectral analysis of the glycoprotein component separated from the sap under treatment to make lacquer (Kumanotani 1981:643;1983:225), (c) SEM photographs of the dispersed polysaccharide particles in the lacquer (unpublished data), and (d) alteration of the visco-elastic properties of the sap in the lacquer-making process (Amari 1980:629).

Drying of sap or lacquer in a humid chamber

The drying of lacquer or sap in the humid chamber is initiated by the laccase-catalyzed oxidation of urushiol into a semiquinone radical (2 or 3 in Fig. 4) accompanied by the reduction of Cu^{++} into Cu^{+} in laccase. Subsequent or continuing oxidation of urushiol needs the oxidation of Cu^{+} in laccase by air (oxygen) accompanied by the formation of water:

Equation 1

Cu^{+} (in laccase) + 1/2 O_2 + $2H^{+}$ = Cu^{++} (laccase) + H_2O

It is easy to understand that the drying begins from the surface of the coated lacquer and is a process determined by the oxygen-diffusion rate. With the progress of drying or film formation from the upper surface to the bottom of the coated lacquer the oxygen-diffusion barrier may become larger but, when the coated lacquer layer is saturated with moisture, oxygen diffusion in the drying lacquer layer becomes easier thus supplying enough oxygen to keep the laccase copper at the active level of Cu^{++} *in situ* after its reduction to Cu^{+} in laccase in the drying process. This is the reason for the use of the humid chamber to complete the drying of the sap or lacquer.

Figure 4. Enzymatic and aerobic polymerization of urushiol.

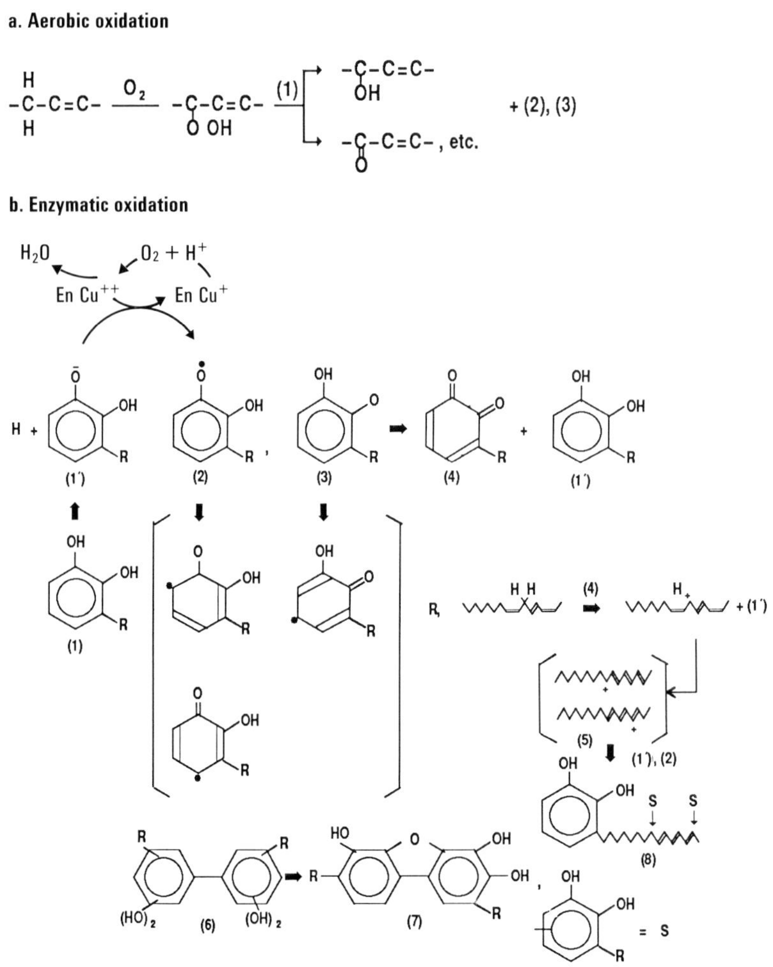

In the drying process, water is produced according to Equation 1. In the initial stage of drying, polymerized urushiol (or the dried skin) formed on the surface retains the water coming up from the deeper layer in drying, with the extraction of the dispersed polysaccharides, making a higher concentration of the polysaccharides in the surface layer of the coated lacquer at the beginning of the drying. The concentration of polysaccharides in the surface layer was examined by applying the ATRIR-spectral technique (Kumanotani 1986).

Furthermore, by the same ATR technique, it is found that in the upper layer of the coated film, rigid structural units made from urushiol, biphenyl and dibenzofuran (6 and 7, Fig. 4) are more highly concentrated than in the deeper layer where the catechol nucleus-side-chain bound structure (8) is more densely packed than the others (Kumanotani 1986).

When isothermal water sorption curves were obtained for the lacquer films in certain conditions (see Fig. 5), the amount of water uptake increased and then decreased with the progress of time, indicating occurrence of the fibrous structure formation promoted by the absorbed water in the lacquer film, which may enhance the barrier to oxygen diffusion into the inside of the grains packed in the films, and contribute to the durability and the surface hardness of the lacquer films.

Polymerization of urushiol

Two pathways are established:

I. Enzymatic polymerization

Urushiol (1) is oxidized in laccase, and the semiquinone radicals (2 or 3) formed undergo C-C coupling with each other or attack the catechol nucleus of urushiol to give diphenyl-type urushiol dimers (6), some of which may be converted to the dibenzofuran derivatives (7) through further enzymatic oxidation.

The semiquinone radicals also undergo disproportionation reactions to give urushiol (1) and urushiol-quinone (4) that may interact with the unsaturated side-chain to produce catechol-side chain-bound products (8). The catechol nucleus of these urushiol dimers can undergo further enzymatic oxidation, to give trimers, tetramers, etc., through laccase-catalyzed polymerization with urushiol or with each other.

II. Oxygen induced polymerization

In the lacquer-making process, the unsaturated side chain of urushiol seems to be oxidized by oxygen and the peroxides formed may undergo a reaction with the catechol nucleus or unsaturated side chain of urushiol to give semiquinone radicals (2 and 3 in Fig. 4) or cross-links between the side chains. In the former case, 2 or 3 then follows the same reaction path as when made enzymatically. Oxidative polymerization, cross-linking or degradation is significant for a lacquer or sap film kept in the air at room temperature or when drying is achieved by baking at high temperatures (110–180C°).

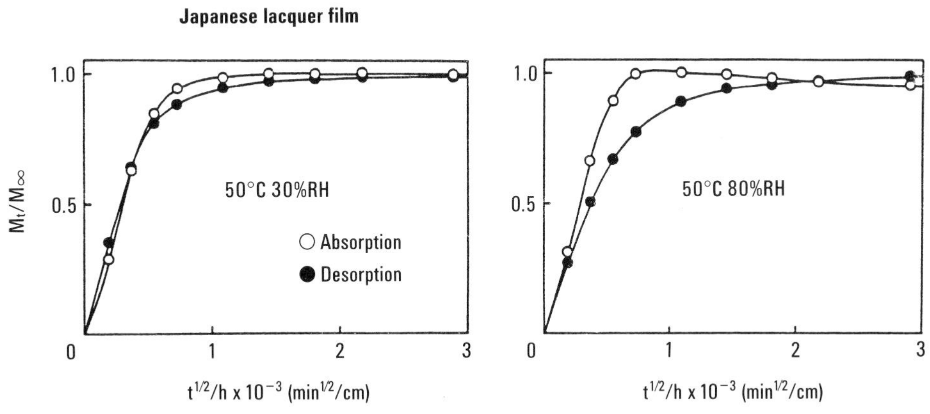

Figure 5. Isothermal water uptake curves. M_t and M are the amounts of water uptake in films at time t and under equilibrium conditions respectively; h is film thickness.

Morphological features and properties of sap and lacquer films

As already mentioned, there is a heterogeneity in the surface layer: a higher concentration of the polysaccharides and the rigid urushiol dimer structure units than in the deeper layer.

It is well-known that degradation of the coated film starts from the surface by an oxygen-induced radical chain reaction, and that polyphenolic compounds are effective as chain stoppers for radical chain reactions. The dibenzofuran or biphenyl unit derived from urushiol may be one of the highly effective chain stoppers rather than the catechol nucleus in the catechol nucleus-sidechain bound structure units.

Based on the results of studying the durability of the lacquer film by dynamechanical property, density, etc., as a function of storage time, as well as from the SEM photographs of the etched lacquer film (Fig. 6), it is suggested that the lacquer film is packed densely with grains 1000Å in diameter, which are responsible for its excellent durability.

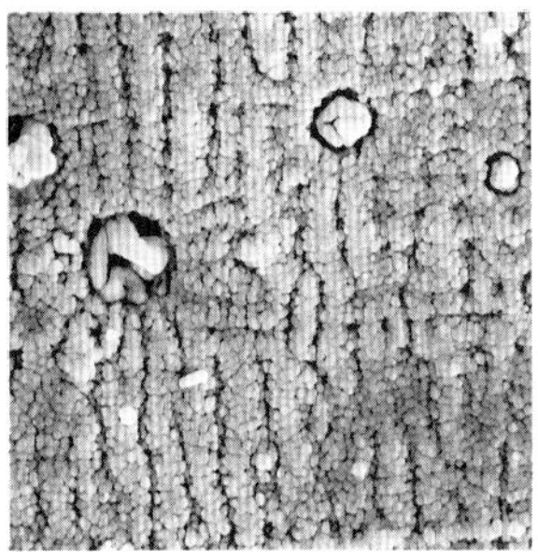

Figure 6. SEM photomicrograph of the etched lacquer film. The inside of the lacquer film is densely packed with grains 1000Å in diameter. A few larger irregular parts in the photomicrograph may come from polysaccharides that were not well pulverized in processing the sap to make lacquer.

Each grain is composed of a wall of polysaccharides with polymerized urushiol inside, "sewn" together with the polymerized urushiol and glycoprotein moiety. The polysaccharide wall, which may absorb humidity, keeps out humidity by the effect of hydrophobic polymerized urushiol linked with the polysaccharides and glycoproteins, so resulting in the polymerized urushiol in the inside of the grain being protected from oxidative degradation (Kumanotani 1981:643, 1983:225).

On the other hand, the sap film is composed of an island (polysaccharides)-sea (polymerized urushiol) texture where the polymerized domain is without any protection from air and is easily oxidized, resulting in cross-linking, followed finally by oxidative degradation (Kumanotani 1981:643; 1983:225).

The more highly concentrated polysaccharides in the surface layer may participate in making a lacquer film with high surface hardness or with high glass transition temperature, forming a good barrier to oxygen diffusion.

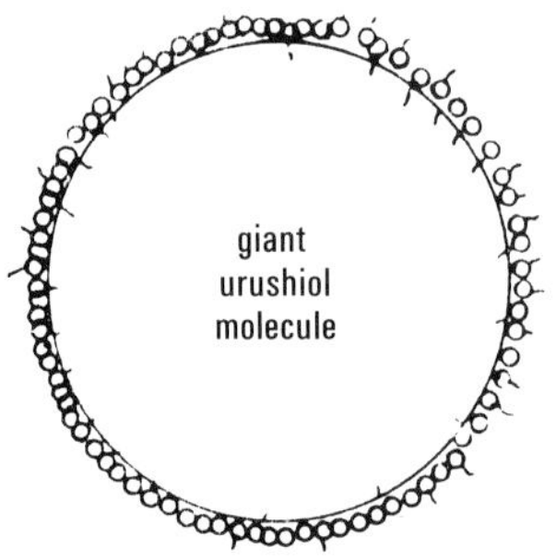

○ polysaccharide and

⚲ polysaccharide associated with glycoprotein

Figure 7. A proposed structure for a grain in the lacquer film. Each of the grains is bound with polymerized urushiol and glycoproteins.

When exposed to sunshine, the lacquer film may absorb sunlight and suffer from photo-oxidative degradation of the urushiol moiety binding each of the grains; the first row of the grains in the films disappears as powdery efflorescence and the second row appears as a new surface, indicating that the lacquer film is degradable out of doors but able to keep the same appearance as the original.

As may be seen from Figure 7, a phase-inversion occurs with progress of the polymerization of urushiol; the high molecular weight urushiol polymers with polar hydroxyl groups on the outside may absorb on their surface the polysaccharides or those associated with glycoproteins or urushiol grafted glycoproteins, so making a grain-structured dried film.

Indoors, it is found that light with wavelength shorter than 360 nm should be removed from the lamps used for the exhibited lacquerware in order to reduce the extent of photodegradation, which is observed as a decreasing degree of gloss of the lacquerware (Araki 1978:1). This measure is understandable, since the oligomeric urushiol isolated from a sample of lacquer showed absorption for light with wavelength shorter than 360 nm.

Water, sorption, and desorption

Lacquer contains not only a trace of water (1-4%, usually 3%) but also hydrophilic components such as polysaccharides and glycoproteins. The glass transition temperature (T_g) of water is below -100°C. Water is expected to lower the T_g of the lacquer film when it is absorbed in the lacquer, and it causes easier oxygendiffusion in films, possibly leading to promotion of oxidative degradation by air (oxygen). Furthermore, desorption of the absorbed water in equilibrium in films brings shrinkage or internal stress in the lacquerware, sometimes introducing cracks in technically reinforced lacquerware.

Table 3. Equilibrium sorption water C∞ (%) and diffusion constant D_{av} (cm^2/sec) for films, *Rhus verniciflua*

Relative humidity (%)	*Temperature (°C)* 30		40		50	
	C∞ (%)	$D_{av}.10^{-7}$	C∞ (%)	$D_{av}.10^{-7}$	C∞ (%)	$D_{av}.10^{-7}$
Sap						
30	1.5	0.78	1.2	1.55	1.3	3.13
60	2.1	0.96				
80	3.1	1.12	3.2	2.29	3.5	4.01
Lacquer						
30	1.3	1.16	1.0	2.52	1.1	5.26
60	1.0	1.44				
80	1.1	1.67	2.9	3.34	3.1	6.25

Based on the measured isothermal absorption and desorption curves for water under various conditions (see Fig. 5), the equilibrium water uptake and diffusion constant of water vapor in films were obtained (Table 3). It became clear that the water uptake is larger for the sap film than for the lacquer, increasing with increasing RH and measuring temperature, and it ranges between 1–4% at 30–50°C and 30–80% RH. Larger values of the diffusion constant are given for the lacquer film, indicating that this film may have favorable properties allowing it to recover an equilibrium amount of water as quickly as possible in response to alterations in temperature and humidity. This is also a very important property for the high durability of lacquer films.

From the T_2 time (spin-spin relaxation time) measured by a pulse NMR for the sap and lacquer films, polysaccharides, and glycoproteins under various conditions of temperature and humidity, it was found that water is also one component of the construction of the film, and an internal stress is expected to occur on water desorption.

Consequently an equilibrium amount of water-uptake is essential in the lacquer films to keep their shape without imposition of stress, otherwise troubles can happen in lacquerware that has been kept for a long time, even though it is strengthened by application of a material reinforcing technique. Of course it should be noted that excess water over the equilibrium amount would be expected to reduce the oxygen-diffusion barrier of the polysaccharide wall of the grains in the lacquer films, promoting degradation of the films by oxygen.

Appendix

Here is a brief description of the man-made technique established empirically on the basis of the results of studies on Japanese lacquer chemistry.

1. Selection of sap

Color, odor, and flow properties of sap have traditionally been an index to estimate the quality of the sap, which should be replaced by the modern analytical method already described for each constituent of the sap. Based on the film formation mechanism studied, dryability of the sap or lacquer and mechanical strength of the dried film seem to be much influenced by the content of urushiol with a triene side chain and laccase, and a balanced composition of the sap like that of the Japanese *Rhus verniciflua*.

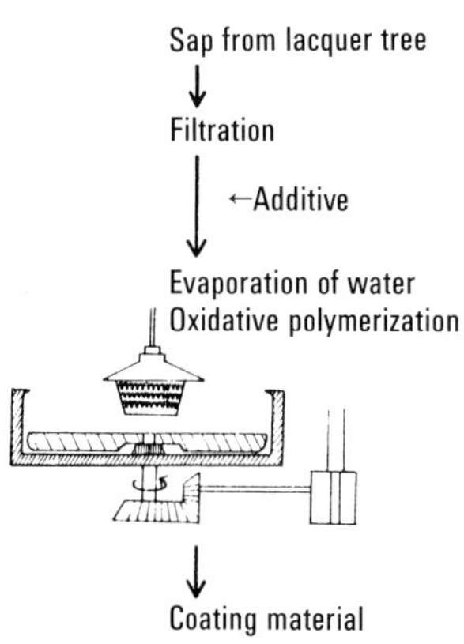

Figure 8. The process of making lacquer from sap, using a vessel equipped with a stirrer.

2. Lacquer making process

As already mentioned, an homogeneous lacquer is made through the pulverization of the polysaccharides, followed by dispersion into urushiol; the design of the wooden vessel with stirrer for making lacquer may be important here. Heating of the sap to make lacquer, caused by friction between the stirrer and the bottom of the vessel (see Fig. 8) may diminish the activity of laccase in the sap under treatment. Too much clearance between the stirrer and the bottom of the vessel results in failure to pulverize the polysaccharides, and yields an inhomogeneous lacquer that may result in defects in the dried films made by the deposited polysaccharides from water droplets in the films (Fig. 6).

3. Skillful coatings

The lacquer artist or craftsman should coat the lacquer thinly on the substrate, for example to a thickness of 5–10 μm, and dry it at a temperature and humidity selected empirically. It was found in the laboratory that thickly coated lacquer needs a longer drying time than a thinly coated one, and before the complete drying of the coated lacquer the dispersing polysaccharides, particularly those with larger sizes or their aggregates, fall to the bottom layer facing the substrate, introducing a larger heterogeneity in the dried lacquer than in that made without precipitation of the polysaccharides. In addition to the surface heterogeneity already mentioned, the bottom layer of the film is rich in precipitated polysaccharides; thus the major, middle part is mainly composed of the urushiol moiety (with less of the polysaccharides), and may consequently suffer from oxidation or degradation like the oxidizable sap film with its separated polymerized urushiol and polysaccharide phases.

An example supporting this laboratory work was that of a lacquer-coated pillar of a shrine which turned white after being degraded out of doors. Moreover, even homogeneous lacquer is found to give dried films with some defects, as already mentioned. To correct these defects a second coating is required. In practice, along with the experimentally observed facts, thin-layer multicoating techniques have been established empirically by craftsmen without scientific knowledge. This is indispensable for getting durable lacquer films.

Bibliography

Araki, T. and H. Sato

1978 "Exhibition lighting and lacquered ware discoloration." *Kobunkazai no Kagaku*:1–24.

Amari, T., J. Kumanotani, and M. Achiwa

1980 "Viscoelastic properties of a sap of Japanese lacquer trees and the lac from it." *Shikizai Kyokaisha* 53:629–664.

Du, Y., R. Oshima, and J. Kumanotani

1984 "Reversed-phase liquid chromatography separation and identification of constituents of urushiol in sap of the lac tree, *Rhus vernicifera*." *Journal of Chromatography* 284:463–473.

Kumanotani, J.

1981 "High durability and structure of Japanese lac and attempts to synthesize it." *American Chemical Society, Division of Organic Coatings and Plastics Chemistry* 45:643–648.

1982 "Chemistry and archaeology of urushi lacquers." *Zairyo Kagaku* 19:88–85.

1983 "Japanese lacquer—a super durable coating." *Polymer Application of Renewable-Resource Materials*, edited by C.E. Carraher and L. H. Sperling. Plenum Publishing Co.

Kumanotani, J., K. Inoue, and L. W. Chen

1985 "The behavior of water in the surface of Oriental lacquers and analogues." *Polymeric Materials Science & Engineering* 52:163–167.

Kumanotani, J., S. Tanaka, and T. Matsui

1986 "Heterogenity in the surface of Oriental lacquer films." *In Proceedings 11 of the 12th International Conference on Organic Coatings, Science and Technology* (Athens):195–201.

Kuwata, T., J. Kumanotani, and S. Kazama

1961 "Physical and chemical properties of coating films of Japanese lac (urushi) (rigidity modulus-temperature relations measured by torsional pendulum method)." *Bulletin of the Chemical Society of Japan* 34:1678–1685.

Lappin, A.G.

1981 "Properties of copper 'blue' proteins." In *Metal Ions in Biological Systems, Vol. 13 Copper Proteins*, edited by H. Siegl, Marcel Dekker:34.

Nakamura, T.

1958 "Purification and physico-chemical properties of laccase." *Biochimica et Biophysica Acta* 30:538–542.

Oshima, R. and J. Kumanotani

1984 "Structural studies of plant gum from sap of the laquer tree, *Rhus verniciflua*." *Carbohydrate Research* 127:43–57.

Oshima, R., Y. Yamauchi, C. Watanabe, and J. Kumanotani

1985 "Enzymic oxidative coupling of urushiol in sap of the lac tree, *Rhus vernicifera*." *Journal of Organic Chemistry* 50:2613–2621.

Glossary

Many of the Japanese words used in this volume appear only once and their meaning is given in parentheses in the text. This glossary contains terms which are used frequently and thus not defined on every occasion. More extensive explanations of techniques are also given.

It must be emphasized that the terminology in this field has not yet been completely standardized. The interpretations in this glossary are drawn partly from the papers themselves and partly from other sources. They should not be regarded as definitive.

The assistance of Professor Kazuo Yamasaki and Mr. Toshikatsu Nakasato is gratefully acknowledged, as well as the "List of Technical Terms" compiled during the 1985 Urushi Study Group meeting by Andreas Burmester and Toshiko Kenjo.

A glossary of specifically architectural terms appears in the article by Yoshio Hasegawa. Historical terms for stages in Chinese *kyushitsu* are given by Hirokazu Arakawa and current Japanese usage by Shogyo Ohba. The names of shells used for *raden* form an appendix to the paper by Yoshikuni Taguchi. Most of this terminology has not been repeated here.

akikusa	autumn plants and grasses, a typical design in Kodaiji lacquers
aogai	shell of abalone used in inlays (*raden*); literally "blue-green shell"
aokin	an alloy of gold and silver
atsugai	shells used for inlay, thicker than *usugai*; literally "a thick shell"
awabi	of the genus *Haliotis* (e.g., abalone or sea-ear)
bengara	a red ocher pigment
bon	a tray
bunko	a document box
chinkin-bori	a technique in which the urushi surface is carved with a sharp instrument and the incised design is filled with gold (see also *qiangjin*)
choshitsu	a technique in which a thick layer of urushi is built up with many coats and a design is engraved on the surface
Chusonji	a temple at Hiraizumi, Iwate prefecture, noted for the richness of its lacquer decorations
Coromandel lacquer	lacquerware produced for the European export trade (named after the East India Company's trading post on the Coromandel coast of southeast India); designs are cut through several layers of lacquer to the wooden base, and the spaces filled with colored lacquer and other substances
danmon	fine cracks on an urushi-varnished surface
dei-e	painting with gold or silver powder mixed with animal glue (*nikawa*)
doroji	an urushi substitute made of animal glue (*nikawa*) mixed with powdered seashells (*gofun*) and finely ground baked clay (*tonoko*)
dozuri	a technique for polishing the kind of urushi surface known as *roiro-urushi*, using a mixture of carbon powder and water, or a mixture of finely ground baked clay (*tonoko*) and vegetable oil
dry lacquer	see *kanshitsu*
e-nashiji	*nashiji* used to make a design rather than used as a background; literally "picture *nashiji*"
fude	a brush; also a technique using a brush, as in *fude-sabi* (using a *makie* brush to apply a soft paste made from *sabi* and water)
fun	filings or powder, usually of metal, e.g., *maru-fun* (round powder) and *sai-fun* (fine powder)
fundame	a fine ground layer of gold or silver powder worked to a smooth, mat finish
gai, kai	a shell
gami, kami	paper
gin	silver, as in *gin-makie*
gofun	powdered seashells used as a white pigment or priming material
gosu	a round box
guri	a technique in which layers of different-colored urushi are applied to an object and a design is carved deeply into the surface to reveal the layered effect; the typical spiral scroll design gives the technique its name, literally "crooked ring" (see also *qulun*)
hagiokoshi	a technique for removing the top coating applied over shell inlay (*raden*) or *hyomon*
hakudo	white clay mixed with a thin solution of animal glue (*nikawa*)
haritsuke	a simplified technique of *hyomon*
heidatsu	thin sheets of metal (thicker than *kirigane*) cut into decorative shapes and applied to a lacquer base; further coats of lacquer are applied and the surface is then polished to reveal the metal
hidori-urushi	urushi in which the moisture is evaporated in sunlight
hinoki	Japanese cypress (*Chamaecyparis obtusa*)

hiramakie	*makie* decoration in very low relief, i.e., only the thickness of the final urushi coating raises it above the surface; literally "flat sprinkled picture"
hirame-fun	flat flakes of metal used for *makie*; literally "flat-eye dust"
homen	a surface prepared by laminating cloth using urushi as the binder
hyomon	thin metal plates (thicker than *kirigane*) are cut into a pattern or design, laid on the urushi, and the surface varnished again with urushi; then the top coating of urushi is polished, revealing the metal design, set flush into the urushi (see also *heidatsu*)
ikakeji	lacquerwork coated with a densely applied fine gold power
inro	a small, compartmentalized container, originally for seals, but also used for medicines and other small objects; the boxes are held together by a cord and closed by a bead (*ojime*); the top of the cord runs through a toggle (*netsuke*) by which the *inro* is hung from the sash (*obi*)
iro	literally "color"; thus *iro-urushi* is colored urushi
japanning	a varnish coating used to imitate Oriental lacquerwork; the term was first published in *A Treatise on Japaning and Varnishing* by John Stalker and George Parker, London (1688)
jigatame	ground coating of *jinoko* and urushi
jikiro	a food container
jinoko	clay that is baked, ground, and screened; coarser than *tonoko* (see also *Wajima*)
jitsuke, jizuke	a process in which *jinoko* is applied with a spatula
jubako	a nest of boxes, or stacking boxes
kaizuka	a shell mound
kamakura-bori	wood carved and then thinly coated with urushi (as opposed to true carved urushi, e.g., *guri*)
kanagai	lacquerware inlaid with gold or silver leaf
kanshitsu	a technique for coating an object using layers of cloth soaked with urushi and pressed over a clay and wooden form that is later removed ; also known as dry lacquer
karamushi	ramie; an Asian perennial plant (*Boehmeria rivea*) of the nettle family
katakata-haku, katakata-nui	fabric decoration in which one surface is decorated with gold foil (*haku*) and the other with gold embroidery (*nui*); the effect is similar to that of *katami-gawari*
katami-gawari	the design or pattern on one half of an object is different from that on the other half—a deliberate contrast
katawagu-ruma	a traditional design showing a wheel half-immersed in water
kawari-nuri	a decorative technique in which cracks are made deliberately in the surface of colored urushi
kebori	hairline engraving
kiji	a wooden substrate
kijigatame	the consolidation of the wooden substrate
kijiro-urushi	transparent urushi without the addition of a drying oil
kin	gold
kingin-e	decoration using gold and silver powder; literally "gold-silver picture"
kinma, kimma	a technique in which lines are engraved on the urushi surface and colored urushi is embedded in the incisions and burnished
kinmakie	a type of gold and urushi surface achieved with fine gold powder applied in the *makie* technique, remarkable for its luster
kintai	urushi applied to a metal substrate

kirigane, kirikane	*makie* design made from tiny pieces of sheet-gold, gold foil, or gold wire (in contrast to gold powder) applied with urushi ; literally "cut gold"
kiriko	a paste made of *jinoko* and/or *tonoko*, water, and raw urushi
kiri-mon	a decorative motif depicting the leaves and flowers of the paulownia tree
ki-urushi	raw urushi
Kodaiji-makie	a style of lacquer decoration associated with the Kodaiji temple, Kyoto, fashionable in the Momoyama and early Edo periods
kokuso	a thick paste of kneaded urushi with various admixtures, typically clay powder and sawdust, sometimes hemp fiber and incense powder, used to cover the wooden foundation before the lacquering process
koshoku	antiquing; literally "old color"
koto	a stringed musical instrument, the Japanese harp or zither
kura	a saddle
kuro	black; *kuro-urushi* is made by mixing transparent urushi with an iron compound
kurome	the process of making transparent urushi from raw urushi, in which the water contained in raw urushi is evaporated with gentle heating
kuro-nuri	to coat with black (*kuro*) urushi
kyushitsu	the sequential application of urushi, from the ground coating to the surface coating
magemono	bent work
magewa	thin strip of wood used in making *magemono*
makie	metallic powder, usually gold, is sprinkled on a wet lacquer surface to form a picture or design; literally "sprinkled picture"
makie-fude	a brush used for *makie*
makie-fun	filings or powder, usually of metal, used for *makie*
makie-hanashi	*makie* that is not polished—the gold powder is simply sprinkled on without subsequent treatment; *hanashi* means "to leave alone"
mitsuda-e	oil painting with pigments bound with oil alone or a mixture of oil and lacquer, with a drying agent
mugi-urushi	urushi with adhesive properties made of raw urushi, water, and wheat or rice paste, sometimes with other admixtures
murasaki	purple; *murasaki-urushi* is a method of making purple-colored urushi
nakanuri	the intermediate coating(s)
nashiji	flakes of gold of irregular size and shape are set at differing angles in a bed of wet urushi and a further coating of urushi is applied; literally "pear skin"
natsume	a tea-powder container, named for the *natsume* fruit, which it resembles in shape
Negoro-nuri	the lacquerware at the Negoro temple was originally red but the surfaces became worn in use and the underlying black showed through in places; this effect was later deliberately imitated
nerigaki	a mixture of urushi and gold filings
nerimono	a paste, typically used for molding shapes
nikawa	animal glue; *sanzenbon nikawa* is a kind of animal glue in stick form
nori	rice glue; *nori-urushi* is raw urushi admixed with rice paste
nunokise	a technique in which cloth is applied to a substrate with *nori-urushi*
nuri	literally "coating," especially with urushi; also used to refer to lacquerware, e.g., *Wajima-nuri*
nuritate-shiage	a final coating; literally "top coating with oil luster"
ohaguru	ferric tannate

okigai	small pieces of shell used like mosaic
qiangjin	Chinese term for incised decoration filled with gold (see also *chinkin-bori*)
qiangyin	Chinese term for incised decoration filled with silver
qulun	Chinese name for *guri*; *xipi*, *tixi* also used
raden	shell inlay, especially with mother-of-pearl
rankaku-nuri	finely broken eggshells applied to a wet urushi surface to form a design; literally "egg coating"
rantai	bamboo woven to form a substrate for urushi; *rantai-shikki* are vessels with a bamboo base
roiro-nuri	a mirrorlike, deep black lacquer finish achieved by the multiple application of thin layers of *roiro-urushi*, each polished to a high finish
roiro-urushi	a form of urushi achieved by adding an iron-based pigment to high-quality raw urushi to give a black color
rokuro	a lathe or a potter's wheel
sabi	raw urushi mixed with finely ground baked clay (*tonoko*), used for the ground layer; other substances may be added
saku	"made by" (as in *fecit*)
sakura	the cherry tree or its blossom
se	a Chinese zither
shikki	lacquerware
shippi-bako	a leather box
shitaji	foundation layer; literally "underneath ground"
shitajigatame	ground consolidation (*nunokise* + *jigatame*)
shitanuri	the base coating(s)
Shosoin	the eighth century Imperial Treasure House at Nara
shu	cinnabar
shumidan	a dais for a Buddhist image; an altar
shunkei-nuri	a technique of applying transparent urushi over wood grain so that the beauty of the natural wood pattern shows through; there are several regional variants
soku	an older term for dry lacquer (see *kanshitsu*)
sozo	clay sculpture technique
suki-urushi	a general term for all types of transparent urushi
sumi-e togidashi	a style of *togidashi* imitating an ink-painting
suri-urushi	raw urushi thinned with a diluent or solvent such as camphor
suzuribako	a writing box
takamakie	high-relief *makie*; a technique in which the design is built up and modeled in a mixture of lacquer and charcoal or clay dust
tebako	a small ornamental box for toiletries; cosmetic box
togidashi	*makie* technique in which the completed low relief (*hiramakie*) design and its ground are covered in further layers of lacquer, which are then polished down until the design reappears flush with the new ground
tonoko	a finer version of *jinoko*; literally "grinding powder"
torii	a Shinto shrine archway
tsubuoki	placing pieces of gold or silver on an urushi surface to obtain a particular decorative effect
tsuikoku, tsuishu	techniques similar to *guri*, in which layers of lacquer—black in the case of *tsuikoku*, red for *tsuishu*—are carved in elaborate designs
tsunoko	calcined deer-horn used in polishing

urauchi	Japanese paper applied to the back of cloth for reinforcement
urumi	black urushi admixed with iron oxide red or vermilion; alternatively, urushi colored with vermilion is further admixed with lamp black
urushi	the sap of *Rhus verniciflua* and other trees of the Anacardiaceae family; also used as a general term for the processed sap
urushiburo	a wooden box kept permanently moist by sprinkling with water, in which lacquered objects are placed to harden; also called *shimeshiburo*
urushi-e	a technique of true lacquer painting, used especially during the Edo period, in which colored pigments are dissolved in urushi
urushikaki	sap collectors
urushiol	a mixture of phenol derivatives, the main component of urushi
usugai	shells for inlay thinned by grinding; literally "thin shell"; their thickness is 8.25 mm per 100 sheets
uwanuri	the final coating(s)
vernis Martin	a varnish finish used by Guillaume Martin and his brother in their Paris workshop to imitate urushi, c. 1730
Wajima	a town in Toyama prefecture noted for its production of ground baked clay (*jinoko*); the clay is formed into bricks about 4 cm in diameter and 1.5 cm high, dried, and baked; the baked clay is ground and classified in order of decreasing particle size: *ichi-henji, ni-henji, san-henji*, etc.
Wajima-nuri	lacquerware using a technique similar to *chinkin-bori*; clay from Wajima is mixed with urushi, giving a smooth strong surface for carving
wan	a bowl
warigai	a technique for breaking shells that are first stuck to a paper backing
washi	Japanese paper
zonsei	a technique in which the outline of the design drawn in colored urushi is line-engraved and the incisions are filled with colored urushi or gold dust